The Builders

THE
BUILDERS

*Houses, People,
Neighborhoods,
Governments, Money*

Martin Mayer

W · W · *Norton & Company* · *Inc.*
NEW YORK

Library of Congress Cataloging in Publication Data

Mayer, Martin, 1928–
    The builders.

    Includes bibliographical references and index.
    1. Housing policy–United States.   2. Urban
renewal–United States.   3. City planning–United States.
4. Construction industry–United States.   5. Federal
aid to community development—United States.   6. Urban
economics.   I. Title.
HD7293.M34 1978   301.36'3'0973   77–17026
ISBN 0–393–08796–4

Designed by Paula Wiener
 2 3 4 5 6 7 8 9 0

Contents

For ELLEN

still the best editor
the best homemaker, too

Part I

PERSPECTIVES

Chapter 1

The Subject

You get too much at last of everything.
Of love,
      of bread,
           of music,
                and of sweetmeats.
Of honor,
      cakes,
           of courage,
               and of figs.
Ambition,
      barley-cakes,
           high office,
              lentils.

. . . Thus far, Aristophanes—but there is an exception to the law of surfeit. As Disraeli once pointed out, and as the great mansions of Newport survive to testify, you can't get too much housing. No populace in the world's history has been so well housed as the modern American. We have twice as many rooms as we have people, and over the past twenty years we have steadily increased the average number of square feet per dwelling unit in the United States. Little inherently nasty housing still exists in this country: more than three-fifths of our seventy-million occupied homes were built since World War II, and outside the core area of the older cities and some depressed rural settings most of what survives from earlier times is clearly "decent, safe, and sanitary," which is what the law requires.

"It would be magnificent if America were to make home ownership a goal of national policy," Michael Harrington wrote in 1962 in *The Other America,* the book that launched the poverty programs of the Kennedy and Johnson years. But he thought such ownership "one of the great myths of American life—for more than half the people do not, and cannot, own their own homes." In fact, unbeknownst to Harrington, three-fifths of American families—more than three-quarters of the families with children living home, which is when this aspect of life is most important—were already living in houses they owned, and the fraction has grown since then, though not much. According to a report by the AFL-CIO in 1975, 77 percent of American trade-union members own their homes. For a period in the mid-1960s, according to a recent study by the Urban Institute, an American family at the median income level (which is lower than the average income level) could buy a standard new house (which is by definition more expensive than a standard older house) at a cost of only 21 percent of pre-tax, perhaps 24 percent of post-tax, income. As individuals, then and now, something more than 80 percent of Americans—more than 95 percent of middle-income Americans—will tell a public opinion pollster they are "satisfied" with where they live. But as a society we want more: we have—we always have—what architectural critic Wolf von Eckardt in summer 1977 called "a chronic and desperate shortage of decent housing."

That's nonsense, of course—except maybe it isn't, because our standards of "decency," legal and psychological, are always rising faster than our production of new homes. Jacob Riis was simply pushed aside when he protested that the poor could never afford to pay for housing that met the minimum code standards the housing reformers of the turn of the century were determined to write (central heating in workers' apartments, for example). Most individuals who are now "satisfied" nevertheless expect to live in "better" housing some day.

Housing is an area of American life where society's reach is supposed to exceed its grasp: that's what this heaven is for. As other societies have grown wealthier, the desire for bigger and better and more private homes has begun to seem less culture-determined and more universal to the animal. Alfred Marshall, the greatest of nineteenth-century economists, predicted it as a *Principle of Economics:* "When the condition of society is healthy, and there is no check on general prosperity, there seems always to be an elastic demand for house-room, on account both of the real conveniences and the social distinction it affords." The cause may be even deeper: "It's like a sex drive," says Lou Winnick, a nervously energetic economist, meticulously enthusiastic, who lays the plans in this area for the Ford Foundation—"to own my own home on my own piece of land."

Statistically, the average American family moves about four times during the course of the marriage, and the move is overwhelmingly "up." Realtor William Edmunds of Orlando, a round young man who runs a nationwide realty training program, says cheerfully that "the buyer always wants more than he can afford, and usually more than what he needs." The American is still Crevecoeur's "New Man," forever moving on; it is the doubtful reverse of the desired coin of social mobility. Pierre Schaefer, secretary-general of the French National Federation of Promoters-Builders, says that the great difference between France and America is that "in France the home is the patrimony, it is what a man will leave to his children, it is *immortal*. In America, a man's home is a consumer good; he uses it like other consumer goods and buys a new one when he is tired of this one."

Home in the form of neighborhood is the centerpiece of what Veblen called conspicuous consumption. "That's the way to live in New York," said a character in Henry James's *Washington Square*—"to move every three or four years. Then you always get the last thing." People's homes are a prime topic of conversation, everything from snotty gossip to envious or self-satisfied wonder at what so-and-so paid only frammis months ago and what the house is "worth" today. The house is, as the real estate brokers keep saying, justifying a 6 percent or higher commission, "the biggest purchase a family ever makes."

At the intersection of shelter need and house-pride stands a stupendous investment. The present value of American homes exceeds

$1.5 *trillion,* more than the total assets of *Fortune*'s thousand larg-est American corporations. Mortgage debt on residential real estate in the United States, which rose by about $120 billion between the end of 1975 and the end of 1977, closely approaches the total in-debtedness of the federal government. It is hard to pin down exactly what we mean by "housing" or "housing costs," but a reasonable guess would be that Americans spend something more than $200 billion a year—over $3,000 per household unit, on the average—for their homes.

The construction industry is the nation's largest, and home-build-ing is the largest category in construction. Real estate brokerage is our largest licensed occupation—we have more brokers than school-teachers—and the total income from real estate brokerage exceeds the total revenue of all American lawyers in private practice. Law-yers themselves make something like one-tenth of their income on matters related to housing, and—as every home-buyer or home-seller grimly learns—there are lots of other occupations that unex-pectedly come to feed at the trough of a real estate transaction. The fee income of title insurance companies, an obscure business if ever there was one, runs $600 million a year. More than half the total assets of our banking institutions—banks, mutual savings banks, and savings associations taken together—are mortgages or construc-tion loans for homes.

Much of this is quite recent. Until well into this century, housing built deliberately for poor people was both cheap and nasty, and the opportunity for lower-middle-class city-dwellers to occupy attrac-tive housing arose only after the development of transportation sys-tems, the end of the historic "pedestrian city," offered the wealth-ier fraction a reasonable trade-off of commuting against space and grassy amenity. Lord Bryce saw the American city of the early 1900s as "a huge space of ground covered with houses, two or three square miles appropriated by the richer sort, fifteen or twenty, stretching out into the suburbs, filled with the dwellings of the poorer."

With rare exceptions, the poor didn't own those dwellings, either: lacking access to mortgage loans, they rented. "Real estate banks" or "land banks" had flitted across the American financial horizons in flocks at various times in the nation's development—the most sophisticated such, and the worst disasters, were the so-called "mortgage guaranty companies" of the 1920s and the Real Estate Investment Trusts of the 1970s—but as Alexander Hamilton had warned from the beginning, such institutions were sure to abort whenever times got rough. "Savings"—or "Buildings"—"and Loan Associations" had sprung up in various localities as coopera- tive ventures by which neighbors helped each other buy their homes (it was unusual for such associations to lend money to anyone who was not a member of long standing); but there weren't enough such to provide sufficient financing at reasonable interest rates. Most mortgages were rather private. "When I was a boy in North Caro- lina," says Robert P. Cunningham, who now runs the inspection programs of the Federal Housing Administration, "the typical hous- ing loan was from one individual to another, the typical term was one year, and the maximum loan was 50 percent of value." There was nowhere near enough such money to enable poor people, or even lower-middle-class people, to buy their homes. Off the farms, at the turn of the century, nearly two-thirds of the population rented.

Even on a rental basis, and even with the necessary long walks for the breadwinner to work and the children to school (only the lat- ter walk was idealized in later nostalgia), spread-out unattractive housing was considered far more desirable than the disgusting "ten- ant houses" a.k.a. "tenements" erected (often enough with the help of loans from the wealthiest families and their churches, which continued to own the land) to house the immigrants forever flowing to the centers of population. The residents of the desert Bryce saw everywhere outside the center of American cities got their chance at decent housing only as the trolley, the commuter railroad, and the automobile successively accelerated the mobility of the upper mid- dle class. And they became *homeowners* only in the aftermath of the New Deal legislation that established the self-amortizing long-term insured mortgage, federally chartered Savings and Loan Associa- tions to write it, Home Loan Banks to shore up these S&Ls, and the beginnings of a federally sponsored secondary market in mortgages.

Even then it took the revolutionary prosperity of the years after World War II to accomplish the change, which occurred so quickly

that the people who lived through it could not begin to comprehend
what had happened. "Between 1940 and 1956," Lou Winnick
wrote in 1958, "the increase in the number of nonfarm homeowners
exceeded the net gain in homeownership in the preceding 150 years
of American history. . . . This is a remarkable achievement, utterly
without parallel in other large industrial countries, and one which
represents a giant forward step in the realization of the Jeffersonian
ideal of a nation of freeholders." Building was part of the story: at
the war's end, three million of the nation's 33.5 million married
couples— 9 percent of all families—were sharing living quarters,
and ten years later the proportion of married couples "doubled up"
was virtually invisible. But the effect of the new housing was much
greater than that. Coupled with the mobility that disturbs M. Schaefer's French sensibilities, the new housing built in the twenty years
after the war affected many more people than its proud possessors,
for it meant that a higher income American was likely to vacate the
home that had been built for him before as much as a quarter of its
useful life had elapsed.

"Filtering" was necessarily (and still is necessarily) the means
by which most people improve their housing. The purchasers of
new single-family houses are moving either from an apartment or
from a house they already own, and it is the resulting vacancies that
enable others to upgrade their shelter. In 1976 more than three
million private homes changed hands in the United States. Urging
the construction of profit-oriented rental housing as the salvation of
the cities in the 1950s, Winnick pointed out the obvious but somehow disliked fact that "every thousand families who can be persuaded to rent new luxury apartments will free rental space for up to
a thousand families who cannot afford new-apartment rentals."

The need to keep the house attractive for subsequent purchasers
and the need to maintain the apartment house for new tenants are
major forces working to preserve and indeed improve the nation's
housing stock. It is when these systems have aborted, usually because the income levels of the newcomers were insufficient to maintain the property, that we get the modern slum, an underpopulated
wasteland rather than the steaming hell of crowded bodies that af-
frighted Jacob Riis.

The proportion of their income that people spend for housing runs from something under 10 percent to something over 40 percent. (Newark makes special loans to people rehabilitating their homes if without such loans their housing expenditures would exceed 65 *percent* of their income.) Everyone seems to believe that the poor spend a higher percentage for housing than the rich, but that may be another area of social criticism where faith transcends information.

Reporting to the President's Committee on Urban Housing in 1968, a General Electric think-tank said they had "searched for systematic relationships between housing expenditure/income ratios and such household characteristics as income, age, family size and race. The outcome is that even within the most finely specified household category there is great variation in this ratio." Four years later, Henry Aaron of The Brookings Institution decided that "average housing expenditures . . . depend on family size and age of head, but the fraction of income spent on housing was almost the same for all income classes." In summer 1977, the Bureau of Labor Statistics calculated budgets for families at the $11,000, $19,000 and $30,000 income levels in the New York-Northeastern New Jersey metropolitan area, and found the lower-income group spending about 19 percent of its income for housing while the other two spent about 25 percent of theirs. But the dispersal around the mean is so extreme—so many families spend much more or much less than the average of their income or demographic group——that meaningful generalization may be impossible.

For two decades after the war, the proportion of income allocated to housing declined on the average for all income classes in America. City planner Martin Meyerson (now president of the University of Pennsylvania) blamed excessive advertising of consumer goods for a growing desire of Americans to economize on their housing and spend the money elsewhere: "This widespread shift in consumer values," he wrote in 1958, "hits hard at every city's struggle to stay viable, for the state of our dwellings and the state of our cities are inseparable. " Dear, dead days, those were, for what Meyerson complained about was in hindsight a joyous expansion of discretionary income, giving increasing numbers of American families the chance to buy the things they wanted as well as the things they needed; we grievously miss this expansiveness now that it's gone.

At all times, of course, lots of Americans, even in the decaying cities, have chosen to spend new-found discretionary income on more housing. After more than three decades of rent control, New Yorkers are probably more reluctant to spend money on housing, and certainly more vociferous about the matter, than people anywhere else in America. But in Greenwich Village and Brooklyn Heights, then Chelsea and Park Slope, now Bay Ridge and Soho and the middle West Side, New Yorkers of all creeds and colors have poured money into the rehabilitation of dwellings (more recently, into the creation of dwellings from the unpromising material of old factory lofts). With and without government assistance (often it seems not to make much difference), people whose appearance and stated attitudes would not lead any observer to characterize them as conservators have made elegant quarters of Chicago's Old Town, Philadelphia's Society Hill, Washington's Georgetown and Capitol Hill and now Mt. Pleasant, Kansas City's Westport, San Francisco's North Beach, Tampa's Hyde Park, Boston's North End—most cities can offer illustrations.

Statistically more significant has been the extraordinary love and care lavished by the lower-middle-class suburbanites of the immediate postwar years on the tract housing denounced when new as instant slums. Robert Wood, later president of the University of Massachusetts, warned in his book *Suburbia* in 1959 that "when housing values new or old are compared, there is a persistent tendency for the central city to have higher residential values and to retain them. . . . The owner of a $10,000 Cape Cod house on a quarter acre lot has little economic incentive to build an extra room or add a garage. These 'improvements' will not only reduce the value of his site in a theoretical sense . . . but they increase his taxes more than they increase the benefits he receives from local public service. . . . In extreme cases . . . the market price of the cheaper houses will fall . . ." So much for the crystal ball of academia. In 1977 on Long Island, the Levittown "Cape Cod" houses that sold originally for $8,000 to $10,000—attics now finished, rooms and garages added, trees and bushes planted and nurtured—were selling

for about $40,000, after more than a quarter of a century of oc-
cupancy.

Housing is one of the areas of life where the blue-collar family
has an edge, especially as the wages of skilled workers rise.
(There's a nice story floating around the bar associations about the
lawyer who gets a bill for a plumbing job in his house. *"Forty
dollars an hour!"* he howls. "I'm a lawyer, and *I* don't make forty
dollars an hour!" To which the plumber replies, "Neither did I,
when I was a lawyer.") Many lower-middle-income American families pay for their home, in effect, by the work they do around a
three-family or four-family building where they occupy one of the
flats and rent out the others. The industry with the biggest number of
workers is undoubtedly Home Improvement: the house is the only
thing you can buy that gets better with use. ("Blue jeans," mutters
a reader of this manuscript; to hell with him.) And the most effective do-it-yourselfer, of course, is the man who knows what he's
doing.

The physical product of the American housing industry has been
severely criticized by (in fact, mostly by) distinguished and sensitive observers. "Well," Saul Bellow recently wrote of his return to
Chicago from a trip to Israel, "we're back, riding through the bungalow belt. Who knows how many brick bungalows there are in
Chicago—a galactic number. There must be a single blueprint for
them all: so much concrete, so many gingersnap bricks, a living
room, dining room, two bedrooms, kitchen, porch, back yard, and
garage. Below, a den or rumpus room. And wall-to-wall everything, and the drapes, and the Venetian blinds, the deep freeze, TV,
washer and dryer, flueless fireplace—plainness, regularity, family
attachments, dollar worries, fear of crime, acceptance of routine.
We ride for twenty minutes through these bungalow blocks, silent,
no need to say what we are thinking: the case states itself."

Indeed, who could prefer these "bungalows" to the handsome
high-rise apartment houses in Vallingby and Farsta, ingeniously
planned suburbs of the Stockholm where Bellow was about to receive his Nobel Prize? Well, the people who live in the bungalows,
now fighting "to preserve their neighborhoods" against thoughtless
batterings by uncomprehending governments and unsympathetic

banks—*and* the people who live in the Swedish high-rises, moving out of their apartments and into single-family housing at such a rate that the vacancies threaten bankruptcy for the government-sponsored cooperatives that built them. That's who.

Don't cheer yet: what's past is prologue. We now must face what George Sternlieb, director of urban policy studies at Rutgers, recently called "the changed realities of housing costs versus consumer means." According to Allan R. Talbot, executive director of the Citizens Housing and Planning Council, "we've built in standards of living we can't afford. . . . We must determine not what is desirable but what is necessary." Getting more specific, Carter B. Horsley of *The New York Times* reports that builders have been asking, "Does every child need his own bedroom . . . does every person need his own toilet? Why wasn't it considered arduous until the nineteen-forties to walk up five flights of stairs?" *Five flights of stairs?*

The years of the great baby boom were 1953 to 1963, with more than four million live births in each—more than four and a quarter million in 1957–58–59. Media pressure to the contrary notwithstanding, the young adult boom that inevitably followed the baby boom has produced an unprecedented rate of new household formation, and each new household needs a home. In 1968, a Presidential Committee estimated the need for new housing in the 1970s at more than 26 million "units." Presidential Committees in 1968 took a highly expansionist view of "needs" (this Committee, for example, found a "need" for 4.4 million additional vacant units; housing economist Frank Kristof suggested at the time that there might be a shortage of enterpreneurs "willing to produce new housing to increase the vacancy supply"). Still, a shortfall of seven million— which is what the 1970s will in fact provide—probably does mean that the nation will enter the 1980s with greater pressure on its housing stock than Americans have seen for a quarter of a century.

Worse: the inflation of the early 1970s and its aftermath—and the rational fear that it will return—have pushed costs more violently in housing than in any other area of consumer expenditure. The problem does not lie in the costs of construction of living space per square foot, which despite large increases in wage rates and lumber costs have actually risen less than the consumer price index as a whole, but in the costs of land, of money, and of government.

In 1948, developed land for housing—bulldozing completed, roads built, pipes installed to bring in water and take out sewage—cost the American builder, on the national average, 11 percent of the price of a housing unit; by 1959, it was 15.5 percent; in 1966, 19.6 percent; in the late 1970s, at least 25 percent, frequently 30 percent of the total bill. ("Of course," says Robert Sheehan of the National Association of Home Builders, conversationally, "You never know what proportion of overhead and profit they're putting on the land cost.") Land costs are still rising, and rapidly, thanks in part to "land use planning," which through zoning and environmental regulation lifts the market price of the land that remains acceptable for residential building and meanwhile adds whopping professional and legal costs while processing is draggingly afoot.

"More and more towns have decided that sidewalks have to be five feet rather than four feet wide," says Charles Rutenberg, a former Chicago furniture merchant who put together a Florida-based building conglomerate called U. S. Homes, one of the first such to be listed on a stock exchange. "No problem—it just adds eighty-eight cents per month for the rest of the homeowner's life. And it doesn't matter to us," Rutenberg adds, bleeding inside but privately, "if the village demands—at a time when cars are getting lighter and speed limits lower—that streets where trucks are prohibited must be built to state highway standards. It's just another $1.70 a month on the cost of the house. . . ."

The real price of housing to a purchaser includes not only the price of the land and the living space but also the price of the money used to buy it, which means, in operating terms to the family, the monthly mortgage bill. From the late 1960s to the late 1970s, the price of the median American home rose from under $25,000 to over $45,000 (the *average* price, pulled up by the $250,000 spectaculars, rose even further). But those figures were more or less in line with the roughly 80 percent increase in consumer prices generally. Because interest rates on mortgages went from 6 percent to 9 percent, however, the monthly mortgage costs on the median house—assuming a 10 percent down payment and a thirty-year self-amortizing loan—went from less than $135 to more than $325, an increase of 140 percent.

There can be truly astonishing gaps between the monthly mortgage bills of neighbors occupying identical houses, depending on when the house was bought. In Levittown, Long Island, for example, a World War II vet who bought one of the first early houses in 1949, for $8,000 with a 4 percent thirty-year mortgage, was still paying the bank $38 a month in 1977. The Vietnam veteran in the house next door, who just paid $40,000 and got a 9 percent thirty-year VA mortgage, pays $321 a month, more than eight times as much, for the same house.

Taxes are up, too, exponentially, the rise in price reflected in assessed value for tax purposes and then escalated by a rise in rates. Between 1965 and 1975, the American homeowner's real estate tax bill rose on the average by 15 percent a year, which means it quadrupled. The operating expenses of housing, concentrated as they are in energy costs (for warmth, coolth, appliances, lighting, and—truly a housing cost—travel to work), rose in the middle of the 1970s even more rapidly than mortgage payments and taxes. At some point, a kind of giddiness sets in, which offers entrepreneurial possibilities. Louis Wolfsheimer, a lawyer fom Baltimore who became Planning Commission Chairman in San Diego, reports overhearing a conversation on an airplane between two Southern California builders:

"Congratulate me, Harry, I've just sold my first $120,000 house!"

"Holy gee, Joe, last time we talked you were building $40,000 houses."

"I still am, but now I get $120,000 for them. . . ."

From a high of nearly 30 percent of the median income in 1950, the annual cost of mortgages, taxes, and insurance for a "standard" new house had dropped by 1965 to only 21 percent of income. By 1977, the figure was over 25 percent everywhere, probably over 30 percent in parts of California and Colorado. Home builders briefly cut back on the house they offered their public—less space, less finishing, fewer appliances—to keep prices from rising too fast; and

they found that the resulting "affordable house" drew few customers. It wasn't a matter of five flights of stairs: one flight of stairs
made a marketing problem.

"We're going to have to go back to the 1930s ratios," says Preston Martin, an elegant former college professor in an elegant office
in the Bank of America building in San Francisco. Once chairman
of the Federal Home Loan Bank Board, Martin is now running his
own mortgage insurance company with a little help from friends at
Sears. "We're going to have to say, 'Mr. Jones, you have to spend
a higher proportion of your income for housing.' " Daniel Rose of
Rose Associates, which builds apartment houses all over the East,
says that "people are going to have a choice of (1) more money; (2)
less space; (3) shoddy building; (4) some combination of those
three."

It must be said that this cry of wolf has been heard before. "Two-
thirds of the population cannot pay a rental or purchase price high
enough to produce a commercial profit on a new dwelling," Edith
Elmer Wood wrote in 1931. In 1958, sociologist Edward Weinfeld
wrote that "nearly 70 percent of American families have incomes
below . . . the amount needed to buy the cheapest of the houses
commonly being produced today." Ten years later, economist Bernard J. Frieden of the Harvard-M.I.T. Joint Center wrote that "The
cost of new housing today prices most American families out of the
market. . . . Ownership of new homes is effectively limited to the
top quarter of American families. . . ." But the fact appears to be
that for the first two decades after the war, about a third of the new
homes built in America, like a third of the new cars, were bought by
families in the lower half of the income distribution. Housing economist Sherman Maisel, a former Governor of the Federal Reserve,
thinks today's beliefs are as inaccurate as Wood's and Weinfeld's
and Frieden's were, that the post-Vietnam pattern will replicate the
post-Korean pattern, with housing costs rising rapidly in the immediate aftermath of the war and then settling back to a slow increase.

Alas, I fear he is wrong: the escalation of wage rates and materials costs that troubled the 1950s were trivial factors beside today's
accelerating rush of land prices, energy costs, the price impact of
the worldwide demand for lumber, environmental restrictions, and

the persistence of inflation (the government has not even targeted, let alone achieved, a reduction in the rate of inflation that would make possible 8 percent mortgage interest rates). The rise in home prices between 1953 and 1965, as The President's Committeee on Urban Housing reported, was mostly the result of a 40 percent increase in the floor area of the median new house, and the introduction of "air conditioning, better thermal and sound insulation, more and better appliances, more tasteful design and landscaping, and many other quality features." No such benefits lie in our path.

What Preston Martin and Daniel Rose are predicting is a decline in the quality of life, not as an abstract formula but as a universally perceived reality. The burden would have to be borne at first mostly by the young, who do not have the increased market value of existing homes as a cushion, and by the community of tenants, who tend to have lower incomes than homeowners. They may take it lying down, but they may not. "People are not inclined to accept this sort of adversity,"says Albert Cole, who was Eisenhower's director of Housing and Home Finance.

In fact, the young householders have decided they do not have to accept it, and they won't. The sacrifice they will make is that the wife will work a few years longer, until inflation brings the husband's income up to the level where he can afford the payments on the mortgage. Take, for example, our Vietnam vet with $40,000 house in Levittown. He earns $250 a week, much below what he would need to pay the $420 a month required by his mortgage, taxes, and insurance on the house. But in eight years, if his salary does no more than keep up with a 6 percent inflation rate, he will be earning about $400 a week. His wife can keep her job till then, and the two of them can participate in political action to hold down the real estate taxes. And if the house increases in value by 6 percent a year, after eight years they'll have a property worth $64,000, and some of the difference between that and the $40,000 purchase price can be borrowed, if necessary, to give the baby a good start in life.

That's the system that worked for the fellow next door, who bought the house new in 1949. He makes only $250 a week himself, but they just bought that great camper for the trip they're going to take this summer with their grandson.

For Donald M. Kaplan, chief economist of the Federal Home Loan Bank Board, "the bottom line is that if most American families had to buy the homes they are living in today, they could not afford them." But for the young couples now crowding the market, the bottom line is that all those people older than themselves are now so well housed because they bought years ago. Couples who once felt they could wait to buy a house—enjoy the relatively easier life of an apartment and save up for the house—are now in a kind of panic to buy.

Since 1967, the Conference Board has included in its periodic public opinion polls a question about the respondent's "plans" to buy a house within the next six months. In July 1977—with the median price of both new and old single-family homes higher by comparison to the median income than it had been at the time of any previous poll—5.2 percent of the families polled, the highest proportion ever recorded, replied that they did indeed have such plans. And what they want, incidentally, is a real house with its own four walls and a piece of land. Rutenberg reports that six out of seven of the purchasers of a U.S. Homes townhouse (defined as a house that shares a wall with a neighbor) say that they regard this purchase as a way-station: within five years, they expect to be living in a free-standing home of their own.

Today's housing market, in short, is creating an ever-increasing pressure group who *need* inflation. They are themselves the losers, because it is expectations of inflation that have pushed the mortgage interest rate so high and made mortgage money sporadically hard to find on any feasible terms. The worst losers, though, are the thirty-odd percent of the population who live in rental housing. If the market value of an apartment house is to keep pace with inflation (and if it doesn't, nobody will wish to own apartment houses), rents must perpetually rise; and the costs of running apartment houses, concentrated as they are in service trades and energy purchases, rise even more rapidly than the inflation rate. Rumblings of resentment and even rebellion can already be heard from the community of tenants—rent strikes, demands for rent control (imposed in Boston, Philadelphia, Washington, New Orleans and El Paso as well as New York), harassment of landlords in courts suddenly sympathetic to tenants, an annual tenant turnover rate that exceeds 25 percent in high-rise elevator buildings and is reaching toward 90 percent in furnished garden apartment projects. Rapidly rising rentals, which can't be got, will be absolutely necessary in the next few years if any ponderable quantity of new multifamily houses are to be built at today's costs. On current trend lines, inflation will within a decade destroy for-profit rental housing in America; and we will have nothing ready to take its place.

We can speak of Americans "owning" their homes only because from early on most states (not all) adopted the Roman *hypotheca* rather than the traditional English *mortgage* (though we kept the English word) as the system by which money could be borrowed on the security of real estate. Under the old English system, title to property was held by the lender; but the borrower had possession and use, which meant that from the lender's point of view the title was a "dead pledge." (The word, pidgin French from the Norman Conquest, shares roots with the "mortal" wound and the "engaged" couple.) In the Roman and American system, title, possession, and use—all three—are held by the borrower, and the lender merely has a claim against the property if the loan goes unpaid. Though this claim is inviolable—a bankrupt who owns his home free and clear may be able to retain it against general creditors through a "homestead exemption" in the bankruptcy laws, but a mortgagee has an absolute right to foreclose on the property specified in the mortgage—the fact that the householder has title, possession, and use justifies his boast that he "owns" his house even if he bought it yesterday on a Veterans Administration mortgage that required him to put down not a penny in cash.

Legally, in Anglo-Saxon jurisprudence, the citizen's security at home has elements of sanctity transcending the fact of ownership. In the eighteenth century, the elder William Pitt proclaimed to Parliament that "the poorest man may in his cottage bid defiance to all the forces of the Crown. It may be frail; its roof may shake; the wind may blow through it; the storms may enter, but the King of England cannot enter; all his forces dare not cross the threshold of the ruined tenement." This has been taken as a gloss on private property, but it is not: the resident of a tenement is by definition a tenant, and what Pitt was saying was that even with the consent of a landlord the state could not enter a man's home against his will. Constitutional prohibitions against search-and-seizure protect tenants and homeowners equally in the United States: the police can stop and frisk a man on the street if they have a plausible reason for doing so, but to enter his home they need a warrant. In Britain today—it is the last nail in the coffin that Her Majesty's governments have slowly constructed in a nonpartisan way to bury the rental housing industry—a tenant cannot be evicted by a landlord for any cause other than non-payment of a state-controlled rent, and maybe not even for that.

Whatever the rights of tenants, however, home ownership is the desideratum and the norm, and not only in the United States. More than half the households in Britain, for example, now live in homes they own, and the proportion has passed 40 percent in France. By and large, modern governments don't want to be landlords any more than private investors do. In Singapore, which has built more new housing per capita than any other country in the world in the last dozen years, virtually all the apartments in the state-constructed high-rises have now been "bought" by their occupants. The East bloc is not immune: Hungary has launched a major home-ownership drive, and the Bulgarians have established a system by which any four families that have a certain volume of savings in the state bank are entitled to (1) a mortgage loan and (2) allocation of materials and labor from the state-operated construction industry for the purpose of building their own quadruplex house. Loek Kampschoër of the Netherlands Housing Directorate says a special factor is at work in the Communist-bloc drive to sell off existing housing: "They've never allocated any money for maintenance, so now they're in trouble and want to get rid of it."

Statistics in the United States have always been muddled by the presence of large numbers of small multifamily buildings where the owner occupies one of the units and rents the rest. Almost 90 percent of our rental units are owned by individuals (as distinguished from corporations or partnerships), and the proportion of the population living in "owner-occupied structures" is probably ten percentage points higher than the proportion of people living in housing they own themselves. Sometimes the differences are extreme: in Boston in 1970, only 27 percent of the housing units but 71 percent of the city's housing structures were owner-occupied. Newark's "urban homesteading" program, initially considered hopeless by academic and foundation consultants because of the city's high proportion of multifamily homes, has got off the ground through the activities of recent immigrants (especially the Portuguese), who in the traditional American way take one reconditioned apartment for themselves and rent the rest to compatriots. It has become increasingly clear in recent years that housing investment is most likely to happen when the investor benefits by it in his living conditions rather than simply on his bottom line or in his tax return. This is especially true in run-down neighborhoods, where only "owner-residence," George Sternlieb writes, "produces the degree of close supervision required for good maintenance." And governments give benefits to homeowners in their tax returns, too—most strikingly where the income tax rates are highest, in Scandinavia and Britian.

Since the Housing Act of 1961 authorized FHA insurance for individually owned apartments in multifamily structures, American lawyers and legislators have been occupied finding ways to extend the benefits of home ownership to apartment dwellers. There are two somewhat different approaches, under the names "cooperative" and "condominium." In the co-op, the tenants are shareholders in a corporation that owns the entire property, including the shareholder's apartment, on which he has a "proprietary lease." In the condominium system, each resident has title to his own slice of air, and a joint interest in "common areas," which may include anything from the incinerator room through clubhouse-tennis-and-swimming-pool complex. The advantage of condominium procedures, in states which have passed the "Horizontal Property Act," is that they give a bank a mortgageable piece of paper for lending purposes.

Apartment ownership plans in subsidized situations may be fake, because the subsidizers (not unreasonably) insist on "recapturing" any increase in the value of the home. In New York, the nonprofit United Housing Foundation, progenitor of the gigantic Rochdale

Village project in Queens and Co-op City in the Bronx (among many others), still forbids ''profiteering'' as it did in its early days as an offshoot of the Amalgamated Clothing Workers Union, with social attitudes derived from the East European Jewish Bund. As a result, the supposed ''owners'' of these projects are in reality tenants who have been compelled to put down a deposit to rent their apartment: when they wish to move out and sell, they can get nothing back but what they originally put in—without interest.

The Foundation does not itself profit by this forced sale because the apartment is immediately made available at the subsidized price to some other family that could not otherwise afford it. And purchasers who have ''bought'' apartments for one-tenth the going price can scarcely expect to be permitted to turn around and reap a windfall by reselling at the market. Still, by denying its customers the inflation hedge which is otherwise a major reason for home-ownership, the Foundation invites the political brouhaha that has wracked Co-op City, where the ''owners'' have refused to pay ''maintenance charges'' large enough to match the rising costs of utilities and the interest on the mortgage, which is held by an agency of New York State. (In fairness, the original Co-op City purchasers were never informed of the enormous construction cost increases that had made inevitable much higher mortgage charges—and thus maintenance bills—than they had been told to expect when they signed their contracts and put down their deposits.)

In this bizarre situation, authentically radical (Trotskyite) leadership has bullied the state government into allocating five to ten million dollars a year of subsidy funds to the sustenance of Co-op City, an almost entirely middle-class community that would not on any social logic be regarded as proper recipients for the state's limited housing help. Since the state's ''Mitchell-Lama'' co-op housing program was modeled on the United Housing Foundation plan, and what is done for Co-op City must be done for the others, taxpayers seem to be stuck with upwards of $50 million a year in subsidies to the ''owners'' of middle-income apartments built with the proceeds of state bonds.

The category ''homeowner'' in the United States thus covers an unimaginable range of situations: the Philadelphian whose Daddy left him a mansion on the Main Line, free and clear; the black couple in Baltimore restoring a row house with a subsidized loan; the confident Californian who moves from development to development every five years, counting on a 25 percent annual appreciation of his house to buy him a better one next time; the struggling Vietnam veteran with a no-down-payment loan on an 1,100-square-foot townhouse twenty miles from Jacksonville; the Seattle furniture salesman with a second mortgage taken to consolidate the debts he incurred after Boeing slowed down; the retired college professor with a small loan to put a year-round heating plant in his cottage in the Berkshires; the shareholder in the subsidized co-op high-rise whose equity is forever restricted to his purchase price. All these and many more are placed by commentators and the Census Bureau in the same analytical set, when the subject is housing.

Nor is "ownership" the only source of confusion. The foundations of housing rest on lucre, on the enormous income streams that entrepreneurs, salesmen, and politicians wish to channel in their direction. From the days when the land speculators laid out the first New Towns of the West—with plans solemnly announced for a Railroad Station here, a Hotel there, a Theater on this lot, etc.— there has been entirely too much money to be made in real estate in America. "Town lots," wrote economic historian Miles Colean, "became the currency of speculation. Repeated liquidation occurred, but the ardor could be dampened only for short periods. Instability was thus built into the urban as well as the rural land structure."

Projects involving the use of land pay out in real values only over a long period of time, but realizable paper profits can come fast: the promotor, the speculator, the developer, the builder, the seller, the lawyer, the insurance agent can and do take their money and run, letting the purchaser and the resident pay later for mistakes. "They don't build firewalls in townhouses here," says John Fitzwater, a young English engineer transmuted to a Clarksburg, West Virginia builder. "It's cheaper to have the buyers pay higher fire insurance."

"Most of my colleagues know how to multiply but they don't know how to add," says Samuel Lefrak, a perfect and perfectly confident caricature of the successful New York businessman, who has built more apartment houses than anyone else alive—he claims to have housed one in sixteen New Yorkers, and probably still owns some 45,000 apartments in the city with an annual rent roll of $100 million. "On Wall Street they have bulls and bears. In real estate, we have pigs."

Eminent domain over land is inextricably bound to sovereignty, and the bundle of rights and uses contained in the apparently simple notion of title to land has bedeviled the Anglo-American legal system since it first appeared on paper in Magna Charta. The United States Government has been involved in this business since before there was such a thing—since April 26, 1785, to be exact, when the Continental Congress adopted the "rectangular survey," by which land in thirty of the fifty states is still officially described, using a grid of "principal meridians," "guide meridians," and "base lines" that overlays the continent. Real estate, unfortunately, is nonfungible—no two pieces are exactly alike—which makes contracts for the purchase or use of land and buildings more detailed and potentially more onerous than any others. Colean lists some questions that must be answered: "Are there easements or rights-of-way that must be honored? Have dower rights been released? Is there any delinquency in taxes? Do undischarged liens of any sort exist? Is there an unexpired lease binding on the purchaser?" There are even more questions if one cares to ask: that's why mortgagees insist on title insurance.

Land grants have been the inducement for settlement and investment on this continent from the first colonies through Title I of the Urban Renewal Act. The value of land, and indeed structure, is in large part a function of the government services that go with it—the creation of official records and courts, the maintenance of navigable waters and the highway system, the police and fire departments and the schools.

Owners of real property always want something from government, and (to put the matter bluntly) they have the money to pay for what they want. Legal fees and insurance commissions related to real estate are the economic foundations of both major political parties and their local leaders throughout the country, and the various

building departments and zoning commissions are the center of corruption in American municipal government. When the federal government gets deeply involved in local allocation of housing subsidies—as it did in Sections 608, 213, 221, 235, and 236 of the
amended Housing Act*—the most predictable result is the moral
decay of numerous middle-level federal bureaucrats in the field offices. But the second most predictable result of federal subsidy is the
construction of a great deal of housing for moderate-income families that would not otherwise be built. Especially, these days, for
blacks.

For housing is the element in American life where race prejudice
bites most deep. Blacks in America have never had a square shake
in access to land or housing. The Homestead Act was enacted in
1862, before the Emancipation Proclamation, and opened the farm-
lands of the West to whites only. Any thoughts that the Freedmen's
Bureau might distribute confiscated and abandoned Southern farms
to the former slaves was quickly scotched by the Amnesty Procla-
mations of both Lincoln and Johnson, restoring these lands to their
former owners.

Jacob Riis wrote bitterly of the Negro's condition in New York in
1890: "The Czar of all the Russians is not more absolute upon his
own soil than the New York landlord in his dealings with colored
tenants. Where he permits them to live, they go; where he shuts the
door, stay out. . . . Cleanliness is the characteristic of the
negro. . . . In this respect he is immensely the superior of the
lowest of the whites, the Italians and the Polish Jews, below whom
he has been classed in the past in the tenant scale. Nevertheless, he
has always had to pay higher rents than even those for the poorest
and most stinted rooms. . . . The reason advanced for this systematic robbery is that white people will not live in the same house with
colored tenants, or even in a house recently occupied by negroes,
and that consequently its selling value is injured. . . ."

Riis thought these conditions would soon improve; instead, they
worsened. In many cities, segregation moved from being a land-

---

* Section 608 was not intended as a subsidy, but became one. The various federal housing
programs will be described as needed in these pages. A quick-reference numerical glossary
of Housing Acts, Sections and Titles can be found in the Appendix.

lord's privilege to public policy: Louisville, Baltimore, Richmond, and Atlanta pioneered the new legislation in 1912 and 1913. When the federal government entered the scene as a promoter of housing, Charles Abrams wrote, "FHA adopted a policy that could well have been culled from the Nuremberg laws. From its inception, FHA set itself up as the protector of the all-white neighborhood. It sent its agents into the field to keep Negroes and other minorities from buying homes in white neighborhoods. It exerted pressure against builders who dared to build for minorities, and against lenders willing to lend on mortgages. . . . Racism was bluntly written into the FHA's official manual: 'If a neighborhood is to retain stability, it is necessary that properties shall continue to be occupied by the same social and racial classes.' . . . It even exhorted the use of a model covenant, providing that 'no person of any race other than _____ [race to be inserted] shall use or occupy any building or any lot, except that this covenant shall not prevent occupancy by domestic servants of a different race domiciled with an owner or tenant.' " Until 1947, such covenants were enforceable in American courts.

Even now, though the gap is narrowing, blacks are only half as likely as whites to own their homes. They rarely buy new houses, and the old houses they buy are almost by definition in "declining" neighborhoods, with all the phrase implies in the deterioration of schools, police protection, street maintenance, and garbage collection. They often pay more than whites have been paying for similar nearby properties, enabling integrationists to "refute the myth" that black occupancy produces lower real estate values; but because their resources are stretched by their purchase, they are unable to maintain what they have bought, and the resulting deterioration of the properties refutes the integrationists and buttresses the myth. Meanwhile, the local welfare department pours all the human sludge of the area into the black neighborhoods and the authorities put the public housing there, because black opponents to the deluge find it hard to protest and even harder to gain attention when they do.

The apartment houses blacks occupy in the cities are most often decayed goods being milked by a fourth or fifth generation of owners to yield the greatest return prior to rapidly approaching demolition (which may be planned in concert with arsonists). These

owners are very possibly black themselves: in his study of the Newark slums in 1967, George Sternlieb of Rutgers found a third of the
ownership black, and the proportion must be higher now. For the
tenant, the ethnic origin of his landlord is usually a trivial concern: I
remember my generous grandmother saying in her accented English
to the refugee who had bought the side-street apartment house
where she lived, ''All the *nice* Jews Hitler killed; why did he have
to miss you?''

The filtering process that brought the benefits of middle-class
housing to poor families has been slowed though not blocked, because the new middle-class housing is in the suburbs, and the chain
reaction expressed by busy moving vans weakens with distance.
When it does reach the city, the result is likely to be panic, landlords
renting to any and all applicants, guaranteeing the speedy decline of
the whole house. In most of the larger cities, black tenancy of public
housing—and blacks, less than 12 percent of the population, fill 70
percent of the public housing units—has been disastrous both for the
residents and for the projects, especially in recent years, as public
housing becomes increasingly a refuge for female-headed families,
demoralized second- and third-generation welfare recipients.

Worst of all, blacks have not been able to use their housing expenditures as a means of saving, of building up equity in their
homes: for the last generation, white families selling homes have
walked away with profits large enough to compensate for inflation,
while black families selling homes have often failed to get back
even what they paid. The ''white noose'' of restricted suburbs
around the central cities is today pretty much history; the middle-income black family today can (and does) get out. But what it leaves
behind are neighborhoods where vacancies and abandonments dra-
matically reduce the possible market price of sound dwellings.

Yet all the statistics and averages and generalizations may conceal more than they uncover, for housing in America is an astonishingly individual, unsystematic, even accidental artifact. Several million blacks live in pleasant apartments and luxurious houses;
elegantly designed new townhouses can be found in tract developments not only in the advertised New Towns like Columbia, Res-

ton, and Irvine but also in and around Houston and San Diego, Denver, New Orleans, Minneapolis. A majority of American homeowners are still paying off on mortgages that carry an interest rate of 6½ percent or less. A builder like Fox and Jacobs of Dallas still offers a three-bedroom house (1500 square feet of living space plus garage) with luxury features (air conditioning, dishwasher, disposal, two baths, walk-in closets, fireplace) for something like $28,000, $1500 down and $250 a month including taxes and insurance. While public housing projects in St. Louis, Boston, San Francisco, and Newark show grievous vacancy rates, the New York City Public Housing Authority (by far the nation's largest, with 550,000 residents) maintains an active waiting list of more than 120,000 families, and in Atlanta forty-year-old Techwood Homes, the nation's oldest public housing project, is annually besieged by new applicants. Rehabilitation of the brick rowhouses of Baltimore is proceeding lickety-split, while the older housing stock of Cleveland and Detroit and the Bronx literally disintegrates.

A few apartment house owners and managing companies operate all over the country, and so do promoters of retirement colonies like Rossmoor and the Del Webb operation (the later concentrated, however, in the sunbelt states). But the nationwide builders who emerged from chrysalis in the early 1970s with a great flourish of trumpets in the stock market have all gone back to the cocoon or down the tubes, and the few far-flung builders who survive have mostly settled back into a small number of localities. ("You buy property in an area you don't know," says Robert Levenstein, the new president of Kaufman & Broad, which continues to operate in five states, "it's with you a *long* time.") Companies as different as ITT, Boise Cascade, McCulloch Oil, GE, Singer, CNA Insurance and American Cyanamid have come a cropper trying to build housing in a number of different markets. Big builders, says Nat Rogg, who used to run the National Association of Home Builders, "are like the dinosaurs whose bodies were too big for their brains to control." As the Real Estate Investment Trusts and their stockholders learned the hard way, it isn't even safe to lend money in markets far from the eye of a guiding entrepreneur.

For a brief period after the war, the major insurance companies became equity-holders as well us lenders in the housing industry, nationwide. Metropolitan Life built Parkchester and Stuyvesant Town, Peter Cooper and Riverton in New York, Parkfairfax outside Washington, Parkmerced and Parklabrea in Los Angeles; and the nation's first major integrated middle-income apartment house— Chicago's Lake Meadows—was put up by New York Life, which also sponsored at Fresh Meadows in Queens what may be the only high-rise apartments that Lewis Mumford ever liked. But they all found there was neither profit nor public praise in such ventures—as Lou Winnick puts it, "They couldn't take the heat." Then they turned to lending, with equity participations in the projects to give them their hedge against inflation, and presently they were looking at great heaps of multifamily housing, apartments for rent and for sale, which they had never expected to own but by God now they did. No central office can know enough to make housing investment decisions all over the United States.

More than two-thirds of the population of New York lives in apartment houses of one kind or another; more than two-thirds of the population of Philadelphia lives in single-family housing. In some cities, neighborhoods have been very durable: what was the best place to live fifty years ago in Baltimore, Kansas City, San Francisco, and Charlotte is still the best place to live today; what was best fifty years ago in Boston, Detroit, and St. Louis is today a slum. Land prices vary astonishingly from city to city: building lots which analysis would find comparable in size, amnenity, and neighborhood status sell in Dallas for less than half their price in Houston. Construction costs are significantly different in different locales: "We can't build our Lincolnshire line around here for less than $55,000," says James Sattler of Toledo's panelized Scholz Homes; "elsewhere, you might be able to get it for $45,000." Even money costs are by no means the same. "Mortgage rates can be different in Springfield and Worcester," says economist Saul Klaman, president of the National Association of Mutual Savings Banks. He can't explain it either.

This is a big, complicated subject in a big, complicated country.

Chapter 2

The Object

You say the word "house" and it means so many different things to different people. One person sees it on an open road in the countryside; another sees a village. One thinks of a farm; another of a cliff dwelling in the urban landscape. Environment is a culture and culture is archetypal; it grows from deep within you, embodies long-lived feelings toward shelter, family, community, and self.

—Moshe Safdie, Israeli architect of
Habitat at Expo 67, Montreal

The diseases of housing rival those in pathology. They include irritations over spatial, physical, and financial limitations. They are involved with neighborhood tensions, the shortcomings of neighborhood schools, transportation, and police protection; lack of proper playgrounds, parks, and open spaces; noise, smoke, smells, smog, drafts, dirt, insects, and vermin. The personal vexations of the housing problem are not only multiple and complex but they defy categorization.

—Charles Abrams, lawyer, housing
commissioner, planner, teacher

The house was at the end of a dirt road . . . which was lined with tall trees and dusty shrubs. . . . The house was fashioned of gray stone, somewhat in the manner of a peasant house, and the outside was not especially lovely. There was a tin roof and a weatherworn wooden porch, with several old chairs set against the wall. "You can't imagine how many rubles I had to pay for this," said Sarkis, "and I will never tell you. But it is charming, is it not? Imagine living in an apartment house when one could live in such a house!"

—Michael Arlen, reporting on his
visit to Soviet Armenia

A few years ago, while he was still a partner in the Wall Street house of E. M. Warburg & Co., unwittingly on his way to the job of New York State Commissioner of Housing (and later U.S. Comptroller of the Currency), John Heimann observed that when they think about housing most people make one crucial mistake. They consider housing a product—a thing, a house—while in reality housing is a service. The physical structure is part of the service, and the most important part, but it expresses the totality of the home about as well as the courtroom and the law library express the totality of legal services, or the classroom and the playing field say all that has to be said about education. "This is a nameless industry for providing shelter," says Arthur Holden, a New York architect who has been active in housing and monetary reform movements for more than fifty years. "Building is the repair section of the nameless industry."

Perhaps 250,000 of the new homes built in American every year are private enterprise in the most restricted sense of the term: they are put up by the people who will live in them, sometimes with the help of pre-cut kits and almost always with contributions from licensed electricians and plumbers, as the law requires. ("Perhaps" is a necessary qualifier for the number: the official figures are based mostly on permits issued by local building authorities, and such permits may represent only added rooms or even swimming pools, not new homes.) This is not, of course, an exclusively American phenomenon: most housing in undeveloped countries is put up without "builders," and in overtaxed countries like Sweden and Finland tradesmen compete for summer work in the upcountry, where people building their own summer homes are always happy to put some unreported cash in the pockets of workers who can help out with the things that take training. There is reason to believe that a similar phenomenon is commonplace in the Soviet Union: while there in 1976, I picked up a poster denouncing managers who permit workers to steal materials from building sites. In America, too, incidentally, the do-it-yourself housing movement may have a help-yourself component: large builders in the Houston area send a truck around every morning stocked to replace what may have been swiped overnight.

Another hundred thousand or so American homes are produced annually for sale by part-time builders. Robert Prendergast, a West Virginia broker and developer who would later become president of that state's Board of Realtors, went off to Dallas to try his luck in 1954: "There were so many speculative houses that had been built by policemen and firemen in their spare time. . . . I didn't find it a bed of roses." Bob Broadway of Charlotte, N.C., speaks of builders "who've got a garage and a pickup truck, and they're in business. The banks—you'd think they'd have learned, but they haven't—will put up $100,000, $200,000. . ."

The distribution list of Sweet's Catalogue Files indicates about twenty thousand full-time in-business homebuilders to produce the other additions to the housing stock, additions oscillating between one and two million units a year. From one hundred (in 1974) to two hundred (in 1977) building firms put up more than a thousand homes a year, and a thousand homes is a big business—gross revenues of $45 million a year, if the average sale is at the median price, pretax profits of $2 million to $4 million, though profits for most builders run on the yo-yo with the number of units, and bankruptcies are by no means uncommon. The hundred largest builders all together account for only a few more new homes than the 250,000 do-it-yourselfers: small operations dominate this industry.

Much American housing has problems of workmanship and of esthetics. Too many builders will use moist lumber rather than lose a day in the schedule, thereby guaranteeing a creaking floor from subsequent shrinkage. Joseph Ciskowski, research director of Jim Walter Corp., says a man would be hard put to find a new house in America where you can lay a builder's square in the corner of any room and not find some space around it—"and nobody knows," he adds, "what the wiring is behind those gypsum walls." A man who sells bathroom fixtures to builders says, "Quality doesn't get you anything with these people: they'll cut your throat for a quarter." Concrete is poured every day from mixes that have not been given a routine slump test—and is cured in temperatures too low or atmospheres too dry to permit proper hydration. Every so often terraces fall off condominiums, and in Boston and Washington entire apartment houses have collapsed because of failures to supervise concrete work. Roofs leak; so do cellars. On swampy land throughout the country, from New York's Queens and Staten Island to Houston, homes are settling into the ground for failure to perform proper soil tests before building.

"With some honorable and infrequent exceptions," architect Jonathan Barnett writes testily, "developers have been selling the public a very inferior product." That's probably overkill, because there are certainly more proud builders than bums. Most builders like to think of themselves as producers, not as profit-centers, and as sellers of dreams. Home ownership, the J. C. Nichols company of Kansas City proclaimed right after the war in a magniloquent brochure entitled *Your Dream House,* is "an adventure in happiness." By the nature of the business, a builder tends to be still in the neighborhood selling his wares when something goes wrong with the previous batch, which gives incentive to care—incentive heightened by the fact that the small builder's name and personal assets go on the note before the bank makes him a construction loan. Still, Mandell Shimberg of Tampa, a former tract developer now profitably engaged in rescuing the government from insured bad loans on subsidized apartment houses, observes that "it is a big problem in housing, to instill a conscience into builders."

Housing design in America is done almost always without the participation of architects. The head of one of the largest home building corporations likes to say that "homes are designed by geniuses and built by gorillas"—but he considers that he himself really designed the homes his company builds. So does Billy Satterfield in Winston-Salem. "I never fool with architects," he says, very contentedly. "There's no way they would listen to me because they've had too much education."

The hostility between builders and architects is deeply felt: "as a young architect," Frank Lloyd Wright wrote, ". . . my lot was cast with an inebriate lot of criminals called builders." The economic basis of the hostility is the fact that the architect's fee is the element of expense the builder can most easily eliminate, turning instead to floor plans and elevations in books and catalogues. (One enterprising publisher offers a Custom Homes Plans Club, a House-of-the-Month service, each house complete with twelve sets of working drawings "drawn to FHA and VA specifications.") Even the apartment houses, which by law in most jurisdictions must be built from designs "sealed" by registered architects and engineers, tend to be put together by the builder from catalogue parts, with the architect collecting his fee for making some drawings and signing his name.

Structural engineer E. Harrison Rhame of Cocoa Beach says that when a builder gets drawings and cost estimates, "one of the first things he does is call in his subs [subcontractors] and says, 'Can you think of any way we can save some money on this?' And the sub always says, Oh, there are some places . . .'" Architect Louis Davis, designer of, among others, the striking Waterside project below the United Nations on the East River in New York, says bitterly (exempting the builder of Waterside), "The end product is of no consequence to these people." Or, incidentally, to a government that promulgated an FHA regulation insisting that "Elements of elaboration, of decorative effect and special equipment constitute *luxury* quality."

Builders say that architects do bonehead things like calling for pipes behind bathroom cabinets that fill the wall cavity. "There aren't many architects around," says Jesse Harris, a Dallas builder, "who know you can't get a four-inch stack in a four-inch wall. When I work with an architect, I make him a present of an electric eraser, because it's a lot cheaper to make changes that way." Workmen bear out the accusation. One of the reasons builders should use union carpenters who have been through the established four-year apprenticeship program, says Donald Danielson of the Carpenters & Joiners, "is that a trained carpenter knows enough to correct the mistakes on the drawings—make the modifications so the plumbing can go in, the wiring can go in." Worst of all, say the builders, architects don't care about the public. "Very few of them are willing to follow it through to the marketplace and see if people want to buy it," says Carson Cowherd of Kansas City.

"I would not give a famous architect a dog house to build," says David Rose of New York's Rose Associates, who has been a builder for sixty years and has employed a variety of architects. "Why not? Because when I build a dog house I'm interested in the dog. He's not." It's unfair (not least because builders can forget the dog, too—David's nephew Fred Rose talks about "all sorts of housing built simply for tax purposes, with never a thought about anybody living there"), but it's by no means false. The architect of a residence to be occupied by people he doesn't know, purchaser or renter, feels torn between his Art and the builder's Mammon, and in that titanic struggle the little household gods get trampled.

Not infrequently, especially in expensive apartment projects, Art and Mammon wind up in cahoots. Reporting a conference on *Who Designs America?*, Eric Larrabee noted that architect "Victor Gruen was reminded of the time when the builder of the Lake Shore Apartments in Chicago, the late Herbert Greenwald, expressed his admiration for Mies van der Rohe, who designed them. 'You know,' he reported Greenwald as saying, 'he's a genius, and not only that, I can build him for $2 less per square foot than any other architect.' Mr. Gruen was very surprised by this, in view of the expensively pure exterior of the towers. 'That's true,' he recalled Greenwald replying, 'but inside he lets me do what I want.'"

Gruen continued, "When you went inside the Lake Shore buildings, it became evident. The elevators were the cheapest thing in Chicago. Except in Greenwald's own apartment, the heating was terrible. The air conditioning was nonexistent, in spite of the completely glass walls—it had to be left out at the last minute, because something had to give—and the people in these apartments had to carry their garbage cans across the halls to the stairways, which would not be acceptable in a public housing project, and generally speaking the inside was shoddy and shabby. Yet you could find enough people who were, interestingly enough, so artistic that they wanted to live in that sculpture even with all the unpleasantness—like, for example, frying and freezing in it and being blinded by the sun."

But when builders do their own designs they tend to replicate what has been selling for others, and they tack meaningless gables onto their roofs or Spanish arches before a tiny front porch. "You have to be a little innovative," said Norvin Knutson of Milwaukee, a bearded blond young man explaining his business to his peers at a homebuilders' convention; "look around and see what other people are doing. . . ." Left alone, most builders will level the hills and fill the streams and lay out gridwork straight to the horizon, because you get the most homes to the acre when the land is flat and the lots are rectangular.

What is needed and what is wanted both vary greatly from one part of the country to another. In the Northeast and especially the Middle West, people like basements under their homes; and these basements, ready to be finished as rumpus rooms or family rooms, may add five or six hundred square feet of usable floor space to an apparently small home. You might as well build a basement in the Middle West, because the foundation has to be dug beneath the frost line anyway. In Texas and California, almost nobody has a basement: soil and climate conditions are such that basement walls would have to be impervious to water beyond the normal capacities of concrete to make the space usable. Jesse Harris of Dallas recently recalled the trumpeted entrance and fast exit from his city of "a builder from the North who was going to capture the market with a full basement—nobody in Dallas was offering a full basement. The first time we had a rainstorm, his customers had full basements, all right." In most of the South, most homes are built on piers, with a crawl space between the floor joists and the unprepared ground that will be masked in better homes by a skirt of siding.

Houses with carports rather than garages are acceptable to the middle-income market as far north as Chicago, but not in California. "People here want garages," said Frank Hughes, who ran homebuilding operations for the gigantic Irvine Ranch south of Los Angeles. "The demand is for the car to be in the living room with them—that's what they think of their cars." In the East, a Formica-surfaced kitchen helps to sell a house, but along the Pacific that's declassé; the fashionable *cuisiniére* and her imitators demand ceramic tile surfaces.

The hard plaster called stucco, which can be impregnated with color before it is applied, is the most practical facing for a house in areas where frost is rare—but builders slap up brick veneers, because lots of people agree with the third little pig (who was able to laugh at the big bad wolf because he built his house of brick). The great bulk of American housing, including low-rise "garden apartments," is built up from a wood frame (this is a necessity in California, where earthquakes shatter load-bearing masonry walls), but in Florida, despite the proximity of the sturdiest American construction lumber (Southern yellow pine, especially Dade County pine), virtually all new single family homes these days are built with lightweight concrete blocks. Nobody remembers why.

"To a large extent," says Steve Messner, who teaches real estate for the University of Connecticut at Storrs, "the kind of house that goes up relates to the local labor market. When I first came here, I wondered, Why don't they have forced air heat, heat pumps, and air conditioning—the way we do it in the midwest? But it turns out they don't have anybody around here who can do the sheet-metal work, whereas any plumber can do the radiators."

Pitched roofs are useful to prevent accumulations of rainwater, and steeply pitched roofs are recommended where there is any danger of accumulating snow—but the most steeply pitched in America are to be found in Southern California, the driest and warmest section of the country, because architects there have had a love affair with cedar shakes. (The roofs look peculiar to visitors from the rest of the country, because there are no gutters at the eaves: not enough water to justify catching it and conveying it to rainspouts.) Thanks to the quest for originality in the California roofs, California builders operating in the most thoroughly unionized building market in the country, where on-site labor is the most expensive, cannot use preassembled roof trusses and must hire carpenters to build up their roofs from rafters in the old-fashioned way. In some California cities, of course, it's the union rules that forbid the use of roof trusses.

Building patterns are localized in part because locations come one place at a time: topography, soil conditions, climate vary from piece of land to piece of land. In general, states have permitted localities to create their own building codes, and even the twenty states that have now adopted one or another of the three proposed national codes usually permit local interpretation of the rules. "A lot of our builders build in all five of the communities in the valley," said Richard Mettler, director of the Home Builders Association of Phoenix, "and you have to get different bids in all five communities. They have a uniform building code, but the interpretations aren't uniform. You complain to Phoenix, they say Tempe or Mesa doesn't know what it's doing. If you got a true uniform code it would probably take the toughest section from each interpretation, so I don't fight for it.''

Technical standards vary for reasons nobody pretends to understand. Jens Holm, who now runs the Jespersen modular building operation in Denmark, spent a dozen years as boss of a Portland cement factory in Allentown, Pa., and remembers that ''we had sixteen sealed silos of cement, made for different specifications.'' Often enough, the building codes reflect not local physical conditions but the relative power of the unions in each state or municipality—New York's prohibition of prefabricated ''plumbing trees,'' and the insistence that each pipe joint be hand-wiped on the site, being perhaps the most impressive illustration. This problem is hard to solve: the British abolished local bylaws in 1965 and instituted National Building Regulations, and the local inspectors kept doing just what they'd done before. ''There's always some little Hitler who wants to make sure he gets his own way,'' says C. D. Mitchell of the British homebuilders association.

''I went to the Merchant Marine Academy,'' says Kenneth Hofmann, who started life as a plasterer's apprentice and now produces about a thousand units a year in the Walnut Creek suburb of San Francisco. He is a stout man with gray hair and a gravelly voice, at home these days behind a big desk, wearing a gray flannel suit with extremely wide lapels and bold stitching. ''I took a course in navigation. If you take *The Practical Navigator* by Bowditch, from beginning to end, that's all there is to the subject. Housing—leaving out the social aspect—well, you could say it all in one-third of Bowditch. It's simple.'' But if it were really that simple, Hofmann would not need the computerized cost accounting that enables him to tell a visitor, for example, that his average house in 1975 absorbed 185 man-hours of on-site work, or that a Bay Area union carpenter that year cost a contractor $32,000 in salary, employment taxes, and fringe benefits, though the man himself took home only $13,212 ''to buy our house.''

As the only manufactured good produced by skilled workers brought specially and briefly to the production site, in an inflexible

sequence that may nor may not move at a real-time pace, homebuilding presents some of the most complicated and sensitive sched-
uling problems in industrial production. But businessmen can fail
to resolve them and survive, because the costs incurred can be
loaded onto the mortgage and paid by the customer slowly, over
thirty years, in depreciating currency.

Archy the cockroach poet once lamented to Mehitabel the hooker
cat that "both our professions are being ruined by amateurs." If
there had been housing people in his circle, he could have included
many more professions. "Anybody who hasn't made a success in
his own business jumps into the real estate business," says San
Francisco's eighty-odd but essentially ageless Benjamin Swig, operator of the Fairmount Hotel and a former chairman of that region's
Federal Home Loan Bank. (Also a sometime partner of high-flyer
Bill Zeckendorf—"smartest real estate man I ever saw in my life";
but it was Swig who walked away with money, like a good builder,
when Zeckendorf crashed.) Housing looks so easy, and so high a
proportion of the players can use other people's money. As a business that rewards genius but not talent, it draws too many untalented
people. And there is inexperience in the work force as well as in the
office: at least 70 percent of the new housing units produced in
America each year are built by non-union carpenters and masons.

Lack of expertise in the supporting and ancillary sections of the
nameless industry does more harm, however, than amateurishness
in the production of the house. "The historical basis for construction," says Gerald F. Prange of the National Forest Products Association, "is that it's almost impossible to do a bad job. The house
that looks like a schlocky job really has the same strong structure.
That's why a guy doesn't worry about whether he has three or four
nails in a joint: he never gets a callback on that, it's always the leaky
basement or the slab that settled in a corner." It will be noted that
Mr. Prange speaks for the lumber industry, and that the examples he
gives of defective construction are both masonry faults. Housing is
like that: full of motes in other people's eyes.

A home, to begin at the beginning, is a hugely complicated
Thing. It houses cooking, eating, washing, dressing and undress-

ing, playing, reading, entertaining and being entertained, sleeping, copulating, defecating. Its esthetics probably do say something about its occupants, because there is always a choice. ("People don't buy a house as an investment," says Jim Guest, a tweed-jacketed former aerospace engineer who transformed himself into the biggest real estate broker in Cocoa Beach, Florida. "It's bought emotionally. If Mama likes a pink bathroom and it's got a pink bathroom, she will buy it. If it has a blue bathroom that a gallon of paint will change, you may have trouble selling it.") Its furniture certainly says something about the residents, and the housekeeping quality says more. Care of lawn and shrubbery, fences and paint will be significant, as homeowners' associations are quick to point out where there are such things. A home has an expressive function, inward and outward; home is where the heart is, and the pocketbook, too.

Still, the practicalities are preconditions. As the early American landscape architect Andrew Jackson Downing wrote before the Civil War, "however full of ornament or luxury a house may be, if its apartments do not afford that convenience, comfort, and adaptation to human wants, which the habits of those who live in it demand, it must always fail to satisfy."

In his 1953 book *The House,* Robert Woods Kennedy described "the human states with which the architect is actually dealing" as *going, being,* and *doing.* To help architects get a grip on the spaces that will be necessary—and "necessary" is the attitude, for Kennedy is designing with rich families in mind, the devil with the costs—he describes the various zones of the house in terms of the activities they will contain. The typology offers five zones:

- public, where guests and others enter, "the buffer between the world and the family";
- social, where "the family gathers by itself and where it is, as a unit, private from the world," though guests too are entertained here;
- operative, for "washing clothes . . . cooking . . . eating . . . child care . . . all of the servants' [sic] living functions;
- semiprivate, where "spaces are considered as the personal property of others . . . , divisions emphasize sex differentiation, and

. . . depending on the circumstances, it can offer embarrassing surprises, delightful moments alone, or moments of significant human interchange'';
* private, for ''sleeping, studying, copulating, excreting, certain kinds of creating, doing nothing, and maintenance.''

For each of the zones, Kennedy considers ten elements: Privacy, Communication, Significance, Noise, Sight, Smell, Activities, Spaces, Location, and Use Frequency. Circulation is important —how do guests get from the front door to the garden? how does mother get from the kitchen to where the children are playing? how does the food get to the table? the laundry to the washing machine? (American domestic design was until recently pretty consistently faulty in circulation patterns, the horror story being the ''railroad flat,'' where residents could get to one bedroom only by going through another—known to Negroes as a ''shotgun apartment,'' because a single blast from a shotgun could kill everybody in the place.) But at the same time, the zones must be reasonably isolated from each other—indeed, the conflict to be balanced by circulation design is between wasted space (which probably implies wasted time) and loss of privacy. The obvious solution is the two-story house, with circulation on the stairway from the public, social, and operative zones to most of the more private areas. And women like stairs: ''One is seen and sees people on stairs in motion,'' Kennedy writes, ''which is in itself fascinating. Variations in height are tremendously potent, subjectively as they produce sensations, and objectively as they create interesting effects.''

But that is only for women who can afford interesting effects. For the others, Kennedy notes, ''Stairs use up about ten times as much energy as the equivalent distance on a horizontal plane. Worst of all, in the crisis situations with which we all must occasionally deal, such as a heart disease or a broken leg, the needless stairs and corkscrew circulations of the average house demand physical and emotional tolls out of all proportion to the flimsy excuses which justify them. To the time and energy elements in this equation, we must add another, that of our essentially precarious balance. . . .''

So Kennedy's solution, when he could get it, was to set the two types of zones on the same plane at an angle to each other, defining two sides of an open space behind. Because the house bent, as it were, in the middle, it could be related to land contours in ways that were impossible for the rectangular box. Kennedy here was essentially (and proudly) following Frank Lloyd Wright, and the heart's ease the American upper middle class has missed through the failure of our architects and builders to adopt such plans can be fully understood only by someone lucky enough to have spent a little time inside the satisfactions of one of the great Wright homes—my own experience was in the glorious house he nestled into a hillside for education professor Paul Hanna at Stanford.

Another problem tackled directly by Kennedy (following Wright) was the need for both active and formal social spaces, and it was resolved by what we would now call—the phrase did not exist in 1953—"the family room." For Wright and Kennedy, as for most American builders in the 1970s, the central feature of this room was the kitchen, to be separated off from the social area by a counter rather than a wall. (Kennedy, like Wright, knew that servants were disappearing; it was just that his clients didn't.) The origins of this room lay in the old farm kitchen, the center of home life in rural America, which Kennedy speculated had been cut down in suburban housing because WASPs were a little ashamed of their uncultured farmer forebears. Wright would have done away with the "provincial squeamishness that made the American parlor," and put all formal entertaining, too, in the family room—as many families do. Kennedy kept a living room, for he was designing for the American upper class, which wants such a facility.

The break between the lower and upper middle class in contemporary homes comes here: in the provision or lack of a separate formal area. Love of the parlor persists in America, however little the living room is used and however funereal it may look, with its textured carpeting and brocaded upholstery covered by transparent vinyl. And there is a strong housekeeper's logic to it, for one of the most tedious activities known to woman is cleaning the kids' junk out of the living room so the adults can "have company." The failure of apartment design to provide family rooms separate from living rooms, even in the most expensive buildings, is among the

late 1940s, quite suddenly, more capacious plumbing to handle the immense rush of water from the toilets when the Milton Berle show ended. . . .

Sweet's Catalogue Files, the McGraw-Hill publishing venture that organizes the American construction industry, offers easy access to ready-made components. "This is in fact how these construction projects are put together," says Irving Gross, New York regional manager for Sweet's. "You don't go to K-Mart; you shop for it out of catalogues." Specifications solemnly printed on architects' construction documents can be frighteningly detailed, earning oohs and aahs from the investor or the sales-oriented developer, but they are usually copied from a catalogue.

We are still far from exhausting the list of skills and occupations. A house sits on a piece of land on soil of knowable plasticity: someone should take samples and analyze. Fellows with theodolites map precisely the contours of the land, and a "slope analysis" spots unbuildable areas. Someone must plat the land and site the houses, keeping drainage areas clear, avoiding south and west exposures in the sun belt, north-facing hills in snow country. (Especially with multifamily plans, there may be unrecognized trade-offs in the solutions to this problem. "We play with cardboard boxes," says the Finnish planner Pekka Sevula, "to get maximum sunlight without too much visibility from one apartment to another.") Houses can be placed to minimize or accept noise from adjacent freeways or railroads. Custom or law often forces a front lawn of considerable dimensions, to the irritation of builders and land-planners. "That twenty-five-foot setback!" Tom Hodges of Little Rock said scornfully to a home builders' meeting. "All the kids can do on that lawn is run into the street and get killed."

The list of support systems stretches out far beyond our horizons. Transportation systems (i.e., roads, for ninety-odd percent of Americans), schools, police, garbage pickup, parks, libraries, shops, theaters, restaurants—all affect people's feelings about where they live. And let us not forget taxes—or the relationship between the amount of taxes paid and the real estate values derived from the quality of the services the taxes buy. The introduction of a sewer system makes underlying land more valuable (though sometimes only in ways not everyone likes, by opening up for development at seven units an acre land that was once zoned for two-acre lots).

The fire department, for institutional reasons, may be the most important service of all. Fifteen or so years ago in Denver, I was struck by the squeezingly high density of housing on one side of a street, with nothing but the local burned-out grass and litter blowing about on the other side. My host explained that the Denver fire department serviced houses only up to that street, and the county on the other side had only some remote volunteer units; thus the insurance companies wouldn't write policies for homes on the wrong side, and the lending institutions wouldn't write mortgages on homes without fire insurance. Nothing, not even a rape, panics a neighborhood so profoundly as suspicion directed against the competence or efficiency of the fire service: note the paranoia in Chicago's Woodlawn section, where the Ford Foundation-sponsored community group believes landlords are systematically torching their properties for insurance payments, with the connivance of the Chicago fire department. Fire prevention codes are separate from other building codes, and they may be separately enforced by firemen, on the grounds that firemen will fear for their own and their colleagues' lives if they pass a violation, while the building inspector's self-interest may run in other directions.

All this, too, is part of "housing."

There is no single "housing market" like the stock market or the meat market, because the units to be bought and sold are big, unique, and immovable.

Housing markets may be segregated by age, family status, location in terms ranging from gross geographical difference (section of the country) to neighborhood distinction (side of the tracks or the freeway, or just "north of the city"), employment patterns, tastes, and income. And by race.

In the aftermath of the boom in the early part of this decade, we won't need new rental apartments in Orlando or Sarasota or Fort Lauderdale for at least five years; we need them now and will get them, because the numbers work out, in San Diego and Houston;

we need them now and won't get them in New York, because the
rents that can be projected for new apartments in New York (even
with rent controls on the old apartments, which pushes up rents for
the new) don't begin to cover the costs of building them and operating them. What can it mean to talk about a national "rental apartment market" or "multifamily starts?"

Young people, both singles and childless couples, will happily
pay for "life style" from Marina del Rey in Los Angeles to the
North End in Boston. Ambulatory "senior citizens" may be looking for a townhouse near a golf course in a "leisure village," or a
"double-wide" in a "mobile home community," or an efficiency
in an apartment house specially built for the elderly, where meals
can be eaten if you like in a communal dining room and there are activities throughout the day. Families with young children want their
own piece of land easily visible from the house, at least two baths
(one private for the master bedroom), handy washers and dryers,
more storage space, and conveniences of layout and equipment. "A
single-family house," Robert Woods Kennedy wrote, "is by and
large a specialized building for the production, care, and education
of children."

Even these obvious categories fail to include statistically significant numbers of individuals. My wife and I (I am statistically significant to myself) brought up our boys on Manhattan Island, and
wouldn't have had it any other way. That's how we were brought
up. Developers of "adult" projects find that their one-bedroom-
and-den units become two-bedroom-with-children units unless they
establish a firm rule against kids. Meanwhile, suburban builders are
bewildered by the number of three-bedroom detached homes they
have been selling in recent years to childless couples or even singles. (Nationally, one house in twelve is sold to a currently un-
married female.) The elderly insist on remaining in the homes they
have occupied for years, and refuse to cooperate with renovation
programs because they're not very interested in the resale value or
future occupancy of the house and don't wish to go back into debt.
They raise the political specter of "gray power" to gain exemptions
from real estate taxes.

Tastes change. Among the signs of functional obsolescence that reduce the value of residential property, Stanley McMichael wrote in 1951, are "unnecessarily thick walls, large rooms, high ceilings, gaudy decorations . . . filigree wooden porches, bays, towers and steeples." Nor is McMichael simply philistine: Lewis Mumford wrote of industrialized ornament (the only kind ever applied in quantity to housing) as "novelties of contortion in wood, unique in ugliness and imbecile in design." But such stuff is pure gold today in San Francisco, Minneapolis, Washington, New York, and Boston.

Every broker winces when he first sees an architectural prizewinner of a house. "The more 'artistic' the house," says Jackson Wells of Coldwell, Banker in San Francisco, "the longer it takes to sell; the more traditional, the more 'homey,' the easier it is to sell." The first customer to come by after that artistic house is listed, however, may be a painter who married money (as lots of painters do), for whom that three-story-high north window means not a chance to freeze but a chance to create.

Builders are always being criticized for producing cookie-cutter three-bedroom, two-bath houses, with the design talent if any being used to make the outsides of identical houses look somewhat different. But builders know in their bones, they can still feel the bruises, that targeting a market for a house is a high-risk activity. Among the established maxims is an insistence that "pioneers die broke." If for some reason a builder misses his target, he may be left with unsold, almost unsalable inventory and the interest payments on the construction loan falling due monthly, like the slow tolling of a funeral bell. And the reason for missing the target may be something as simple as the arrival on the market of a bunch of homes from another builder, aimed at the same prospective purchasers. "There's been a feasibility study," says Jack Pullen of the Federal Home Loan Bank in San Francisco, "and the study said there was a need for two hundred condominium units around a tennis club. Five lenders each financed construction of one hundred units, being prudent, building for only half the need. In this business, each builder and lender works in a vacuum."

At their least ambitious, American homes today are spectacularly more comfortable than those of only forty years ago. At that time,

Abraham Goldfeld, a pioneer manager of housing projects for the poor, accepted "the belief current in housing circles that 1½ persons per room is acceptable"; today the standard of crowding *internationally* is anything more than one person per room, and Americans expect better. Recalling his early days in the homebuilding business in North Carolina, Robert P. Cunningham, now supervisor of FHA's Minimum Property Standards, remembers "a skeleton of a house, often with an unfinished second floor—the cliché was that the second story would be completed in time for the daughter's wedding. Pine floors were put in initially, with the hope that later you'd put in an oak floor. There was no hot water except for a jacket on the back of the kitchen stove, which was a wood stove; and even when there was circulating hot water, it was through a 'laundry stove' rather than a real heater. There weren't more than three or four homes in my town that had automatic heating—we called it the Iron Fireman. But in all things, the luxury of one time becomes necessary to marketability at another time."

Relatively few of each year's new housing units—ten to fifteen percent of the total—are purchased under government insurance or guarantee programs. Jim Walter Homes, which has produced more low-income houses than any other "builder" during the last fifteen years—over 120,000 units, scattered through rural America mostly south of the Mason-Dixon line—wants no part of FHA regulations, because a good piece of its business is the partly finished or "shell" home to be completed by its blue-collar purchaser, and also because the company makes its profits from finance charges well above the interest rates customers would pay on FHA mortgages. Most "low-end" new housing, however, is controlled by FHA rules, because some potential purchasers would find it hard to borrow without government insurance of the mortgage and it's worth the builders' work to keep that group in the market for the product, even though most sales will turn out to be on "conventional" (i.e., nongovernmental) mortgages.

FHA property standards must demand more house than the average citizen enjoys at the time the insurance is written. Cunningham's point about luxuries becoming necessities is a matter of future practicality, not nostalgia. An FHA-insured mortgage may run forty years, and because the insurance lies on the house a prudent insurer must ask not "What do people want to live in now?" but "What will people wish to live in a generation from now?" It would be unAmerican not to assume that they will wish something better. The FHA is commanded by law to be prudent and condemned by situation to be American.

Meanwhile, every urban locality adopts what are called Minimum Housing Standards, to protect its residents against unscrupulous builders or landlords, shoddy construction, deterioration, and assorted dangers to health, safety, or comfort. The first such code was adopted in New York in 1867 and did no good at all: far more important was a private prize competition to design a "best tenement," run by a magazine called *Plumbing and Sanitary Engineer* in 1878, and won by the horrendous "dumbbell" design, a building to occupy about 80 percent of a 25 by 100-foot lot—four apartments to a floor, fourteen rooms, ten of them with windows on airshafts. Eighty-two thousand of these were built in the city before further construction was outlawed in 1901; according to Roger Starr, Housing and Development Administrator for the city in the mid-1970s, thirty thousand still survive.

With the "new law" of 1901, government in New York took the plunge into the enforcement of housing standards substantially higher than the existing condition of low-income housing. Strongly backed by Theodore Roosevelt, who as governor of the state had shepherded it through the legislature, the law was a triumph of middle-class political Progressivism. It led to Lawrence Veiller's National Housing Association, and for the first time placed the power of the state behind the demand of poor tenants for minimally decent housing.

Except that it wasn't their demand, and the state couldn't make the law stick. Even the easy part of it—the new standards for multifamily construction—was widely violated. As secretary of the Tenement House Commission, Veiller "discovered only fifteen out of 333 tenements under construction in Manhattan which did not violate the law in some respect, and none in the Bronx, Brooklyn, or Queens." Building as the law demanded would drive rents above the capacity of the tenants to pay. Jacob Riis worried that "our

tenement house reform was taking a shape that ended to make it impossible for anyone not able to pay [up to] $75 to live on Manhattan
Island.''

Even Riis did not anticipate the range of consequences, recently
described by Roger Starr: ''The fact that the new buildings were
somewhat more expensive than the older buildings led their owners
to believe that they would have a difficult time marketing them. As
a result, they tried to make them more attractive to prospective
tenants by including more than the mere legal minimum in amenities . . . [which] added to the cost of construction and operations.
. . . One has merely to compare the skimpy services, the smaller
room sizes, the absence of closet space, the tiny refrigerators, the
unwillingness to include the cost of heating in the rent which characterize European apartment houses with their New York counter-
parts. And it is equally characteristic of American apartment houses
that when they no longer can attract members of the socioeconomic
groups which they were built to serve in the first instance, their deterioration is so rapid that it leads to abandonment at the end of a relatively short period of time.''

Meanwhile, because no new housing was being built for the poor,
crowding persisted and even worsened in the ''old law'' tenements.
Overcrowding, the Tenement House Commission found, ''produced a condition of nervous tension; interfering with the separate-
ness and sacredness of home life; leading to the promiscuous mixing of all ages and sexes in a single room—thus breaking down the
barriers of modesty and conducing to the corruption of the young,
and occasionally to revolting crimes.'' But only cruelty could result
from enforcing the density codes against people who had no other
place to live. Besides, historian Roy Lubove writes, ''Much of the
overcrowding in tenement apartments resulted from the practice of
accepting boarders and lodgers to supplement family income, and
especially to help meet the monthly rent bill. . . . If the Board of
Health really enforced the laws against overcrowding, many immigrant families would have suffered a serious loss of income and
been hard pressed to pay their rent. Objectively, from the point of
view of health, comfort and perhaps morality, the Committee was
justified in demanding that the law prohibit overcrowding in tene-

ment apartments. Yet from the point of view of the immigrant any restriction on his right to accommodate boarders seemed a cruel imposition. . . . Such institutions as the boarder and the sweatshop . . . were, in the final analysis, a stage in the Americanization process. Like Tammany, another undesirable institution in the eyes of the reformers, the boarder and sweatshop helped the immigrant to survive and adapt to American life.''

Ever since those days, the code of Minimum Housing Standards has expressed a terrible, and terribly purposeful, hypocrisy. The standards are set where the middle-class leaders of public opinion believe they should be, and having thus "assured" a higher quality of housing for the poor, the leaders turn their attention elsewhere. A few activists clamor for rigorous code enforcement, some of them even accepting a future when, as New York's Judah Gribetz and Frank Grad wrote in 1966, "the cities will become the landlords of a great mass of uneconomical, deteriorated buildings." According to a Deputy Commissioner of Chicago's Department of Development and Planning, efforts at strict code enforcement were a prime cause of housing abandonment in that city's Woodlawn district from 1960 to 1970, when population declined from 81,000 to 52,000. Mostly, however, the codes simply "require," as housing consultant Anthony Downs puts it, "relatively high minimum quality standards for all households, since that does not 'undemocratically discriminate' against the poor. Then we deliberately fail to enforce these standards so as to accommodate reality." The Boston Redevelopment Authority bitterly notes "a dilemma: how can code inspectors *look* effective (to avoid public criticism), without *being* effective, forcing removal of marginal but habitable dwellings from the housing stock." Meanwhile, however, the standards *are* enforced in the suburbs, keeping out poor people who cannot afford the costs.

Still, over the long run it is probably preferable for housing codes to express hopes rather than realities. No doubt the value of the codes, as of other social prescriptions, has been diminished in recent years by the growth of deliberately "symbolic" legislation which everyone knows is a sop to pressure groups, creating rules that will be administratively subverted even if judicially supported.

Some of the popular new code amendments, with their heavily environmental flavor, are ludicrously inapplicable to real life. "Instead of systematically improving things," says Michael Werner, who resigned as building commissioner in St. Louis to work at the Building Officials and Code Administrators headquarters in Chicago, "you're systematically creating chaos. What worries me is that we're going to draw our definitions so everybody's living in substandard housing." This, too, is not new: James Marston Fitch wrote some years ago that "by the end of the [nineteenth] century, a house without central heating, bath and lights was definitely substandard, even though the vast majority lay precisely in that category." Without the unrealistic codes, we would be much farther back on the road to providing "decent, safe, and sanitary" housing for all those who can in fact discipline their lives sufficiently to keep their homes qualified for such descriptions: to date, our experience has been that this generation's minimum housing codes really do fit the next generation's minimum housing.

Unfortunately, what Americans want to buy when they go home-shopping is not primarily the quality of the structure or even the attractiveness of the "amenities." What they want first of all—what they will spend their money for; what they will demand even before they will let anybody show them a house—is the neighborhood, the place to live.

Part II

THE EXPLODED METROPOLIS

Chapter 3

Situs

If you can tear yourself away from the games of the Circus, you can buy an excellent home at Sora, at Prusino, for what you now pay in Rome to rent a dark garret for a year. And there you will have a little garden . . .

—Juvenal

Every public opinion poll gives the same result: the typical Finn wants to live in a private house on its own plot of land, beside a lake, in the middle of Helsinki.

—Ake Granholm, architect, National
Housing Board, Finland

The value of any parcel of land is determined by three factors. The first is location. The second is location. And the third is location.

—message reputedly found in a cleft
stick in Plymouth Plantations,
c. 1683

Southwest Kansas City is probably the most successful large-scale suburban planning project accomplished within city limits in the United States. "Few people in the world, or in America for that matter," wrote Andre Maurois, "realize that Kansas City is one of the prettiest cities on earth. . . . Why? Because one man wished it so, and insisted upon it." The man was Jessie Clyde Nichols, a barefoot boy from Kansas who had done graduate work at Harvard and later founded the Urban Land Institute. He came to this abandoned area—home of a garbage dump, a battered old harness track,

57

a brick kiln, and a farm where Belgian hares had been raised for butchers—from experience as a builder of clapboard three-story houses on forty-foot lots. This was not very distinguished housing then, and it is now totally obsolete, but it is still viable lower-middle-income housing because the family company, headed these days by Miller Nichols, a vigorous American-Gothic over-seventy son of the founder, has kept the area neat and the planting trim and the corner shops in business through a long chain of financially weakening households. Construction of those houses was completed in 1912, and that year Nichols began accumulating land in what he would later call the Country Club District, though after the earliest years there was not in fact any golf course or even clubhouse associated with the development. Eventually, Nichols would develop 7,500 acres of land—almost twelve square miles—of southwest Kansas City. He started building in 1922.

Nichols' centerpiece was the first automobile-oriented shopping center, all Spanish-style architecture with little towers and red tile roofs, arches, and (especially) fountains, plenty of places to park on sunken lots behind ornamented brick walls, but also grass with sculptures on it. This is five miles from the center of Kansas City. Most buildings were two stories high, with second-floor offices especially for dentists and doctors, whose presence was deliberately encouraged by the construction of a hospital virtually next door. There was no paved road out from downtown, and when the city wouldn't build one—arguing that Nichols or no there would never be enough traffic to justify it—Nichols paid for the paving himself. He also installed, and his successors still operate, a private water department and a nongovernmental fire department, plus a bus service "primarily," Miller Nichols recalls, "for maidservants. Any time you can get rid of a fire department or a bus company," he adds, "do it."

In a sense, the Country Club district was privately zoned. Nichols sold land for apartment houses around the edges of the rectilinear shopping streets in the flat valley to provide customers for the stores, and then began building and selling lots on the surrounding hills. There were three rules: no right angles, follow the contours of the land, and preserve the trees. "Even now," Nichols says, "every new house in one of our areas has three trees; if they're not on the land, we plant them." The land was sold subject to highly restrictive covenants, rquiring driveways and garages, substantial setbacks from the street, foundation planting, and (of course—wouldn't be America otherwise) racial restrictions on resale, these now voided by the courts. In addition, every purchaser would have to join what Nichols called a Homeowners Association, which would take care of the garbage collection and street cleaning in the area, and police the architectural conservatism and property maintenance of all purchasers.

By the late 1920s, when the movie theater opened, southwest had become the most fashionable section of Kansas City, and its newly built high school was among the most ambitious and admired college-preparatory public schools in the country. Nichols himself soon abandoned building and concentrated on the more profitable labors of land development, selling off his parcels with streets, sewer lines, drainage, and water in place. The Country Club District spread like a stain over the rolling hills of this pretty country, and quite naturally passed through the adjacent Kansas-Missouri border without paying too much attention to it. One of the streets in the development was called State Line Road—different from the other streets in that they curved and it was straight, and also in that the cars going north were in Missouri while the cars going south were in Kansas. Much the same houses, on much the same lots, stood on both sides of the road.

The houses on the two sides of State Line Road still look alike, a little large for contemporary America but lovingly tended: a gracefully aging rich suburb. But in fact you can pick up the homes on the Missouri side for ten and twenty percent less than the homes on the Kansas side: the market is much stronger in Kansas. If you bring up your children on the Missouri side, you live with the prospect that one of these days the federal court will order your children bused to the unhappy slums of downtown Kansas City for their schooling; but if you live on the Kansas side you're safely in Shawnee Mission School District, with a high school now rated better than Southwest, and with absolute safety, because there is no judge with a jurisdiction that extends to both sides of the line. Whatever may happen

elsewhere in terms of a court-ordered interchange of students between central city and suburb, the residents of Johnson County,
Kansas will not be touched. (Though, oddly enough, a civil rights
group in Kansas City took a crack at it in 1977.) That's worth cash
money. So virtually all the new homebuilding in this rather prosperous part of the world is located on the Kansas side of the line.

I watched one of these things play itself out near my home turf in
the early 1960s, when I was chairman of a local school board that
had jurisdiction over the huge Stuyvesant Town development in
Manhattan. This collection of tax-break-subsidized lower-middle-
income apartment houses (for upwardly mobile populations) was
served by three public schools, two of them below Fourteenth Street
in an area increasingly black and Puerto Rican, one on Twenty-first
Street near Gramercy Park in a zone that also included the unsubsidized upper-middle-income Peter Cooper project. And as time
passed the families with public-school children concentrated in the
northernmost zone, which gave access to the highly desirable P.S.
40. When we tried to do something about ever-increasing segregation, we were denounced for violating the principle of the neighborhood school: the line in the middle of the project had divided it
into different "neighborhoods." If law or custom had permitted differential rents for the identical apartments in different sections of the
project, Metropolitan Life could have made a bundle in the north.

A prime determinant of the value of a larger home, then, is the
quality—more accurately, the reputation—of the schools. "People
without children are the prime sources of demand for housing in
Washington," says George Peterson of the Urban Institute. "The
townhouses in Northwest have gone up the most in price; the single-
family house with three bedrooms has gone up the least." Where
schools are reputed to be excellent, families with school-age children will outbid others for available homes; where schools are reputed weak, a family with children is likely to lower its bids for
home purchase or rental to leave money for private school fees
(where such institutions exist).

School district is one of many trade-off items, like a swimmng
pool, a larger piece of land, a view, a golf course, easy access to

convenient transportation. Martin Meyerson and colleagues from the citizens pressure group ACTION wrote some years ago that a consumer's housing purchase buys "a package of related goods and services: with the house go schools, churches, shops, visual environment, places to stay, neighbors, status attributes, a municipal administration, a journey to work . . . and even an orientation toward cultural, social and commercial activities—in short, a way of life." True enough; but then the authors go on to say, "The consumer may wish to spend a relatively large share of his income for some items in the package and a relatively small share for others. He must buy the package as a whole, however. . . ." And here the academicians oversimplify, leaving out the fact that anyone with a middle-class income can choose among packages in which different features are salient.

The significance of changes in individual features is most apparent in apartment houses, where absolutely identical units will carry different rents according to how high up they are. The shopper for a detached home often finds himself choosing between, say, a house near the water but in a hollow where it's going to be hot and a house of roughly the same size and equipment up on the hillside, where the views are glorious but you can't walk to anything. Very large developments like Westlake Village outside Los Angeles and Montgomery Village in Maryland will stress the environmental quality of the homeowner's ambience; at the Irvine Ranch, the emphasis is on activities and "life style"; at Winter Park near Orlando, it's the Europeanized restaurants and shopping of an amended Main Street. Some people like greater privacy, others are looking for extended opportunities to "neighbor," a sociologist's verb that works. Real estate brokers plugged into computerized multilisting services can play these games with customers in real time, punching into the computer specific changes in requirements and watching the machine feed back different lists of houses, locations, and prices.

About 40 percent of Americans live today in suburbs, as against less than 30 percent inside the old city boundaries. Suburbs vary vastly from one to the other, covering just about the entire income and social status spectra (there are even some poor black suburbs). Specific neighborhoods tend to be homogeneous because people similarly situated tend to share similar tastes. "Levittowners," Herbert Gans wrote of his sojourn among this mysterious tribe of New Jersey roots-droppers, "wanted homogeneity of age and income—or, rather, they wanted neighbors and friends with common interests and sufficient consensus of values to make for informal and uninhibited relations. Their reasons were motivated neither by antidemocratic feelings nor by an interest in conformity. Children need playmates of the same age, and because child-rearing problems vary with age, mothers like to be near women who have children of similar age. And because these problems also fluctuate with class, they want some similarity of that factor—not homogeneity of occupation and education so much as agreement on the ends and means of caring for child, husband, and home."

Roughly one American household in five moves every year, renters much more often than homeowners, homeowners in Florida and southern California more often than homeowners elsewhere. About two-thirds of all moves are within the same metropolitan area. Despite what the planners say, "convenience" is almost certainly the least important reason to move in any society where automobile ownership is widespread: reporting on residents' attitudes toward "housing estates," a team from the British Housing Development Directorate noted that convenience "emerged as being fairly unimportant: that is, some tenants were satisfied with the neighborhood even though they considered the location of the estate inconvenient—and vice-versa. This finding was contrary to the opinions of the housing managers. . . ." (Convenience is more important to the rich: more than 70 percent of the tenants in the expensive new high-rise apartments in downtown Boston gave "convenience to work" as a "very important factor" in their decision to rent here.) Density counts: lots of people move to get "more room," both in terms of the floor space of their own dwelling and in terms of the neighborhood population per acre. There is also a minority who like high-density living and will put up with a great deal for the pleasures of the city.

Meyerson and colleagues reported that neither physical conditions nor distance from work or shopping pushed people out of their homes, "but some families will change neighborhoods, despite satisfactions with their dewellings, if the social characteristics of their neighbors become obviously different from their own." Studies of Jews in the 1940s and 1950s, Negroes in the 1960s and 1970s have consistently shown that members of minority groups who want an integrated environment define integrated as being 50 percent their own group. But if the two groups are of different sizes and everyone wants at least 50 percent of his neighbors from his own group, the mathematically stable solution to the game will present a high order of segregation and much greater crowding of the smaller group. Anthony Downs stresses the importance of the "absentee market," the potential buyers here who decided to look for homes in some other neighborhood; their actions will be motivated by what economist Kenneth Boulding once called "the image of the image."

The heart of the matter is probably that the suburbs for the past generation have been a better buy. "We looked for a house in San Francisco," says Charles Slatzkin of the real estate development firm of Garson Bakar. "The cheapest we could find was $80,000, and you had to put $20,000 more into it to make it livable. We got the same house in Piedmont for $50,000, and you get sun in the summer and the schools are better." That's San Francisco, of course; across the bay in Oakland, you can get city houses cheap; but then it's the wrong neighborhood.

Because they have sought image, compatibility, and space—and have normally found some component of all three—people who are moving within a metropolitan area feel they have gone "up" and tend to be content after each move. Newcomers, sociologist John Zeisel says, are "more easily satisfied with whatever they have than are housing occupants who have been living in the same place for a long time." Unfortunately, most comment about housing and people comes from the great height of upper-middle-class dissatisfaction with one's own life, solaced a little by the tears of pity that can be shed on the poor ants swarming far below. Mostly, these tears rain acidly on the denizens of the "ticky-tacky boxes" in the suburbs; indeed, a lot of social criticism of American housing can be described as, in Scott Donaldson's words, "a venomous attack on a place in which seventy million Americans now live." Robert Wood printed what is surely the ultimate criticism, of "inadequate, overpriced homes in suburbs inhabited by directionless people who do not know they are unhappy.'' He wrote these words, amusingly, while a resident of Lexington, a Boston suburb: ''My professional opinion,'' he noted in the introduction to his book, ''should never be confused with my personal tastes.'' He didn't know he was unhappy, either.

The interesting fact is not that suburbanites do not know they are unhappy, but that significant elements of satisfaction in one's home can be found in urban areas almost everyone not living there would consider desperate. ''Looking at neighborhoods they had left a decade or two before,'' political scientist Edward Banfield writes, ''suburbanites were often dismayed at what they saw—lawns and shrubbery trampled out, houses unpainted, porches sagging, vacant lots filled with broken bottles and junk. To them—and, of course, even more to the scattering of 'old residents' who for one reason or another remained—these things constituted 'blight' and 'decay.' To the people who were moving into these neighborhoods from old tenements and shanties, however, the situation appeared in a very different light. Many of them cared little or nothing for lawns and had no objection to broken bottles; they knew, too, that the more 'fixed up' things were, the higher rents would be. What mattered most to them was having four or five rooms instead of one or two, plumbing that worked, an inside bathroom that did not have to be shared with strangers down the hall, and central heating. To the least well-off, 'blight' was a blessing. They were able, for the first time in their lives, to occupy housing that was comfortable.''

No doubt being rich is better (much better) but it isn't the end of the world to be poor. ''Slums, they call us,'' said a woman in Milwaukee, explaining her vote against one of the nation's earliest urban redevelopment projects, in 1947 (it failed). ''Why, that's a terrible word—those are our homes, our shrines. We live there.'' Charles Abrams, a New Dealer to his fingertips and a great warrior in the cause of better housing for the poor, once looked back at his own childhood for the benefit of a class he was teaching at Columbia, and said, ''For the family of a pickled-herring vendor in a tenement apartment on the lower East Side amid neighbors and friends who care about them and share their troubles and triumphs, and with all sorts of shops and lively activities right at hand, such a tenement, despite the bathtub in the kitchen and the railroad layout of the place and the insufficiency of sunshine and fresh air, may be (I say, *may* be, mind you) a better, richer housing situation than anything that city planners and bureaucrats are apt to provide as a substitue.''

This, too, is a question of neighborhood—*not* amenities (we make our own amenities in the end), but neighborhood. As the unrich have proved on The Hill in St. Louis, in South Boston, New York's Astoria, Chicago's Back-of-the-Yards, Baltimore's Patterson Park, a place can be kept desirable as a home for many years despite the low-moderate income level of the residents. Barring racial nastiness, only three things will quickly change a residential area: (1) if the citizens are more afraid of the criminals than the criminals are of the police; (2) if learning conditions in the schools are so bad that the life chances of the children are diminished; and (3) if the district becomes so fashionable that it provides an intrusion of what Jane Jacobs called ''cataclysmic money,'' and its present occupants find they can no longer afford their homes. The first two will empty an area of everyone who can get out, and fast; the third will rapidly produce an all-but-complete turnover of population.

Let us not forget (please) that while tenants are a restless lot, the average homeowner moves only once every ten or twelve years. Left alone, people stay put; and they would prefer to be left alone. For reasons too deep to explore—related to the forgotten fact that each of us can be alive only here and only now—thoughts about home bring out the conservatism in almost everyone. We are a mobile people but we want ''roots,'' knowing that rooted life does not move, resenting the flux on which our lives are remorselessly carried. The people who get used to it—Alvin Schorr cites studies of construction workers who live in a succession of trailer camps with solid family structures ''as if geographically stable''—are in families who feel they have a ''real home'' elsewhere to which they return when jobs are scarce.

Time, unfortunately, waits for no man. New technologies make factories dysfunctional; transportation patterns change; structures age; other neighborhoods, for one reason or another, draw off the family next door and ship in a replacement. If the area as a whole is economically successful, the process of change will be faster, not slower. Land economist A. M. Woodruff of Hartford writes of an "iron law" that dictates clearance and reconstruction whenever "the land under a building is worth more without the building than the combined value of land and building for any purpose to which the building can be put." The law fails only when—as in the old western mining camps, inner-ring St. Louis and the South Bronx— the value of *any* structure on the land is perceived as less than the cost of building it.

People have been driven from their homes since the beginnings of urban settlement, sometimes by those who said they were enemies, sometimes by those who said they were friends: many pious Catholics were evicted to enable Bernini to build the great piazza before St. Peter's in Rome. But the modern city, built as it is on the specializalization of labor, is different. Since the industrial revolution, it has been possible to use the land with an *intensity* never before achieved. By the iron law, the grandest mansion any robber baron ever built cannot stand if the land is suitable for an office building, a department store, or even a very large apartment house. And the apparent cost of retaining grassy areas, parkland, or vacant lot grows monstrously as the value of the surrounding property escalates. Cities and suburbs both have grown most rapidly not during the youthful years of "sprawl" (a pejorative word that usually turns out to mean "early settlement") but during the adolescent years of ruthless infill, when the areas between the scattered sprawl sites were built up, and structure, population, and value weighed ever more heavily on the land.

Because modern transportation can bring so many people past a single spot in a given day, it is not even necessary to build significantly on a piece of land to increase its value far beyond anything it could command for residential or recreational purposes. Lyle Fitch and Ruth P. Mack of the Institute of Public Administration report that "in the Philadelphia area . . . land appropriate for filling stations was selling at one time for the equivalent of $100,000 an acre while other land in the same area was selling for $2,000-$3,000 an acre.''

This is quite a different phenomenon from the mobility of, say, the early days in New York, described in 1857 by a committee of the New York State legislature: ''As our wharves became crowded with warehouses, and encompassed with bustle and noise, the wealthier citizens, who peopled old 'Knickerbocker' mansions, near the bay, transferred their residence to streets beyond the din; compensating for remoteness from their counting houses, by the advantage of increased quiet and luxury. Their habitations then passed into the hands, on the one side, of boarding house keepers, on the other, of real estate agents; and here, in its beginning, the tenant house became a real blessing to that class of industrious poor whose small earnings limited their expenses and whose employment in workshops, stores, and about the wharves and thoroughfares, rendered a near residence of much importance. At this period, rents were moderate. . . .''

The Knickerbockers, in short, did not make money out of their move: economic advance had made the neighborhood less attractive, and they left. But observe what happens next, in the same report:

''. . . the rapid march of improvement speedily enhanced the value of the property in the lower wards of the city, and as this took place, rents rose, and accommodations decreased in the same proportion. At first the better class of tenants submitted to retain their single floors, or two and three rooms, at the onerous rates, but this rendered them poorer, and those who were able to do so, followed the example of former proprietors, and emigrated to the upper wards. The spacious dwelling houses then fell before improvements, or languished for a season, as tenant houses of the type which is now the prevailing evil of our city: that is to say, their large rooms were partitioned into several smaller ones (without regard to proper light or ventilation) . . . and they soon became filled, from cellar to garret, with a class of tenantry living from hand to mouth, loose in morals, improvident in habits, degraded or squalid as beggary itself.''

The question of what could be done was not so easy, for it seemed to many that the iron law had been enshrined in an even more iron Constitution, which in its Fifth Amendment proclaimed that nobody could "be deprived of life, liberty, or property, without due process of law" [a clause made applicable to state government as well in the Fourteenth Amendment], "nor shall private property be taken for public use, without just compensation." Restricting the use that could be made of a land parcel was clearly a form of "taking.' So long as it was the rich and their favorite institutions (the churches) that owned the appreciating land—and in the United States until well into the nineteenth century land could not be owned by corporations—there was little pressure to test the Constitution and see just how iron it really was.

Then the economic vitality of New York City got too big for its britches. The owners of the fashionable shops on Fifth Avenue, recently established, found that the land near their emporia had become more valuable to the proprietors of the city's burgeoning clothing factories than it was for commercial use. Massing the Progressive forces that had been writing housing codes for the previous twenty years, Ward McAllister's Four Hundred pushed through the nation's first land use law. They called it "zoning."

The original New York ordinance, a short six pages long, distinguished only among residential, commercial, and industrial uses, "apparently," writes lawyer Richard Babcock, "because the drafters doubted the validity of a classification between single-family and multifamily uses. (A more notable example of the clouded crystal ball is hard to imagine!)" Presently a New Jersey lower court stated precisely the rationale of the system: "Conserving the value of property and encouraging the most appropriate use of the land." The idea of "appropriate" had replaced the idea of "profitable," and zoning was on its way. In 1926, the U.S. Supreme Court went vigorously along, justifying the zoning out of apartment houses from an area by an extension of the old common-law doctrine of nuisance: "A nuisance may be merely a right thing in the wrong place, like a pig in the parlor instead of in the barnyard. . . . Under these circumstances, apartment houses, which in a different environment would be not only entirely unobjectionable but highly desirable, come very near to being nuisances.''

Into the 1950s, the courts gleefully extended the permissible uses of zoning authority. Wisconsin accepted a village's exclusion of a Lutheran high school: ''The presence of the school will lessen the taxable value of nearby homes and will deter the building of new homes in the area.'' Even architectural character could be controlled through the zoning laws. In those swinging phrases that gave his opinions resonance (such as, ''one man, one vote''), Justice Douglas, speaking in a case that involved condemnation rather than zoning, proclaimed that ''The concept of the public welfare is broad and inclusive. The values it represents are spiritual as well as physical, esthetic as well as monetary. . . . If those who govern the District of Columbia decide that the Nation's Capital should be beautiful as well as sanitary, there is nothing in the Fifth Amendment that stands in the way.''

All sorts of restrictions and requirements flowed out along the expanding boundaries of zoning:

Noisome use, obviously—glue factories, airports, and the like.

Commercial use, to prevent big stores and office buildings from inundating areas where the streets can't take the traffic.

Density in terms of land coverage and housing units (number of homes per acre, setbacks from the road, side lots, etc.).

Height and bulk, to preserve light and air and views (the ziggurat skyscraper with its setbacks was an early creation of zoning; today's flat-top is a creation of later zoning, which trades off the release from setback requirements against lesser use of the entire parcel on which the structure sits).

Sewer availability, for obvious reasons.

Off-street parking, at first required to get the cars off the streets (which is still pretty much the story in Europe, incidentally: the subsidized Danish housing at Farum Midpunkt was built under an agreement with the locality that each housing unit would be served with two garage spaces underneath the building, so visitors as well as residents would get their cars out of the way). In America, the environmentalists have now decided that where there is no place for cars there won't be cars, which is true (there won't be people either, but that's not the environmentalists' mission); and now a good deal of zoning for center cities prohibits the provision of parking places.

Esthetic compatibility (even something as vague as "architectural appeal"), approved as a zoning criterion by the Wisconsin Supreme Court because "*Anything* that tends to destroy property values of the inhabitants of the village necessarily adversely affects the prosperity, and therefore the general welfare, of the entire village."

Topographical maintenance, to prevent builders from leveling the hills or building on steep hillsides to endanger the stability of the soil; more recently, to stop the idiocy of building on flood plains.

Hotels and motels, to place the transients away from the settled residents.

Et cetera, very nearly ad infinitum. Recently cities have even used their zoning authority to restrict the whorehouses and smut shops the Supreme Court for some reason has ruled cannot be prohibited *per se*. Virtually all the land use planning that is done in the United States is done under the rubric of the zoning laws.

Examples of triumphant zoning are hard to come by, and most of them are European. (Lewis Mumford says the whole idea is Venetian, citing the cemetery island, the glass-blowers' island, the shipbuilding and armaments area for the Arsenal, and the Lido, "a seashore pleasure resort: a recreational precinct.") European zoning did not seek to separate workplaces from residences (indeed, Mumford is especially pleased with Venetian practice in part because it "minimized the wasteful 'journey to work' "); but the Europeans did set height limitations on buildings, recently and disastrously removed in Paris and London, and might set standards for architectural compatibility, portion of a site that could be built upon, etc.

In recent years, European zoning has been especially effective in preserving historical structures and in maintaining "greenbelts" of farmland around central cities. Thus the growth of Copenhagen has been channeled along five "fingers" of land extending out from knuckles made by transportation nodes in the city, and the Dutch, in the most crowded country in Europe, have kept farmland in production to the well-defined edge of intense urbanization. (Thomas Stamm of Buowcentrum, the building research organization, explains that "People want to see a cow when they go for a walk after dinner.") But in Europe as in America, the conflict between private ownership of land and public zoning powers creates intense political disputes. The Europeans do not have a Fifth Amendment to protect them from their governments, but they do have the leverage of elections, in which intensities of feeling influence the results. The Dutch government, in fact, fell in 1977 on the issue of land use proposals. The British are on their third attempt in two decades to tax away all the profit from the conversion of farm land to residential or industrial use; the first two such laws were repealed. The 1976 Community Land Act—requiring that all land for development be sold to local authorities at a price they set, for resale to developers at the market—also will not survive a return of the Conservatives to office.

The state power that underlies zoning laws is the authority to issue or refuse building permits, and the officials who write and enforce zoning codes are therefore always in a position where they are preventing landowners from doing what they would find profitable or pleasant to do. (In America, the zoning commissioners are typically housewives, storekeepers, and doctors; and building codes are written to a significant degree by city councilmen rather than by engineers. "They are not knowledgeable," says New York housing lawyer Eugene Morris, "and they've got nothing else to do. Nobody can educate them and make money.") The result is that the codes tend to grow ever more detailed and more rigid—in Mumford's Venice, where the Grand Canal is a glorious mishmash of four centuries of architectural style, the zoning policemen of today ban the construction of a home designed by Frank Lloyd Wright for a plot just off the Canal, on the grounds that its style would clash with its neighbors'. And because nearly all zoning codes for already built-up areas permit the continuation of an established "nonconforming use," there is a burr of apparent unfairness ever present beneath the saddle.

In America, the zoning structure itself is built on an ideational swamp, and however much the courts may prop it up, sections keep sinking. Zoning is, to begin with, transparently exclusionist. In a little "Citizen's Guide to Zoning," the Rockefeller Brothers Task

Force on Land Use and Urban Growth tells the potential member of
a zoning board, ''You would be surest of your decision to protect
something nice. ('That's a pleasant neighborhood, and it shouldn't
be changed.') If development within the nice area was uniform, so
much the better: you could just require new development to be like
what was already there. . . . The zoning process is at its best in
those places where the dominant public need is protection of an area
already substantially developed. This is no surprise, since protection is what the originators of zoning had in mind.''

Once past this easy zoning decision, however (the Report continues), ''you would have much more difficulty fashioning regulations'' Indeed. There are, the Report continues, three options:
''(1) 'When in doubt, let 'em do it' . . . (2) 'Pretend that you
know best' . . . (3) 'Wait till you see what happens.' ''

But the task force finds that none of the options yields satisfactory
results. If the zoning board says that the entire farming borderland
of the town is open for residential development, the result may be a
little knot of houses here and a little knot of houses there—
''sprawl''—with greatly increased costs for publicly provided services.

If the board says that development of stated variety can occur just
here and nowhere else, ''you may have created something like a
monopoly in developable land . . . if you had decided that the town
needed one shopping center and had foolishly designated only one
site for it on the zoning map, you could well have driven the land
price so high that no developer could touch it.'' (Unless, that is, the
developer knew in advance that this is what you were going to do,
and picked up the property inexpensively ahead of time. The Metropolitan Council of the Minneapolis-St. Paul region is going to fall
apart on such issues long before completing its arrogantly conceived
master plan—by which, Council director John Boland has said,
''every owner will know everything about his plot; developers will
know a certain plot of land will be used for a grade school in
1986.'')

If the zoning board decides to wait and see, administering what
are quaintly called ''floating zones'' and deciding every application
ad hoc, ''there are problems—by far the most serious and fun-

damental—caused by a lack of trust in the decision-making process and the decision-makers themselves. . . ." In the less genteel language of Professor Donald Hagman of UCLA Law School, "At the local level, where most land use controls are still exercised, land use decisions provide the main opportunity for local government corruption."

Classical "exclusive use" zoning, moreover, has come to look more and more like a disaster for the urban centers, with their dead downtowns at night and dead residential areas by day. The notion that an apartment house is somehow cheapened by ground-floor stores, which was received wisdom from 1920 through 1960 or so, always seemed ludicrous to people who knew Paris (which is where the high-income apartment dwelling got its start: for years good apartments were advertised as "French flats"). Jane Jacobs caught the public imagination in 1961 with a grandiloquently stated case for "mixed zoning" and perpetually lively streets; and it is surely true that if the appeal of the urban life is its diversity, then zoning, with its implications of uniformity, is not going to help the cities.

What has most troubled most critics of zoning, however, is the "easy" decision to maintain a neighborhood at its present standards, for by definition the standards of a middle-class suburb are beyond the purchasing power of poor people, many of them black. "Once you put the lid on housing," says Professor Charles Meyers of Stanford, "the poor get poorer." Anthony Downs says that 99.2 percent of the vacant land zoned for residential use in the New York suburbs is reserved for single-family housing, and that over half the residential land in Connecticut requires at least a one-acre lot per new house. When RCA was unable to get a zoning variance to build a new corporate headquarters on 270 acres in New Canaan, Conn., it sold the land finally to a developer who announced he would put up 125 $225,000 houses on the site, on two-acre lots, which is what the law demanded. Thus is the environment protected.

In sum, Downs claims, unnecessary zoning and building codes in the suburbs make the minimum housing unit there two and a half times as expensive as it need be. That's probably high; but in my own travels in 1974–77, for what it's worth, every reasonably attractive new house I saw for under $30,000 was being built in an area—outside Dallas, Tampa, St. Louis, Salt Lake City—where there was to all intents and purposes no zoning. The worst problem may well be that the fundamental capriciousness of zoning (and environmental) regulation makes homebuilding so risky an occupation that only the most rapacious developers, generating very high profits on the successes to compensate for the cost of the abortions, can survive.

Every so often the courts stir in discomfort, throwing out racial restrictions, or zonings that seem designed purely to preserve one body of commercial interests from competition by another, or rules that require somebody's property to be kept as open space (a "taking" for which the zoning body must pay if that's what it wants to do: Southern California Edison has been awarded several million dollars by a court to compensate for a San Diego zoning change that put a tract the company owned in a no-development category). Several public interest law groups have brought court cases against restrictions that prevent high-density development in the suburbs, but the arguments available are limited, and the lawyers' choice among them may have been unfortunate. Fighting a restriction to five hundred new units a year in the San Francisco suburb of Petaluma, for example, the lawyers based their case on a Constitutional right to travel, a far-fetch rather contemptuously dismissed by Justice Douglas; insisting that Eastlake, Ohio, could not require referendum approval of a zoning change, the lawyers argued that such procedures were an illegitimate delegation of legislative authority to the public, provoking from Chief Justice Burger an irritated reminder—in which Justice Thurgood Marshall joined—that under the American system of government the legislature derives its powers from the people, not vice-versa.

The Supreme Court of Pennsylvania has ruled that "Zoning provisions may not be used . . . to avoid the increased responsibilities and economic burdens which time and natural growth inevitably bring," and that of New Jersey has proclaimed an affirmative duty of that state's higher income townships to make land available for low-cost housing. But the New Jersey court then decided that it can't write new zoning laws to replace the old and thus cannot effectively interfere with the localities except on a case-by-case basis;

and by the time the legal processing is through on a case-by-case basis a developer's costs have risen to the point where he can't afford to build moderately priced housing on that land anyway. "Zoning litigation is expensive," Richard Babcock writes, "because it is conducted almost exclusively through the use of expert testimony . . . by planners, appraisers, traffic engineers, and other experts in the particular line of business involved in the case. Bona fide experts are expensive, and even the phony ones cost something."

Moreover, judges detest this sort of litigation. Babcock quotes as an epigraph to a chapter an opinion of the Michigan Supreme Court, "This is another zoning case." As presented to the courts, zoning challenges are essentially battles between rich adversaries—on the one side, the builder and the landowner whose property will appreciate; on the other side, the householder whose home is worth more if the demand for housing in this suburb is met by a restricted supply and whose taxes will go up if the developer is permitted to build. (Palo Alto, California, commissioned a study of the potential tax impact of new housing in the foothills west of town, and found that it would be cheaper to buy the land for park use than to educate the children of the people who would live there, even if they built expensive homes and paid the appropriate taxes. So Palo Alto got first a new bond issue to pay for the acquisition, and then a new park.)

Each side of the dispute seeks to involve a much wider interest in his behalf; passions run high, and arguments are conducted in the nastiest possible terms. Mary Anne Guitar, in a book significantly entitled *Property Power,* denounces at great length the greed of "the spoilers" and their allies. "Overnight," she writes, "they develop a social conscience and wonder piously about the young couples who won't be able to build modest houses if land is zoned for large lots." Then: "there is little concern for the people who paid their own hard cash for a place they thought would be decently preserved."

John Fischer, for example, former editor of *Harper's Magazine,* proclaimed himself "radicalized" by the threat of development: "The Survivable Society will no longer permit a farmer to convert his meadow into a parking lot any time he likes. He will have to understand that his quick profit may, quite literally, take the bread out of his grandchildren's mouths, and the oxygen from their lungs. For the same reasons, housing developments will not be located where they suit the whim of a real estate speculator or even the convenience of the residents. They will have to go on those few carefully chosen sites where they will do the least damage to the landscape. . . ." But it is quite impossible to imagine a democratic society in which the demands of the landscape will be put before all others. It's a funny kind of radicalism—maybe one should call it "radicalism for the rich," modifying Charles Abrams' famous aphorism—that places the view from a twenty-acre estate over the needs of the people in the cities to get out once in a while, or even (as in the fights over the construction of power lines) to cool off with air conditioning in the summer.

The problem the court has is that both sides are almost always right, especially in their denunciation of the other. Though warriors like Guitar will suggest that really the poor are better off if they can't move to the suburbs, because that will force the government to rebuild the cities (where most poor people are going to live anyway), the conservationists really don't care about the housing of those not already on the ground. And though the builders and their allies in the civil rights movement will speak of Planned Unit Developments that do no harm to the environment, they are really unconcerned about the cherished amenities of the existing suburb—and indeed are occasionally motivated by a desire to pull the temple of the suburbanites down about their ears. Race hate and class hate (sometimes generational hate, as the young lawyers of the poverty program seek to get even with their parents) are easily mobilized; and such tensions are supposed to be controlled by political rather than judicial process.

And there is some question as to what can be accomplished, even in the legislature, so long as building permits are issued and regulations administered by local governments. Massachusetts established a state board of appeals with power to override local zoning codes, and in twenty-two of the first twenty-four cases brought before the board the petitioners overturned local restrictions on the use of their property—but only two of these victories have as yet produced any housing (and in one of the two the apartment house was built in an industrial zone, leaving its supporters wondering what they had been fighting for). In New York, the Urban Development Corporation was originally authorized to override local zoning restrictions and build multifamily housing anywhere in the state; but when UDC announced plans to use these powers in Westchester County the state legislature amended the act and removed the authority.

Even after a municipality is compelled to grant a zoning variance, there are all but endless ways that an unwanted project can be stopped—little elaborations of the building code, sewer requirements, a great range of environmental objections. An interesting example of the imagination that can be employed in these blocking actions is a regulation introduced in 1976 in Prince George's County, Maryland, outside Washington—an area still traumatized from court-ordered school busing—where a builder now must provide special apartments for handicapped people (double-wide doors for wheelchairs, railings, special bathroom equipment, etc.) amounting to at least ten percent of the units in any new multifamily construction. Coupled with the small market for such apartments, the cost of building them should guarantee a halt to multifamily building in the county.

During the last few years, however, a number of well-to-do suburban communities have been amending their codes to permit multifamily housing, and have actually been requiring builders to provide some percentage of units for moderate-income and even low-income people. The purpose is not to release a safety valve for the cities but to assure that municipal employees—teachers, policemen, firemen—can live in the town where they work. Among the towns with such laws, listed in the November 1974 issue of *Planning*, are Cherry Hill, N. J., Lewisboro, N. Y., Lakewood, Colorado, Eden Prairie, Minn., Fremont, Cal., and Arlington, Va. (The first one of these laws that came up for appellate court scrutiny, by the way—from Fairfax County—was thrown out by the Virginia Supreme Court as an illicit "taking" by the zoning authorities.) At this writing, HUD is proposing to deny community development funds to suburbs that are inhospitable to subsidized housing (which the present law may in fact command: lawsuits in Connecticut and Long Island have cut off some federal aid to suburban communities); but many suburbs are beginning to feel that they lose more than they gain, quite apart from questions of exclusion, when they get involved with federal grants.

For the time being, "snob zoning" in the suburbs produces two effects: it gives a little more vitality to the efforts at rehabilitation in the cities, which we shall consider shortly; and it drives increasing proportions of lower-cost building farther out from the cities. Fifteen years ago the resulting "urban sprawl," however distasteful to planners, was looked at more or less benevolently by others, the benefits being less expensive housing for working-class families, while the costs were only a slightly more rapid than "planned" metropolitan expansion. Now, with the drop in the birthrate and with stabilization of the cities probably no more than ten years away, with the edge of urban settlement already so far out, and with public policy committed to reducing the use of gasoline, governmental incentives and disincentives to influence the placement of new housing may be both more desirable and more effective.

But the attitude that government really decides what will happen, which seems to be a kind of infection that spreads quickly among people who attain public office, will make intervention at best ineffective and at worst counterproductive. Population flows and technological advance force change, and except in a handful of wealthy and ferociously organized enclaves (New York's Forest Hills Gardens, San Francisco's St. Francis Wood, Southwest Kansas City, Baltimore's Roland Park), efforts to maintain city or suburb as it is can show no more than temporary and misleading success. The energy in the system is that of the builder, the buyer, and the lender, not that of the government; frustrated in its asserted path, that energy will find outlets through the interstices of any regulation. Policies that impact on costs—land, money, labor, material, and selling costs—will have the most pervasive influence, and can also most easily do harm rather than good. The stretches of abandoned housing in our larger cities, for example, offer opportunities for recycling urban land that will be lost if government through redevelopment programs supports the price of that land at levels now demonstrably too high. Governments are at their least attractive, in

every way, when their aim is to protect people who have made losing bets from the consequences of their mistakes.

Politicians must also understand that land policy and social policy
are very different questions, often likely to be in conflict. To take
what can scarcely be a controversial example, Olmsted's plan to
build Central Park in New York was extraordinarily enlightened
land policy, but at the time it required the eviction from the site of
some thousands of squatters who were the poorest of the poor. The
brilliantly accomplished social policy of encouraging home ownership by the lower middle class in the years after World War II was
abusive land policy and recognized as such from the beginning.
Because the buildings are with us for such a long time, planners
must also consider the possibility (indeed, the likelihood) that what
looks right now may look terribly wrong in ten or fifteen years.
That's not something planners like to do.

Chapter 4

New Communities

Columbia will begin this summer and is scheduled for completion in twelve to fifteen years. We believe it will be a beautiful, an intensely human, a lively, effective city. . . . It does what most of us have known all along was doable—to plan a city intelligently for the people who will live in it. . . . A great tide runs. If we harness it to the tools we have forged for developing and redeveloping our cities, we will revolutionze our urban civilization in our lifetimes. We can wipe out the suffocating oppressiveness of slum and blight and sprawl. We can replace the nobodyness of the massive formless city with the somebodyness of communities that make a man and his family important.

—James W. Rouse, builder and developer (1965)

Town planning has now become a sort of dumping ground for every difficult and unresolved problem such as the birth-rate, the social equilibrium, alcoholism, crime, the moral of the great city, civic affairs and so forth.

—LeCorbusier (1924)

No amount of planning will ever replace dumb luck.

—sign on the office wall of C. James Dowden, executive director, Community Associations Institute

Suppose you could start from scratch, that instead of zoning to control the changes being wrought in a living community, you could take inert land and structure the community on paper before the peo-

ple came. There is something here for everybody who makes plans:
for the Utopian with his vision of the new mankind created by new
surroundings; for the fascist with his belief in an imposed order; for
the investor who can sell the industrial and commercial zones at immense profit because he has brought the workers and shoppers to
live just here and nowhere else. Indeed, this is how the big developers make their money when they make it: from the Irvine Ranch
south of Los Angeles to Reston, Va., from the Shimbergs' Town &
Country area outside Tampa to Lefrak City in Queens, the gain for
the builders derives from the stores and offices and industrial sites
they will rent rather than sell, with value appreciating steadily as the
development grows.

The early "planned communities" of Saint-Simon and Fourier
and Robert Owen were all centered around work. The first to be established on a permanent and successful basis was probably the Cité
Ouvrière in Mulhouse, France, built by textile manufacturer Jean
Dollfus (with money borrowed in Switzerland because the interest
rates were lower); by 1870 it included three thousand little houses,
nearly all of them still in use after a hundred years and not a slum,
though today's residents are inconvenienced by the lack of space for
automobiles. In England, what Colin and Rose Bell called "cold-
blooded philanthropy" built excellent residential communities tied
to the sponsor's factory: Lever's Port Sunlight near Liverpool and
Cadbury's Bournville near Birmingham, still happily occupied, are
the classic cases. America has its own version, modeled in part after
Cadbury's (the two were in the same business), at Hershey, Pennsylvania, where the amenities include a theater, an ice-hockey stadium, elaborate swimming facilities, a splendidly equipped school
for orphan boys, several golf courses, and an elegant resort hotel.

Ebenezer Howard, an English court stenographer inspired by
reading Edward Bellamy's *Looking Backward,* saw that this sort of
thing was good economics as well as humane, and that it did not
need to be tied to a single productive enterprise. A "garden city"
(Howard's term) of forty thousand or fifty thousand people, radiat-
ing out from a Crystal Palace for highbrow entertainment and mer-
chandising, could pay for itself by recapturing the increased value
of the land. The town would be a hole in a doughnut of farmland

that would feed the population and that would be held inviolable. Industrial sites, suitably insulated from both the town center and the residential neighborhood, would provide jobs for all. A single demonstration of the scheme, Howard argued in 1898 in his book *Tomorrow,* would make the hideous great cities start shrinking, as both businesses and people fled to happier surroundings. Within five years, he had his demonstration at Letchworth, forty miles from London—but it was not until fifty years after that, and then in a greatly changed state and force-fed by the government, that Letchworth reached its planned target of thirty thousand population.

Immediately after World War II, England put money and authority behind Howard's ideas. In 1944, as part of the great burst of wartime creativity that produced the modern welfare state, Sir Patrick Abercrombie's Greater London Plan proposed a group of New Towns of sixty to eighty thousand people with industry enough to employ them, located about thirty miles from London on the other side of a permanent greenbelt. Less than a year after the end of the war a working committee headed by Lord Reith, the great snob who built the BBC, had reported back with specific details for the administration of such foundations, and in 1946 the New Towns Act set Britain on a course of deliberately draining people and industrial vitality from London (and, later, the other great cities) for the construction of a better life in the countryside. The problem as then perceived was to handle the "overspill" from the cities. Now, when the age of overspill is ended and the Greater London Council feels that the city is fighting for its life (the process of urban abandonment has begun, fifteen or so years behind the United States), posters in the Underground still tell the traveler how much better both living and business are in the New Towns.

As of the late 1970s, thirty-three New Towns were in operation in the British Isles, housing more than two million people (of whom a million were imports from the cities), providing jobs for two hundred thousand. The capital to launch them, say $5 billion in 1967 dollars, has been provided by the national government on low interest sixty-year loans that are slowly being repaid by the development corporations, which are autonomous public agencies. (Actually, the costs to the national government are much higher, because housing subsidies in the New Towns are enormous, totaling over a quarter of a billion dollars a year.) In the early years, when there were great shortages of both modern industrial capacity and housing in England, New Towns worked like a watch: housing units were made available only to employees of the immigrating companies, which could easily recruit the skilled workers they wanted because housing was available.

Sociologists noted that the arrivals from the London slums were not very happy, missing their pubs, the extensive comadre relationship of the London streets, and their relatives—much more than Americans, the British live in extended families. With a strong assist from television, which made men more content to stay at home at night, these dissatisfactions eased. By 1968, 47 percent of the residents of Harlow, one of the largest New Towns, had blood relations living in the Town outside their own household. "Brother and sisters-in-law, sisters and brothers-in-law," says a PEP pamphlet, expressing a British view of society never heard in America, "are coming to live in the same town. A real community is indeed being created."

Some aspects of the British New Towns are indeed very attractive. Arriving at Stevenage, the first to be designated in 1946 but still not complete, the traveler by train walks on an interior bridge through an immense new "leisure center" finished in 1975. To one side, the glass walls of the bridge open onto a view of a huge indoor artificial-turf bowling lawn, where young men and women are most decorously (and skillfully) throwing black balls at a white one; to the other side, one looks down onto the cleanest little cafeteria and snack bar in England. Elsewhere in the ingeniously designed multi-level shed are squash courts, basketball courts, two theaters, and God knows what else. Beyond the leisure center is the famous "pedestrian precinct," actually a low-rise shopping center with cars parked around the periphery, not so unlike many American shopping centers of the 1960s as the planning profession might wish you to believe, and not so clean as the leisure center. Walkways curl under the roadways to separate the people and the traffic, following designs pioneered by the American Clarence Stein in Radburn, N. J. (Nobody in England, however, seems to have noticed that the children in Radburn go some distance to play in the streets rather than on the luxurious greensward.) The common areas between the row houses, through which the walkways wind, are grassy and imaginatively planted.

The housing itself, unfortunately, is abominable by continental European, let alone American, standards; the only thing to recommend it is that virtually all the homes (90 percent) are single-family. Densities are twelve to sixteen houses to the acre, but for some reason—either bad planning in land acquisition or bad procedures in land preparation—the lot costs to the corporation are more than $5,000 (in depreciated pounds) per house. The ground floor offers a box of a kitchen (without stove or refrigerator: the tenant supplies her own), a living room perhaps 140 square feet in size, exposed radiators, a closet with a hot-water heater nearly filling it, a hallway straight through from back door to front door, vinyl flooring, and stairs to a second floor with a windowless bathroom and three tiny bedrooms offering a poor quality wood floor and no heating facilities whatever ("everyone carpets and uses an electric fire"). Exterior and interior walls are load-bearing masonry, which means all the shapes and sizes are locked in forever. (And forever is not far off it: Colin Jones of the housing section in the Department of the Environment says airily that "if you build in masonry there's no reason why a house shouldn't last 350 years"). The cost to the development corporation for this house in 1976 was $45,000.

Of course, things could have been worse: Leonard Silkin, the first Minister of Town and Country Planning, proposed in a speech in Stevenage in 1950 that dwellings would share a kitchen and everyone could eat out of a communal pot. But things could also have been better, for it turns out that even at $45,000 a house the prices are not sufficient to pay the costs of the amenities package, and when the leisure center opened in Stevenage everybody's "rates" went up by about $30 a year to service the bond issue.

Not surprisingly, the officers of the town government and most of the upper-level people from the factories in industrial park live in the village of Old Stevenage about half a mile away. (A few of the people who work in Stevenage live in the original Letchworth, only a few miles up the road.) Also not surprisingly, the original plan to mix the social classes on the same blocks has been abandoned, and the better houses within the New Town stand on the hillside as they always did, seculae seculorum.

New Towns came to America with Rexford Tugwell's Resettlement Administration in the New Deal, and three were begun: Greenbelt near Washington, Greenhills near Cincinnati, and Greendale near Milwaukee. At the groundbreaking for Greenbelt, Franklin Roosevelt called it "an experiment that ought to be copied by every community in the United States." Nobody was building factories in those days, and the American version of Ebenezer Howard, publicicized by Patrick Geddes and Lewis Mumford, was not so hot on factories anyway. (As town planner and historian Hans Blumenfeld once pointed out, Mumford never remembers to mention that "there can be no good living without making a living.") So the Greenbelt towns, placed in the path of suburban expansion, grew up into dormitory suburbs, as did, on a higher social level, Clarence Stein's small-scale (eight hundred homes) development at Radburn.

Interest in New Towns revived after the war, stimulated by the intellectual prestige of the British reports and by the visible triumph of what is still far and away the most successful venture of this sort—Tapiola, outside Helsinki, tall trees and lakes, a mix of fairly high apartment houses, low-rises, garden apartments, and single homes, individually and together a monument to what was for one blazing generation the greatest design tradition in Europe.

The lessons learned from Tapiola were various. The first was that the job could be done by an entirely private enterprise and essentially by one man, in this case a visionary ex-lawyer named Heikki v. Hertzen, who took for himself the title of President and Planning Director. The second was that a socialized attitude toward housing (Tapiola did achieve mixed-income neighborhoods, though its occupancy is now predominantly middle-class) was entirely compatible with individual ownership (90 percent of the apartment units are co-ops). Another was that the relatively skimpy living spaces of Finnish subsidized housing (the two-bedroom apartments run about seven hundred square feet) were salable to people who could afford more, if the ambience was made sufficiently attractive. A last lesson was that the thing could make money: "Good planning has proved to be good business," Hertzen said in 1973. "It is the best possible investment. . . . The difficulties in the way of modern town and community planning are not of an economic nature." Two elements, however, may not have been sufficiently considered by imitators—the serious housing shortage in the Helsinki area at the time, and the fact that Hertzen's cost for raw land was only 1 percent of his total investment, while his cost for finished land (including not only streets and sewers, but also a plant that generated both steam heat and electricity for the entire project) ran to only 7 percent.

In the America of 1960, the New Town idea had great resonance. The Census Bureau was extrapolating from the huge birth cohorts of the late 1950s to a population at the turn of the century that would require a hundred new cities and a doubling of the number of urban homes. In California a number of very large landowners undertook developments of more than 2,500 acres each; Edward Eichler, who had been a home-builder himself and then chairman of the first Governor Brown's Advisory Commission on Housing Problems, called them "community builders." Outside Phoenix, Goodyear Tire commissioned Victor Gruen to develop a large free-standing community on the Litchfield Park properties near the factory where the blimps are made.

The largest of all was Columbia, Maryland, between Baltimore and Washington, the work of James Rouse, a balding, avuncular figure in horn-rimmed glasses with a slow smile and a casual manner. A successful builder of shopping centers from Baltimore, Rouse had made it on his own (after finishing night law school, he had actually been a lowly employee of the newborn Federal Housing Administration in the middle 1930s); in 1962, he began a long, secretive march through the farming community of Howard County, near Baltimore, to pick up from 328 owners 14,000 acres (22 square miles) on which he would build his New Town of Columbia.

Rouse let his cat out of the bag with the announcement of a planning committee drawn mostly from the social sciences. The expertise represented on this "work group" was entitled Public Administration, Family Life, Recreation System, Community Structure, Economics and Housing Market, Education, Health Systems, Housing, Local Government, Traffic and Transportation, and Communication in the Community. Several members, including the chairman, Donald Michaels, a social psychologist, were drawn from the Institute of Policy Studies, Mark Raskin's New Left think tank in Washington.

The social purposes of Columbia were admirable and have been in some part achieved: there are, for example, three hundred units of federally subsidized housing scattered in groups of sixty through five "villages" of Columbia, indistinguishable in appearance and feel from the townhouses and low-rises around them. "Value-contouring"—placing homes so that the income level of the occupants changes gradually as one moves down the road—has made possible a mix of homes in each village. A Family Life Center (very important to the academic committee) counsels people on their problems. Concerts fill the summer air at the Meriwether Post Pavilion (the first structure at Columbia actually completed). There are indoor tennis courts.

Columbia's Community Association (which everyone must join) taxes residents at a rate of thirty-seven cents per $100 of market value, maintains the roads, parks, and playgrounds, collects garbage, gives fire protection, and absorbs the deficits of a bus system that in 1976 carried 350,000 passengers—"enormously emancipatory," says Rouse, "to old people and kids, especially kids." But the bus costs fifty cents a ride (ten cents if the passenger is part of a family that has paid the $225 annual fee for "the package," which includes membership in the golf, tennis, and swimming clubs). In the original planning, there were to be separate rights-of-way for the buses ("In one of the most euphoric moments of the work group discussions," planner Morton Hoppenfeld reported, "it was proposed to ban cars from Columbia altogether"); but the bus roads were never built, and the projected every-five-minute bus was every thirty minutes at last report.

The lake is pretty, the contours of the land and most of the trees have been preserved, as have the streams in wooded valleys that meander through the villages. The shopping center and its cars are unobtrusive, placed in a hollow with a walkway over the main road to the office buildings, and inside the shopping center the world is rather gay—Rouse is expert at this: he redeveloped the marvelous Faneuil Hall complex in Boston. Here and there in the little village shopping areas one even finds a Greenwich-Village-y restaurant or shop of the kind Holly Whyte predicted could not be sustained in a New Town. The town is integrated: Rouse claims 18 percent black, which seems high, but the representation is much more than token—and, significantly, houses owned by blacks sell to white purchasers as easily as houses owned by whites sell to black purchasers. (Leonard Downie, Jr., of *The Washington Post* is outraged that "teenagers have segregated themselves into one all-black village recreation center and another that is all white"; but it is hard to see what planning or even goodwill can do about that.) Though nothing is architecturally imaginative, nothing is wildly out of place with the word "new," and there are also none of the abominations—the triplex attached townhouses built to look like one big mansion, or the nutty Down East fishing village implanted like a stage set on the bare shores of the Pacific—that deface California's Irvine Ranch.

Most purchasers of homes in Columbia are simply delighted, not least because they've profited by their investment. "My daughter," Rouse reports ruefully, "bought a townhouse here in 1972 for $21,000, and sold it at the end of 1976 for $40,000." That returns some of the investment in Columbia to the Rouse family, but not much; the fact is that to the developers, the New Town has been a disaster. Rouse had sold the idea of Columbia to his board and to Connecticut General Life Insurance by means of what was called internally "the Green Book," an economic model of Columbia that predicted the developer would double his money every five years. (Risk-free, too: Rouse once said in a speech that he had told Connecticut General, "the very worst that can happen to you is that you get rich slowly. Who can get hurt owning fifteen thousand acres of land midway between Baltimore and Washington?") This model was then adjusted several times by economist Robert Gladstone of the working group, as planning came closer to execution, and in June 1964 Gladstone submitted "Working Paper #16," a chilling final projection that showed profits of only $10 million total on an investment of $100 million over fifteen years. At this point all the plans were rejiggered to reduce the allocation of land to open space, increase the density of housing and the share of both commercial and industrial zones; the numbers were somewhat arbitrarily brought back to where Rouse said they had to be; and in November 1964 the first shovels of dirt were turned.

Ten years later Connecticut General threw in as much of the towel as it could, writing off losses of $21 million. Another $3 million or so apparently was written off by Manufacturers Hanover and Morgan Guaranty, the original bankers to the project; and as of the end of Rouse's fiscal 1976 there was an additional $27 million of losses that had been capitalized. Rouse in fact controls today only 15 percent of the voting stock in Howard Research and Development (the development corporation for Columbia); Connecticut General has the rest, and wishes it didn't. Though Rouse has bailed out of the cost of the community facilities (the Community Association has sold $15 million in bonds to repay him; interest on those bonds is of course taxable), and though the prices at which he now sells land and would have been unimaginable in 1964 (early in 1977, a 9,000-square-foot lot for single-family use, less than a quarter of an acre, was priced at $14,500, and a pad for a townhouse at a density of ten per acre was priced at $6,500), there is no way that he and Connecticut General can hope to recoup the lost income from the years when the capital they sank into Columbia earned nothing.

No doubt creative accounting and selective amnesia will make it look okay at the end to the enthusiasts who write about architecture and city planning. From my point of view, too, Columbia is a success, precisely the home they wanted (rich suburban with a social conscience and a contemporary flavor) for the people who live there. It is not, however, the sort of success that would encourage rational others to go and do likewise.

The proposition presented by the unfolding history of Columbia (and of Robert Simon's parallel Reston, Va., on the other side of Washington) was an offer no red-blooded American government could refuse: high purpose, great publicity, and an unlimited supply of others to blame if (rather, when) things went wrong. In 1968,

Title IV of the Housing and Urban Development Act empowered
the Cabinet Department of that name to extend assistance of various
kinds to "private new community developers," and in 1970 Title
VII, the Urban Growth and New Community Development Act, put
Washington into the New Towns business in a big way. As of July
1974 there were supposed to be fifty-five government-assisted New
Towns under construction in the United States. The basic support
tool was to be a federal guarantee of debentures sold by the developers, plus a flow of direct grants, plus preferential access to funds
appropriated under the Urban Mass Transportation Act of 1964, the
Airport and Airway Development Act of 1970, the Public Health
Services Act, the Library Services and Construction Act, the Land
and Water Conservation Fund Act of 1965, the Housing Act of
1961, the Housing and Urban Development Act of 1965, the Federal Water Pollution Control Act, the Consolidated Farmers Home
Administration Act, the Higher Education Facilities Act of 1963,
and the Public Works and Economic Development Act of 1965.
Also "technical assistance."

By the end of 1974, HUD had guaranteed debenture issues for
thirteen projects and had "recognized" two others (Roosevelt Island and Radisson, projects of New York's Urban Development
Corporation, which was selling tax exempt bonds; the law forbade
HUD to guarantee tax exempt instruments). By the end of 1976, six
of the thirteen had in effect gone down the tube, with HUD's own
New Communities Development Corporation acquiring title from
the initial developers; and five of the other seven were unable to pay
the interest on their bonds. Only one looked like a viable entity: The
Woodlands, near Houston, and that was viable only because its
sponsor, George Mitchell of Mitchell Energy Corp., didn't much
care whether he made money on it or not. (In late 1976, HUD gave
Mitchell a little more money as a bribe to plan just a few moderate—or even—please!—low-income units on the site.)

In a talk at a seminar in Columbia in fall 1976, James F. Dausch,
head of the New Communities Administration, started from the assumption that "294 million in loan guarantees may not be recov-
erable." He could see only two possible "strategies":

"Strategy One: Take All Losses Now and Liquidate All Title VII Projects . . .

"Strategy Two: Recapitalize and Reorganize Potentially Successful Projects. Extricate HUD From Others."

The first strategy, he said, had four disadvantages:

"(1) HUD will lose total financial investment, all social/economic opportunities, alternative to sprawl.

"(2) Short-run impact on budget will be substantial.

"(3) All private new towns will be hurt by adverse publicity.

"(4) HUD credibility will be hurt."

It is hard to avoid the feeling that if the second of these four could have been removed from the list, the Department would indeed have walked away.

Except that it couldn't. Richard Karp has detailed the progress of the most complete flop on the list: the New Town of Gananda, New York, twelve miles from Rochester, originally planned for 50,000 people in 17,700 homes on 5,842 acres. Between April 1972 when an HUD guarantee commitment of $22 million established the project and late 1974 when Gananda Development Corp. reported itself unable to meet interest payments, a total indebtedness of almost $45 million had been incurred to create values even charity could not estimate as high as $10 million. A grand total of two model homes had been built on the site; one had burned down, the other had actually sold.

The development company or its affiliates had borrowed $4 million from Chase Manhattan, $3 million from The Lincoln First Bank of Rochester, and $1.2 million from the Northern Ohio Bank (which went bankrupt even before Gananda defaulted). Most serious of all, $11 million was owed on mortgages to the farmers who had sold Ganada the land, and under law the federal government's claim to any surviving values would have to come first, wiping them out, if HUD simply foreclosed. Adding insult to injury, the farmers would also become liable for the $800,000 of real estate taxes Gananda had not paid. HUD would have to accept a loss not of the $22 million it had guaranteed, but of something much closer to $35 million.

The New Town of Newfields, Ohio, seven miles from Dayton, was scarcely any better. In thirty months, the developers there had run through an $18 million government-guaranteed bond issue and other money to build thirty-seven homes, "an Olympic-sized swimming pool," a community center, and "a nine-acre lake" (everybody built a lake, even Gananda). "In-depths studies of the Newfields project," an HUD press release announced in October 1976, three years after issuing a guarantee the law had said should be an "acceptable risk" to the federal government, ". . . showed the project to be financially and economically infeasible as a full-scale Title VII new community. . . ." Congressmen and other politicos from the locality pressed HUD to prevent harm to the area, which meant no part of the $18 million could be recovered.

At the New Town of Flower Mound near Dallas, the New Town of Jonathan near Minneapolis, and the New Town of Riverton on the other side of Rochester, HUD has taken over admitted failures but bravely announced that new developers will be found to make the New Towns a reality. But a year after the announcement that Arlen Realty would do Riverton no contracts had been signed: Riverton was going to be a loser, and Arlen wouldn't move without an open-ended commitment from HUD to make good the losses.

There is no shortage of explanations why everything turned so sour for the New Towns, and HUD is still digging up new ones. Almost $5 million was spent in 1976 on fees to consultants to audit and fussbudget and try to figure out what happened:

Explanation One: Congress did not follow through: $168 million in grant money had been authorized for the first three years to supplement the guarantees, but only $25 million was actually appropriated; after six years, including money from new appropriations and discretionary funds allocated here by the Secretary of HUD, the total grants awarded had reached only $70 million. (Nobody seems to have calculated, however, how much poured down these sluiceways from all the other Acts to which New Towns had preferential access.)

Explanation Two: The huge overbuilding of conventional suburbs in 1971–72, followed by the great bounce in interest rates in 1973–74, crippled everyone's planning. It is hard to make sense of the New Town of Shenandoah near Atlanta when suburbs even nearer are awash in unsold condos.

Explanation Three: The Act, forbidding assistance until after a project "has received all governmental reviews and approvals required by state and local law," sank developers in a quicksand of rural greed and obstruction. "When we first went into this program," says Melvin Margolies, deputy administrator of the HUD New Communities office, "we didn't give as much attention as we should have to local towns and counties. We dealt only with the developers who owned the land. We've learned you can't shove a new town down people's throats."

Explanation Four: It was all thievery: too high a price paid for the land to people who would later express their thanks in some tangible way, monstrous fees for lawyers, architects, bankers, brokers, and publicists in the development group, sweetheart deals with contractors and union leaders, etc.

Explanation Five: Stupidity and its expenses. Texas builders love to tell the story of how the developers of the New Town of Flower Mound brought down consultants from the Museum of Modern Art in New York (more than once) to advise on the putative arts and crafts program for the Community Center that was never built. The Davis-Bacon Act, which requires the Secretary of Labor to certify that everybody is paying union wages on federally subsidized projects, drives the costs of homes in any New Town way above the price of comparable dwellings nearby in the 80 percent of the country where housing construction is nonunion. At Cedar-Riverside, the New–Town–in–Town in Minneapolis, government-subsidized lawyers blocked construction of the necessary high-rise apartment houses by alleging an inadequate environmental impact statement; after three years of paying interest on borrowed money without any opportunity to use it, the development corporation gave up the ghost and let the left hand of the government pick up the losses its right hand had created. An anonymous "housing analyst" quoted in *Planning* said that nobody "had any notion where the money was going. HUD's auditing section was, and it still is, a rat's nest. . . . If you're going to guarantee 40 million bucks and you don't treat it as an investment, you're going to run into problems."

Explanation Six: Initial failure breeds more failure. "The assumption," says Margolies, "was that people would be willing to pay 10 percent more for the amenities. Now with all the bad publicity people won't even pay the going rate—they want a bonus for being pioneers."

Whenever there are so many reasons for a disaster, the suspicion rises that the idea was no good from the start. A study in 1976 by James Landauer Associates reported that "virtually all new towns [including the hundred-odd without government support] have a present asset value between 20 percent and 33 percent of their book value." At the root of the problem is the fact that New Towns ask too many people to buy with their house too many things they do not want, or at least do not want at that price. "We're not selling the house *à la* an airline seat," says Scott Ditch of the Rouse organization. "We're selling the destination. We're selling a *system* of life support; a *system* of recreational facilities, not just a few handball courts; a health *system,* not just a few hospitals; a school *system,* not just your school. The problem is that the payment has to come in the price of the house."

In the 1960s, the huge Janss/Conejo development in the San Fernando Valley sold what the ads called "the sporting life . . . play 18 holes of championship golf after breakfast, ride a horse after lunch, take kids for a hike in the afternoon and finish off with a few games of bowling." Foster City near San Francisco boasted of its linked lagoons and the pleasures of boating thereon. But when interviewed for the Eichler study on community builders, only 16 percent of the purchasers at Foster City said they thought they'd go boating on the lagoons, and only 7 percent of the purchasers at Janss/Conejo thought they would make much use of the golf course. The importance of the "amenities package," Eichler decided, was not that it offered pleasures but that it provided "symbolic investment protection for the residents."

There is indeed a community that will gladly pay for wooded streams and concert halls, separations of vehicular and pedestrian traffic, large shared open spaces, but it is overwhelmingly an upper-middle-class community, and most of it already has a satisfactory home. *"The most important factor in the purchase decision to the buyers of the less expensive homes,"* Eichler writes (italics in the original, *"seems to be space and construction quality."* If the builder outside the New Town, with no amenities to pay for, can offer more space and better quality for the same or a lower price, which is not difficult in areas where federal sponsorship requires union wages (not to mention processing delays) that can otherwise be avoided, the number of customers for the New Town may be significantly restricted. The New Town developer who succeeds will be the charlatan who can sell smoke as a branded product called "life style."

The economics of the situation work against New Towns in multiple ways. The federally guaranteed projects all contemplated building periods of twenty years and up, which means that large parcels of land must be held inert after purchase, eating up interest payments and tax payments far greater than any likely income from farming them. The amenities package must be put in place first to draw the customers, and this expense, too, accumulates interest costs until it can be paid off by the sale of later houses at higher prices. Moreover, the New Town developer, unless it is a state agency, must finance the parks, community centers, and playing fields with taxable paper.

"We build amenities," says Leon Weiner of Wilmington, Delaware, proprietor of a Planned Unit Development almost large enough to qualify as a New Town. He is a forceful third-generation builder, small, bald, cigar-smoking, an almost perfectly round face, blue eyes behind gold-rimmed glasses. He is generally considered the most effective president the National Association of Home Builders has ever had. "We build recreation facilities, open space, on the theory that people will pay for it. What we mean is that society is trying to impose its views of what a community should be.

"Because our project was planned comprehensively rather than as a series of forty- or fifty-acre parcels, we paid for wider streets and an arterial system, for sanitary sewers, we had to deal with a drainage system that would otherwise have been built by the public. We provided a clubhouse, which people could later have bought for

themselves. We made the homebuyer pay at high mortgage interest
rates for what could have been built later with tax-exempt bonds. I
think we were wrong.''

In most of the New Towns and larger PUDs, disputes between the
early purchasers and the developer grow increasingly bitter within
the community association, because their interests conflict. The developer cannot get his profit out until he sells the last 10 or 15 percent of the project, but the community on the spot sees those last
sales as the newcomers who will overcrowd the amenities they
bought to enjoy.

''There's no doubt in my mind,'' Rouse says wistfully, ''that the
gross economic modeling of Columbia would have proved out to
completion—that the amenities *were* free, the communitity facilities, the transportation, the lakes. . . . The economics of New Town
development have not dealt successfully with the value-creating
process.'' In practical terms, though, what Rouse is saying is that
the last purchasers, because the values have become tangible, will
pay much higher prices than the first purchasers to bring the developer out ahead. This process has indeed been playing itself through
in Columbia, too late to save the investors. The result, however, is
to squeeze out the moderate-income families who cannot possibly
pick up a tab of this size; and Rouse did honestly care about helping
those families: he was not in this venture just to make money. What
he demonstrated was that the social costs of the model outweigh the
benefits.

Americans tend to be vain, to claim the biggest this, the best that,
the worst of the other. Given the dimension of the failure of our
government-sponsored New Towns, you would think we had a clear
track to the superlative in this category, but we don't. The French
Villes Nouvelles, planned on a much larger scale than ours—
Marne-la-Vallée was until recently supposed to house 500,000 people on about thirty square miles of terrain beginning only twelve
miles from the eastern edge of Paris—have been a correspondingly
greater drain on that country's economy and a more severe test of
the patience of national and local political leaders and bureaucrats,
road and railroad designers and builders, pioneering residents, in-

dustrial planners. When the French make a decision, they stick to it, and these New Towns will indeed be completed. What remains uncertain is whether they will be occupied.

Each French New Town (five are being developed near Paris and four elsewhere in the country) is by our standards an assembly of full-scale Towns—fairly widely separated centers that may have as many as 100,000 residents each, linked through considerable stretches of open countryside to one grand center with major urban facilities. (These get cutesy names, like "Agora" and "Piazza," comme chez nous.) France being a country where farmers tend to live in small towns rather than on their farms, pieces of land that size require the inclusion of existing communities and the creation of large new governmental entities, each an Etablissement Public d'Aménagement. For the plans of the EPA to go forward, the government must provide an immense infrastructure—roads, railroads, water and sewage, electricity and telephone. Like the American government, the French government does not build and operate housing for anyone but prisoners, military, and lunatics, so negotiations have to be carried on with the local nonprofit associations that sponsor workers' housing and the private builders that will provide the less subsidized housing for the better off (there is no completely unsubsidized housing in France).

Aspects of the French approach are much admired by some American commentators. "The French can put in industry," says Melvin Margolies of HUD's New Communities office. "We can't." But that's only what the laws says: life is more complicated. "In France," says Richard LeFebre, a young engineer who uses very fancy four-color brochures to help him sell industry on locations in Marne-la-Vallée, "there is very little mobility. The *chef d'entreprise* is afraid to lose people, afraid to lose customers."

Robert Donzet of the Ministry says, "We can use blackmail: if you want to build in Paris, you must *also* build in new towns." Then it turns out that only the government-controlled banks and insurance companies, and the American multinational corporations which always need favors, are susceptible to such pressures. The government has been able to push into the new towns the grandes écoles, those unique combinations of Ivy League colleges and secret societies that produce the future leaders of French corporate and public bureaucracies. But the economic benefit of a grande école is probably less than that of the community colleges the American New Towns have routinely attracted, certainly less than that of the full-scale University of California implanted on land given by the developer in Irvine.

Leonard Downie admires ''a strong French national new town law [which] authorizes the creation of a new local entity for each new town site—an organization that swallows up existing suburban political bodies and territories whether they like it or not.'' One does not, in fact, swallow up established communes that way in France. Bulding permits still must be issued by the local Syndicat Communitaire d'Aménagement. ''In the early years,'' Serge Goldberg, director of the EPA at Saint-Quentin-en-Yvelines wrote in 1974, ''some ambiguities remained on the division of responsibilities between the EPA and the SCA.'' Mais, oui. Michel Rouselot, director of the EPA at Marne-la-Vallée, says that the local SCA for two years held up the extension of the A-4 superhighway that was to be the lifeline of his New Town, insisting that the Ministry of Education build a new school for the commune as a recompense for the nuisance. Blackmail works both ways. The road won't be in until some time after 1980, when very large stretches of the Town will be complete. The result is that wherever *here* may be, you can't get to Marne-la-Vallée from here: the twelve-mile trip from Paris is forty minutes of excruciating slogging behind camions through small-town centers. The extension of the subway line hasn't got built either.

Some laws turn out to be inviolable. The planners not unintelligently wanted their secondary schools in the centers of the towns, where the transportation routes converge. But you can't open a new drinking establishment within 1,500 feet of a school door in France, which means no cafes. The belligerently imposed separation of walkways and roadways destroys the small shopkeeper who sets up on the walkway; it isn't only that he doesn't get business—he gets lonely, never seeing anybody. In general, says Mme. Bouret of the Ministry, ''the problem with the New Towns is the lack of animation.'' This is a problem, incidentally, in all countries: the people drawn to New Towns are normally two-income families. During the day both mama and papa are at work, and the children are in school or at the day-care center (especially in France, where three-fifths of all children between ages two and five are taken care of in the lavishly equipped and stongly staffed state-run écoles maternelles). During the weekend, everybody piles in the car and gets to hell out.

Moreover, the administrators of the EPAs do not necessarily have a free hand in siting their housing. Saint-Quentin-en-Yvelines (always "SQ") is built around the existing industrial commune of Trappes, an automobile-building center and a stronghold of the French Communist Party. The general principle of French New Town development has been a fifty-fifty division between high-rise workers' housing in HLMs (the initials stand for moderate-rent habitations) and private construction. The commune of Trappes insisted that at least 70 percent of all housing in its borders must be HLM; "and now I have to tell them," Goldberg says (in English, by the way), "that the housing for them will be at least 90 percent HLM, because nobody wants to invest there, the image of Trappes is such . . ." Rents in the HLMs run to about $150 a month for a two-bedroom apartment, and the tenant's employer, under the French system of mandatory housing finance, must contribute about $10,000 in the form of a loan to the HLM at 1 percent interest.

Meanwhile, elsewhere in SQ, several single-family tracts of a kind quite familiar to Americans—one developed by Levitt, one by Kaufman & Broad—have sold out fast at prices of $75,000 and up (partly financed by low-interest loans from the government). Those purchases are helping to pay for the HLMs: Goldberg, who has had some discretion in this area, sells New Town land to the subsidized housing projects at a loss of about $2,500 per apartment, and makes up the losses on his sales to private builders.

Goldberg controls the choice of architect for the high-rise HLMs in Trappes, and many of them present façades of some grace and charm, the boxiness broken up by geometric surprises, in great contrast to the truly horrid prisons the Trappes municipality had put up before the New Town arrived. (There is some payment for these surprises inside, like a terrace oddly entered from the narrow angle of a trapezoidal kitchen; but the apartments are larger than the French norm, because the New Towns as experimental venture benefit by construction bonuses from the state.) In general, Goldberg has had trouble with architects, as have all the New Town managers: "These days," he says, "they are all interested in their schools rather than in the plumbing. They are all amateur sociologists, and I don't think they are very good at it. I tell them, "Let's do what we do well . . .' " Those New Towns that have run architectural competitions (Goldberg won't have them, because he doesn't trust the juries) have wound up with extravagances (inverted pyramids with towers rising from them, for example) that call attention to themselves but solve no problems.

In some ways the most interesting part of SQ is one that was there before the New Town was designated. Goldberg and the Ministry never mention this section and will not show it to a visitor, though in 1976 almost half the actual resident population of the New Town lived there. This is Verriere Maurepas, home to 25,000 people in apartment houses and single-family homes, all built as a private enterprise by one developer, Jacques Riboud. An engineer by training, Riboud spent the better part of two decades in the United States—working first in the oil fields of Texas and then, during the war, at the Pentagon designing tanks—before coming home with his American wife. He is a solid man with grave gestures, very interested in the world, and by no means a laissez-faire fellow. He is, indeed, shocked by the unplanned way housing has been created in France: "It is incredible to build five million houses without even the start of a doctrine to say where they should go . . ."

The literature for Verriere Maurepas is classic New Town, but I think one should catch the French flavor:

A VERRIERE MAUREPAS, les habitants des maisons individuelles du Groupe Riboud ont à leur disposition et pour leurs commodités DES ECOLES MATERNELLES, PRIMAIRES, SECONDAIRES, UN HOTEL DE VILLE, UNE GENDARMERIE, UNE PERCEPTION, UN CENTRE COMMERCIAL, DES MAGASINS, UN QUARTIER D'ARTISANS, DES CAFES, UN HOTEL RESTAURANT, DES GYMNASES, TENNIS, PISCINE, UN CENTRE CUL-

TUREL, UN MARCHE FORAIN, UNE EGLISE, UN
FOYER DE RETRAITE, UN GROUPE MEDICAL ETC.
. . . VERRIERE MAUREPAS, C'EST AUTRE CHOSE!

Note from this list the artisans' shops, the police station, the
open market, the town hall (built by Riboud and given to the town).
Riboud is not attempting to give people the world of the future,
which they may or may not want; he is offering in new surroundings
and new facilities a world not unlike the one the French over two
thousand years have made for themselves. The New Town in his
mind does not give its developer either the right or the opportunity
to solve the world's problems; it gives him a chance (profitable one,
if the state will leave him alone) to meet the legitimate demands of a
population.

Also it's *interesting:* "The problem of a new city is a thousand
problems all at once." Riboud manages them separately. In a period
when housing for the elderly was being placed out on the edges of
all the New Towns around the world, to give the old folks the
benefits of views, trees, and grass, Riboud (over the dead bodies of
two ministries) placed his in a high-rise built on top of the super-
market: "Old people want to be where life is, and it's convenient
for mother to leave the children with grandmother while she goes
shopping." Instead of the usual circular or square shopping center
(giving equal access for shoppers parked all around it), Riboud built
a "stringbean shape," so people can walk along it as they would a
village street, with brick-textured paving—and low-rise apartment
houses over all the shops, as in a real city. To minimize the presence
of the cars in the parking lots before and behind this area, Riboud
used the spoil from the rest of the building process to raise the whole
section a dozen feet above this flat part of France: then people look
over the cars and don't notice them so much. The slope is planted
with a prickly ground-cover Riboud found rather than with grass, to
reduce maintenance and keep the kids from rolling down it. To
avoid the loss of vitality that seems to characterize all the New
Towns, Riboud put aside a large space behind the middle of the
stringbean for an open farmers' market, terribly crowded with
pickup trucks, tents, and people three days a week, shoppers from
all over the sterilities of SQ—from the HLMs in Trappes and the

fancy houses of Coignières—and he pays for cleaning up the mess after every session.

One section of Verrière Maurepas is a collection of several hundred minuscule town houses, twenty to the acre, each about seven hundred square feet of living space on two floors, with a tiny garden behind a fence against the narrow street (no parking: that's done in boxes around the perimeter of the area). These houses with stucco façades and slate roofs and wooden picket fences were built and sold in the late 1960s at prices under $12,000, which people who worked for Riboud in service capacities could afford. They and the little plots they stand on are tended with love; it's a pleasure to walk by them. The pitched-roof flavor is reminiscent of southern French villages: it is, Riboud explains, an *"urbanisme provençale.* To reduce the problems of a new city, you must give people something that reminds them of the place where they grew up."

Riboud looked around him at the fenced-off gardens that surround a little *place* in this section, and pointed out that the doors in a few of the fences were much lower than the rest of the fence, adding variety to the visual aspect. "They were all like that when we opened, and I was outraged when I found people were putting in high gates instead, because they weren't allowed to do that. But if the gates were low, the dogs could jump them." Riboud shrugged. "You cannot control what people do . . ."

For Riboud, the great villain of the French New Towns was LeCorbusier, whose "radiant city" left a legacy of sterile design and dictatorial "planning." (Riboud has written a book denouncing LeCorbusier.) But LeCorbusier reached Riboud's conclusions in his own way. Told of the mess the residents had made of his simple design for Pessac, adding rooms and changing window styles and painting new colors, LeCorbusier said, "You know, it is always life that is right and the architect who is wrong." If the people who planned the New Towns in all these countries had been capable of learning that, the idea might have been made to work. But it seems to be an idea that draws the wrong people.

Chapter 5

In the Cities

A city, too, like an individual, has a work to do; and that city which is best adapted to the fulfillment of its work is to be deemed greatest.

—Aristotle

To listen to our urban developers, there is something sinful in this trend toward decentralization that threatens the city's land values and tax base; it should be reversed! It is time for these modern King Canutes to understand that this trend will never be reversed and that anyone who tries to do so is bound to come to grief. The densely crowded agglomeration of the nineteenth century with its concomitant, the fantastic skyrocketing of urban land values, was a short-lived passing phenomenon caused by the time lag between the modernization of interurban and intraurban traffic; once this time lag was overcome, it was bound to disappear forever; and few will regret its passing.

—Hans Blumenfeld, town planner and historian (1948)

In the middle of the city grows the Tree of Life and the gates of it shall never be shut.

—Isaiah 60:11

About fifteen years ago, I participated in a one-day meeting at the Minneapolis Institute of Art, to which the five speakers were supposed to bring suggestions for keeping the middle class in the cities. It was then a very unfashionable subject, and I can't remember that any of us had any very good ideas.

What I do remember vividly from the meeting is something said not by a member of the panel but by an irritated member of the audience. Minneapolis, he complained, was becoming intolerably overcrowded. When you went to the Mississippi for ice fishing in the winter, there were so many other people out there you had to hunt around for a place to pitch your tent. What were we going to do about that? It took us a while to realize he wasn't joking, and to note that there was really nothing at all that could be done for him, for the word ''city '' implies shared facilities, and as the country grew more prosperous there would be more people sharing them, even in declining cities.

The number of confidently expressed theories about the origins and growth of cities seems to be limited only by the imagination of the people who express them. Contradictions are the norm. Eliel Saarinen tells us that medieval cities lacked straight streets because ''the medieval time was cruel and full of dangers, and during the frequent enemy invasions the streets of the town were the battle-ground of attacks and counterattacks. Matters being so, why should the enemy have straight and convenient roads along which to enter?'' Colin and Rose Bell note that ''towns put up by soldiers'' in England after the Norman Conquest ''have straight streets; if the inhabitants grow fractious, they can be put down more easily, and if an enemy takes the town, he finds it more difficult to defend. . . . From Alexander to Napoleon, soldiers have plumped for the tidy grid. . . .'' (The relatively few military implantations in the United States, incidentally—Colorado Springs is perhaps the most obvious—show not only straight streets but extremely wide streets, wide enough for a cavalry troop to ride in parade order.) ''Wherever the city is viewed primarily as a fortess,'' Hans Blumenfeld writes, ''The circle is regarded as the ideal form.''

North European cities grew with long, narrow lots, Blumenfeld reports, because that was the agricultural pattern: long, narrow ''hides'' for the plowman. Latin cities grew in square blocks because fields were plowed crosswise, once in each direction. There were occasional wide streets in medieval times to serve as fire-breaks. Renaissance planners divided cities into discrete neighborhoods to make a layout matching the functions established by the craft guilds, which thereupon broke down. "Ways of shaping city form," writes MIT's Kevin Lynch, "arise by custom or, more rarely, by creative innovation. These models include not only large-scale ideas . . . but also many smaller elements; superblocks, grid-irons, or U-shaped bays of row houses. Once applied and shown to work reasonably well, and sometimes even if never successfully applied, they exert a compelling force on future plans, by virtue of their very simplicity and decisiveness in the face of uncertainty. Their power lies in their usefulness to the designer. . . . We notice that the plan for a new port city in Africa looks suspiciously like a student scheme for an American 'new town.' "

The city's prime function is that of the marketplace. The market facilitates, then encourages, division of labor, which in turn extends the market. Necessarily, the city imports the food sold in its market, which means it must generate a surplus of manufactured goods or salable services (not infrequently, government services) to trade for what to eat.

The city requires a multiform transportation system, a category that includes not only the obvious road and rail services but also water supply, waste disposal, and energy transmission channels. If the only water supply is the well or the town pump, the implications for the pattern of the city's growth are clear; also for its health. Most people forget how recent many of these transportation facilities are. The water truck and the honey wagon were common sights on the streets of American cities into the nineteenth century. At the time of the Civil War, Blake McKelvey reports, no American city had a public water supply system that reached half its households, and cities the size of Rochester and Milwaukee "relied entirely on private wells and water carriers." Today in the largest city in the world, Tokyo, nearly half the population lives in districts not served by a sewer system. This infrastructure is expensive to build, requiring considerable economic surplus from the manufacture of goods and services; it is also expensive to maintain.

Manufacturing the city's economic surplus requires a great supply of labor, which must be housed. In the earlier years of the growth of American cities, there was no public transportation cheap enough for workers to use, and the pedestrian city crowded up beyond modern imagining. Present-day Calcutta is less jammed than the New York of the turn of the century, when one ward achieved a density of more than 350,000 people to the square mile. (At that density, the entire current population of the New York metropolitan area would fit onto Manhattan Island alone.) Reporting to the government of Los Angeles in 1925, the consulting firm of Kelper & DeLeuw declared the urgency of building a subway or elevated system immediately, because only rapid transit "enables the city to expand [geographically] and prevents undue concentration of great masses of people in a small area." Like so many other consultants before and since, they seem to have left something out.

Until this century, a high proportion of the economic production of the city was carried on at, or in conjunction with, people's housing: people lived behind or above their shops, with their apprentice employees in the same building. In New York during the heaviest period of immigration and the worst time of crowding, the workshop and the home were often one, as pieces of the garment-making industry were allocated to family work-groups, which then carted their finished work through the streets by hand to the next site. Cigar makers, too—it was a big business in New York in the late nineteenth century—worked at home. In France up to twenty years ago, it was considered a hardship to work so far from home that one couldn't get home for the midday meal, which was dinner; the kids came home from school, too.

Though the argument is clearly one about chickens and eggs, it probably makes more sense in the United States to say that the industry followed the people. The city fathers thought so: municipalities made great concessions, including cash donations, to lure railroad builders, and gave the trains the freedom of the city. In Europe, where city centers were already settled and monumentalized before the industrial age began, the railroad yards were some small distance from the center of town; the tracks never ran along the banks of the Thames or the Seine or the Tiber as they did along the Hudson, the Ohio, the Mississippi, Lake Michigan, and Lake Erie. Thus European industry stood a little farther out from the political-commercial center of the European cities, and though every city had its rookeries, industrial workers were more likely to live out among the factories, in the rude and nasty suburb. For the French it started as a ''fauxbourg,'' a false city, and the street names changed where the city walls used to be. The Oxford English Dictionary gives ''suburban'' as ''having characteristics that are regarded as belonging especially to life in the suburbs of a city, having the inferior manners, the narrowness of view, etc., attributed to residents in suburbs.'' By contrast, the American College Dictionary offers the nonpejorative ''pertaining to, inhabiting or being in the suburbs of a city or town.''

In the United States the cermonial values associated with the city center were minimal: government was a dirty game, and religion was plural. The railroad dominated downtown. Industry was seen as the life of the city (even in the old cities formed for reasons of transshipment and commerce); and labor, instead of spreading out into the industrial suburbs as it did in Europe, was crowded near the center. There was neither rank nor title: land was in the hands of people who expected to make money on it. ''A rapid advance in real estate values,'' McKelvey notes, ''provided the foundation of credit.'' It did not make sense for the better-off householders to hang on; they sold at a profit and went off where the air and the streets were cleaner. Presently, starting in 1870, the horse cars made it possible for merely salaried people to follow as the new immigrants, willing to live in conditions their predecessors found intolerable, jammed into the tenements. The foundations of the plural American democracy were laid in the transfer of ownership of residential property—at a price—to the more ambitious, sometimes more grasping, of the newcomers, the urban equivalents of Jefferson's yeomen, who were pretty grasping too.

Thanks to the crowding, the city was hospitable to small enterprise. A large labor pool trained by compeers was available for seasonal employment, shipping facilities geared to handle a volume of small orders were available, both services and supplies could be bought as needed rather than kept on the shelf at the proprietor's expense. Robert Weaver lyrically extols a survivor (more or less), Manhattan's garment industry: ''Those involved are small operators, and a tremendous number of services have to be provided for them. People who can repair sewing machines and people who can give various other specialized services are required. These services are supported jointly. Also, if one man runs out of green thread, his competitor will lend it to him, because the competitor knows that if he runs out of green thread, he wants to be able to borrow it. The result is that neither has to keep enormous inventories, thereby cutting down on overhead. Of course, too, there are 'trimmings jobbers' who stock small items . . . which they will sell in small quantities.'' For the producers of nonstandardized products, from advertising campaigns to couturier clothing, the city gives indispensable access to idiosyncratic customers.

But the income from such a social organization cannot be restricted to its actual operators: because the profits rest so completely on propinquity, the owners of the land seize a large share. It was not an accident that the most original and important social theory thrown up in the United States was Henry George's Single Tax. His information base was wrong (it was not true that land ownership was tending toward monopoly in America) and his central formulation was a simpleminded equation (Product − Rent = Wages + Interest). But nineteenth-century economics had a weakness for such stuff—Marx is awash in it—and George's insight was (and is) correct: that it is bad for a community when a high proportion of the value added by the work of the people and the ventures of the entrepreneur can be taken by the owner of the inert factor, land. Urban crowding yielded too great a return to land: "Where each house was once surrounded by a garden and orchard," George wrote, "a lot of twenty feet front now carries family upon family, living, on top of each other, in tiers. . . . The ownership of square feet now enables [a man] to live in luxurious idleness on the toil of his fellow citizens.''

Into this economic context came the elevator. Multistory "lofts" arose to house manufacturing operations and warehouses. Department stores, a novelty from France, grew in downtown districts, their operations facilitated by cash registers, telephones, mass advertising, and eventually the pneumatic-tube system that permitted immediate control from a central office. In 1869, in what was the most livable neighborhood of New York, the one most like London,

in the heart of a triangle of Squares (Union, Madison, and Stuyvesant, with Gramercy Park next door), there came the first American
apartment house; and now wealthy people too could live in tiers and
help escalate the value of the underlying land.

The fact that construction technology facilitated vertical expansion hid for a long time the equally significant fact that patterns of
land ownership and valuation had made it all but impossible for
urban industry to expand horizontally. The manufacturing base of
center cities became increasingly vulnerable to possible change in
the organization and functioning of business firms. Economies of
scale in financing, marketing, and production—plus some heavy-
handed "rationalization" of industries to eliminate competition,
hindered but not stopped by the antitrust laws—reduced the share of
output represented by the smaller enterprises that needed the shared
services of center city. Make the total market big enough, and variety in essentially standard products can substitute for custom-
made—especially if one sells for much less than the other. Between
1900 and 1910, nonfarm gross product in the United States rose
about 80 percent in value (52 percent in constant dollars), but the
number of business firms rose by only 29 percent. As electric power
supplanted steam power, more and more industries moved to continuous production processes (i.e., assembly lines) that required an
expanse of space on one floor. The enormous growth of interurban
trolley services from 1890 to 1910 greatly extended the catchment
area from which businesses could draw their work force—and, of
course, produced an increasing dispersal of workers' housing, row
houses, three-flats, small apartment houses marching to the horizon
along the lines of transit.

All these developments were masked in the 1900s by economic
growth that proceeded at a rate of 6.5 percent a year compounded
(4.5 percent in constant prices) and by the heaviest immigration the
nation has ever known, more than ten million new arrivals in ten
years. Center-city real estate values kept rising with the skyline. But
between 1899 and 1909 every one of the nation's thirteen largest cities except Detroit saw a decline in its share of the total manufacturing employment of its area.

The 1920s were a quasi-miraculous period when the GNP grew at a rate of 4.8 percent compounded, employment grew at a rate of 3 percent compounded, and prices remained stable. (The figures are taken 1922–1929, to eliminate the distortion of the deep depression of 1921; they would be even more impressive taken from the lower base.) Immigration was cut back by law, but the farmers' sons still flowed to town. The cities seemed to have developed what could be called a stable pattern of growth. Philadelphia, St. Louis, San Francisco, and Boston joined Detroit as cities that saw manufacturing employment rise as a proportion of the metropolitan total; Baltimore, Cleveland, and New York held relatively even with their surrounding region.

The pattern was a radial city with commerce, retailing, and more industry than people now remember downtown; an inner ring of low-income housing, a middle ring of industry, and then expanding spokes of less dense settlement out along the trolley lines, the commuter lines, and the ever lengthening paved roads. Hans Blumenfeld explains the logic: "White-collar workers who can find a market for their skills exclusively in the business center may choose their residence at any place on the outskirts from which they can commute; but the families of unskilled workers, in which several members sell their labor at frequent, unsteady jobs anywhere at the outskirts as well as in the center, must seek a central location from which any part of the region can be easily reached."

The apparent security of prosperity in the 1920s led to a great overvaluation of capital assets, which it was assumed would gain in earning power forever. Real estate, as the most durable of these assets, was especially inflated, and because the leverage on real estate purchases is so high (that is, the purchaser borrows so great a share of the price) it is peculiarly susceptible to disaster if prices turn down. During the decade 1920–1930, residential mortgage debt tripled; and it is a measure of the 1930s collapse that in 1940 mortgages on residential property were still 20 percent below the 1930 peak, though by then GNP had turned and was up 10 percent from 1930. The cities stood littered with unoccupied industrial facilities, office buildings, and housing—which, presently, the war economy filled up.

Our cities came out of the war, then, apparently rich but really rotted. Important work had been done on the infrastructure during the Depression—there were new roads, bridges, airports, dams and power plants, sewer facilities and parks—but the industrial, commercial and residential facilities were very much as they had been fifteen years before, only fifteen years older. The banks having had no incentive to foreclose (alternative investments for the money, if they could have got it, were yielding only a percent or two a year through much of the 1930s), bankruptcies after the initial panic had been less widespread than was believed: to a large extent, the land was still in the hands of the many owners who had held it before the market laid its egg. Though it wasn't very profitable now—rent control was all but universal right after the war—urban real estate seemed to individual investors, insurance companies, and banks to be an excellent place for money. After a war that had brought huge deficits to the government accounts the nation was swimming in money.

In fact, urban real estate was a terrible place for money. The conveyor belt and the fork-lift truck were much more efficient than the elevator for materials handling, and the old loft building could never again be competitive as a place for large-scale manufacturing or warehousing. Trucks could service suburban locations for less than the cost of inching through downtown traffic; indeed, they could service much of the city itself more economically from a suburban location. In the 1960s, the businesses most dependent on efficiencies of distribution moved out of center city: bakeries, breweries, the newspaper plants, even the wholesale food markets departed—in Paris and London, where the Halles and Covent Garden were shut, as well as in American cities. Factories and department stores in suburban locations could offer workers and shoppers parking for their cars that downtown could not match. There were bureaucratic termites in the structure: "less onerous regulation and lower building costs in the outlying communities in a metropolitan area," Miles Colean wrote in 1950, "contribute to the dispersal of population and business from the central areas and to the consequent loss of their property values."

Even when a manufacturer wanted to expand on his old center-city site, because it was convenient to customers or a specific labor force or specialized transportaion, he often found he couldn't. "The assembly of a city site was a formidable operation," economist Raymond Vernon wrote in 1959. "As the city developed, most of its land was cut up in small parcels and covered with durable structure of one kind or another. The problem of assembling these sites . . . required a planning horizon of many years and a willingness to risk the possibility of price gouging by the last holdout. Moreover, once a site was acquired, razing costs alone could easily run on the order of $50,000 an acre in current dollar terms. All told, the value of the site could amount to twenty or thirty times more than that of an equivalent area in a developed suburban location."

In fact, the departing industry did not spend much less for land in the suburbs than it would have spent in the city, and wound up with higher land costs than expected, because real estate taxes are higher in the suburbs (a Minneapolis study in 1970 found relocated factories paid a real estate tax bill per worker hour triple what they had paid in the city). But they got greater comfort now, parking spaces for the work force, and assured room for expansion later. And the receipts from the sale of the urban property greatly reduced the out-of-pocket expense.

Residentially, too, much of what had been built in the cities was obsolete: the wiring couldn't handle air conditioners, the plumbing couldn't manage washing machines, the kitchen was too small, there was only one bathroom and no convenient place for baby carriage or bicycle, there were roaches and rats. After fifteen years of insufficient maintenance, with wage rates up and rents held down, the apartment houses were deteriorating. Popular opinion, some years ahead of the planners, held that apartments were unsatisfactory as a place to raise children. And government policy, then responsive to popular opinion rather than to planners, was promoting home ownership.

"The VA mortgage," Louis Winnick wrote more than twenty years ago, "was probably as effective an instrument for narrowing the market for new, family-sized rental units as was ever devised." Mortgage insurance or guarantee, by lengthening the time for which lenders would write mortgages, greatly reduced the monthly cost of owning a home. In the mid-1950s, mortgage rates were about 5.5

percent and the average new home price was about $12,000. On a fifteen-year self-amortizing mortgage with no down payment, the monthly payment for such a home would be $98.05; on a thirty-year mortgage, it was only $68.13—a reduction of more than 30 percent. Amortization and equity accumulation started slow on a thirty-year mortgage, of course—but they were zero on a rental. Nonsubsidized rental housing simply could not compete: "Easier credit terms," Winnick wrote, "not only altered the balance between renting and owning, but indeed made owning *the only* form of tenure that many families desiring or forced to seek new housing could afford."

The photographer Louis Schlivek, writing a memoir of *Man in Metropolis,* remembers why he left New York: "We could find nothing to meet our needs in any neighborhood we cared to live in at a price we could afford. Instead, what we did find, poring over the real estate pages of the paper, was ad after ad urging us to buy a house in the suburbs. We weren't interested in living in the suburbs, and had not planned on buying a home, but the terms made us rub our eyes in disbelief. It was impossible to resist at least going out to look. Imagine, a six-room house with a yard of its own which could be 'carried'—amortization, taxes, insurance—for a monthly payment lower than the rent on our one-room apartment."

Once again, the dimensions of the problem were concealed by the onrushing prosperity of the country. Industrial growth was concentrated in the larger plants that sprang up in the suburbs, but there was enough left over to permit the maintenance of the very high urban land prices that allowed the established companies to sell out of the cities at a profit. The housing shortage was acute after fifteen years of depression and war: in the city as in the countryside, builders could rent or sell whatever they built. And the last wave of immigrants was rolling into the cities, this one from the Southern farms, from the depressed Negro sections of the Southern cities, from Puerto Rico, and (legally or otherwise) from Mexico. These newcomers quickly filled—indeed, overfilled—the spaces left vacant by the departing middle class and the upwardly mobile working class.

Incidentally, the notion that the arrival of black and Hispanic populations created a new situation in city neighborhoods simply cannot be sustained by the evidence. *All* newcomers over the years had provoked the same reactions. Writing of the "ethnics" whose willingness to reinvest in their homes is now looked upon as the cities' best hope, appraiser Stanley McMichael wrote as late as 1951 that their settlements "sprang up, in many cases, on the ruins of once high-grade home districts occupied by earlier American inhabitants. Infiltration of a few foreigners promptly started an avalanche of changes. Within a very few years a once high-type residential section became a worn-out, 'blighted' area. Large structures were cut up into many units or became cheap boarding houses. . . . Some buildings were occupied until they collapsed, burned down, or were demolished upon orders of the health department. An appraiser sent into such a district to evaluate a property raised his hands in anguish, made as good a guess as he could, and fled."

Like McMichaels' foreigners, the Negroes and Puerto Ricans who crowded into the cities in the 1950s heightened what would otherwise have been depreciating values of urban dwellings, both tenements and formerly single-family homes that could be cut up. "Owners of such properties," the National Urban League says, "were able to realize a greatly increased income without being forced to invest in any substantial improvements." But the brief profitability of these slums did not draw entrepreneurs: "There were practically no building additions to the supply of low-rent housing—as had occurred in the tenement construction periods that responded to the great European migrations." The lenders withdrew, leaving the market to be bid up by a crazy patchwork of greedy immigrants, cynical syndicators, amateur mortgage-writers. As the professionals and some of the academicians realized, these newcomers would be the last wave. When they got the chance to move out of these noisome slums, the homes they left would stand empty.

If city densities are to be lower—a development that has been considered desirable by most observers for at least 150 years—there will be housing abandonment; no way to avoid it. Lowered densities will be—indeed, already have been—conducive to the development of separated nodes of urban activity: commercial, industrial, cultural, governmental, all four. The real question is not how to pre-

vent abandonment, but what to do with the vacated land. A number
of answers are possible, covering the entire spectrum from additional parks to revived manufacturing to the establishment of new
residential colonies conveniently situated to the entire spectrum of
economic activity on the periphery of the city. But all of the possible answers require that somebody take the losses implied by the
reduced densities, and the resistance to accepting these losses will
be all but universal in the community. The urban infrastructure itself—the utility and transit systems—must be written down in value
if it is to serve a smaller population. As the center city declines, this
expensive infrastructure, to quote George Sternlieb and James
Hughes of Rutgers, "becomes much more a hindrance than a positive asset."

Somehow, the thing can't be allowed. The city has assessed the
now-abandoned property for tax purposes, and even if taxes are not
being paid it wishes to maintain the pretense that the tax base is
there. The asserted value of local land is what supports the municipal borrowing structure. The local banks and savings associations
that still hold some of the mortgages want to pretend if they possibly
can that this paper still has value; and the bank examiners, politically pressured and frightened of the damage that could be done the
bank, will go along.

"Land prices become institutionalized," says John Heimann,
"because they're in everybody's mortgage." After all, even without investors, a purchaser for the property may be found. "The only
real *cash* buyer of slum tenements," George Sternlieb reported in
1966, "is the public authority, either in the person of urban renewal
or highway clearance." Fine: we'll bring in the government. From
the throats of the bankers, the property owners, the leaders of the
construction unions, the local officials rises the call for government
intervention, for urban renewal.

Chapter 6

Urban Renewal

Thenceforward, instead of a flattened-out and jumbled city . . . our city rises vertical to the sky, open to light and air, clear and radiant and sparkling. . . . Imagine all this junk, which till now has lain spread out over the soil like a dry crust, cleaned off and carted away and replaced by immense clear crystals of glass, rising to a height of over 600 feet; each at a good distance from the next and all standing with their bases set among trees. Our city, which has crawled on the ground till now, suddenly rises to its feet in the most natural way. . . .

> —LeCorbusier (1924)

Slums and their populations are the victims (and the perpetuators) of seemingly endless troubles that reinforce each other. Slums operate as vicious circles. . . . Our present urban renewal laws are an attempt to break . . . the vicious circles by forthrightly wiping away slums and their populations, and replacing them with projects intended to produce higher tax yields, or to lure back easier populations with less expensive public requirements. The method fails. At best, it merely shifts slums from here to there, adding its own tincture of extra hardship and disruption. At worst, it destroys neighborhoods where constructive and improving communities exist.

> —Jane Jacobs (1961)

There are many important issues in the urban renewal question, but there is one which is both the most important and easiest to understand. The local government must have the power to take by force the private property of one man—his home, his land, his business—with the intent of turning it over to some other man for his private use and personal gain. It is on the acceptance or rejection of this principle that the fate of urban renewal rests. . . .

> —Martin Anderson (1965)

"Let me give you an example of the opportunity we have taken in the St. Louis area," James San Jule told a meeting in San Francisco in 1969. ". . . This development I have largely been responsible for in St. Louis is called Mansion House. It is right downtown. It is right on the river. It is right next to that magnificent Saarinen Arch. Five years ago where Manson House is there was the worst slum area of the city of St. Louis, right downtown. After nine o'clock at night the only things that moved were rats as big as cats, and you didn't dare walk around down there because you took your life in your hands. Today, Mansion House is built. It is three magnificent twenty-nine story towers and a great complex of commercial and other facilities. It is highly successful: we have nearly 1,500 people living there now. It's approaching 85 percent occupancy. It is the pivot point of the rebirth of downtown St. Louis. St. Louis is no longer the dead old lady, gray city, that it was, because one man had guts enough to say that a Mansion House should have been there. He was not even a builder, he was a lawyer, and without a great deal of money. He had the guts, he had the imagination, he had the creativity, to go out and get the equity financing and to build it. Incidentally, to make a lot of money out of it. . . ."

Six years later, Robert Olson, a doctoral candidate at Washington University in St. Louis, wrote a dissertation about the workings of the Urban Redevelopment Corporations Law in Missouri. "Mansion House," he reported, "was the second application of the URC Law in the St. Louis CBD [Central Business District]. . . . The total building cost at the completion of the 1,248-unit upper-income apartment complex is listed by the Plan Commission at $52 million. The development was partially financed by a Federal Housing Administration insured $36 million loan, and the original equity capital was supplied by nationally prominent businessmen, including Henry Ford II. . . . It was a bold attempt to bring the upper middle class back to live in the inner city. However, the project was not successful in attracting residents, and vacancies have averaged around 50 percent since its opening. The original private investors have been able to use the complex as an enormous tax shelter, allowing them to avoid paying a total to $26 million in personal income taxes to date. The mortgage payments are not current, and the ultimate loser is the Federal Government, which insured the $36 million original loan. In addition the City of St. Louis has abated property taxes amounting to $400,000 a year for the ten years of the project's existence. . . .''

Mansion House has since been forced into bankruptcy, despite the conversion of one of the ''magnificent twenty-nine-story towers'' to hotel use (over the angry complaints of the town's other hotels). Vacancies have been much reduced by cutting the rents below the break-even point. Because two creditors filed conflicting claims in conflicting jurisdictions, a good deal of testimony has been taken and it appears that someone may indeed have made ''a lot of money'' out of Mansion House, but it wasn't the investors.

''Urban Redevelopment'' was Title I in the Housing Act of 1949, and it was designed to clear slums and to pick up neatly what were in fact the two great burdens the older cities faced looking into the postwar era. City land had been cut up into parcels too small to meet the needs of modern retail, commercial, industrial, and maybe even residential use; and the time and probably cost of assembling a site of sufficient size was enough to discourage most builders and to justify an unholy return on investment for the few who persisted. Under urban renewal, a city could move to condemn large tracts of skid row and inner-ring slum (and anything else the city wished to declare ''blighted'') and make them available to private builders. A rather odd, perhaps overly clever provision of the act gave it a housing reference: the land had to be used for housing either prior to its acquisition or subsequent to the completion of the program. Thus existing housing could be torn down to make way for stores or office buildings or factories; or existing stores, etc., could be torn down to make way for new housing. What couldn't be done was to replace one set of business uses with another set.

The second great burden on the cities was the sheer cost of land, made even higher than it otherwise would have been by the need to demolish existing structures before erecting new ones. Federal grants were made available both to acquire the land—for whatever prices the local courts might determine were the fair market prices before the announcement of the urban renewal program—and to clear the buildings. (More than half the federal money spent on urban renewal in the first dozen years went to buying property—i.e., into the pockets of landowners.) The grants in effect covered two-thirds of the cost, with the other third to be picked up by the municipality—but the municipal contribution could be in labor or materials, or even in the form of a pledge to build a school or plant a park in some section of the urban renewal site. Land would then be made available to private builders at whatever price reduction was necessary to make their projects economically feasible. (On the average, the developers paid the local authority about a third of what the land had cost the government.) Loans to finance the new housing and commercial structures would come mostly from the private market, encouraged by government guarantees, but for projects that worried the private lenders, inexpensive direct federal loans at the Treasury's own borrowing rate would be available.

In terms of commerce and the attractiveness of downtown, there seems no question that urban renewal has been a benefit to the cities. San Francisco's Embarcadero Center (to take the best first), Philadelphia's Penn Center, Pittsburgh's Golden Triangle, Denver's Mile High Center, the civic center area in Brooklyn, Boston's Government Center, Bunker Hill in Los Angeles—the list is all but endless. If St. Louis got the fiasco of Mansion House, it also got the Saarinen Arch, the greatest public monument ever built on this continent; a fancy stadium; and a large collection of office buildings to fight back against the new downtown in suburban Clayton. There are significant arts centers in Milwaukee, in Atlanta, in New Orleans, in Minneapolis (where the astonishing Federal Reserve Building is also built on an urban renewal site).

In terms of housing and the overall livability of the cities, however, urban renewal was a disaster. Something like a million people were, in Martin Anderson's loaded but accurate phrase, "forced to pack their belongings and leave their homes." The sociologist Herbert Gans lived in the West End of Boston as a participant-observer during the latter part of the nine years that its demolition and replacement was being planned, and explained poignantly the failure of the community to organize in self-defense: "some felt sure until the end that while the West End as a whole was coming down, *their* street would not be taken."

Between 1949 and 1968, some 425,000 units of housing, nearly all of them home to low-income households, were torn down by urban renewal projects, and only about 125,000 new housing units were built on the sites—well over half of them homes for the rich and the upper middle class. The law had provided "that there be," as a House of Representatives committee report puts it, "a feasible method for the temporary relocation of families displaced from the project area and the permanent provision of decent dwellings at prices and rents within the financial means of such families." That this requirement was not taken terribly seriously was the gravamen of a good deal of complaint in the 1950s and 1960s, and most of the complaint was true. Robert Weaver, while Kennedy's administrator of the Housing and Home Finance Agency, answered the critics defensively, noting that "in this Nation some 12 million families change their place of residence each year. Those forced to move because of urban renewal are but 2 percent of this total." It was not a strong response. Roger Starr defended more effectively with an attack: "If nothing is done for the family living in the dark, cramped, dirty, neglected buildings, with vermin, rats, sleazy plumbing, and the unconquerable dirt, we may be postponing the fears engendered by proposed change, but we do nothing to mitigate the ever-present, grinding reminder of inferior status which a shoddy home represents."

Even granting all that, a great deal depends on where the change leads the family. No very good records were kept of where the expellees went (only half of them applied for relocation assistance), but occasional studies by official bodies did indicate that most of them secured better housing than they had previously occupied (Weaver said only 8 percent of the evicted families had wound up in substandard housing)—at a considerable increase in the share of their income they had to devote to their homes. Surveying ten unofficial studies in 1972, Chester Hartman noted that in New York City the average rent increase for people displaced by urban renewal was 36 percent, and that in Detroit, while the average increase was only 16 percent, a fifth of "the relocatees" had wound up paying more than half their income for rent.

"Anyone who sees improved housing as the primary goal of

urban renewal programs,'' Hartman concluded, drawing on these unofficial studies (which drew a much uglier picture than the government's), ''can only be shocked by [the] results. Redevelopment projects are leaving 30, 40, and 50 percent of displacees in substandard conditions while destroying substantial numbers of decent low-rent units. Overcrowding is being only marginally relieved, and in some case exacerbated. General neighborhood conditions often have deteriorated or become less satisfactory, thus canceling out any gains in individual housing conditions. This is unacceptable public action, if not public policy. Further, once the overall stock of low-rent housing—decent or otherwise—diminishes, landlords tend to allow their properties to deteriorate and municipal services tend to decline. What may be rated as standard housing at the time the displacee moves in may be seriously substandard within a matter of two to three years.''

The damage may have been even greater than that. When a few blocks of slum dwellings were demolished in 1938 to build the approaches to New York's Triborough Bridge the rents of the remaining tenement apartments in the area rose by about 25 percent. By reducing the supply of low-rent housing, urban renewal probably raised the rents of all poor people in the cities where it occurred. Where dislocations were heavy, it became profitable for landlords in aging but still sound sections of the city to cut up existing units and put them in the market for slumdwellers, spreading the slums as well as the dwellers.

Small businesses, the kind that help people escape from poverty, were severely damaged: something more than 100,000 were displaced, and 35,000 expired of the shock. These were neighborhoods where bankruptcies of small businesses were common anyway, of course, and were likely to increase as the area ran down: a study by University of Chicago argued that ''the liquidation rate in Hyde Park-Kenwood [the university's home territory] was not excessive. . . . It was not as high as it might have become if renewal had not taken place and Hyde Park had experienced the same transition as Woodlawn [the university's neighbor to the south]. . . . Viewed dispassionately, dislocation at worst hastened the inevitable for a few.'' But *which* few may have been more important in this context than these authors were willing to admit; and their conclusion that the drop in total enterprise was caused primarily by the inability of newcomers to start businesses in the area is more deadly to the city's future than they realize, for a successful city—and by extension a successful neighborhood—is an incubator of businesses. It should also be noted that in evicting local businesses to gain clear title to land for redevelopment, the urban renewal authorities shamelessly victimized their owners, reimbursing them only for the value of their future leaseholds and giving them nothing for the value of the business itself even in situations where eviction meant certain death.

Perhaps the most destructive aspect of urban renewal, however, was the long delay between the announcement that an area had been adopted for this treatment and the actual construction of anything new on the site. Homes and businesses could stand in limbo for years, rapidly deteriorating until they were torn down; and then the land might stand empty for more years through the interminable processes of proposal, hearing, federal approval, court action, design, and financing. It was great for lawyers but not for anybody else. Rather than increasing the values of the adjacent property, urban renewal through decay and vacancy on the site eroded the values of its neighbors. And urban renewal promoted and legitimated the custom later known as "redlining"—marking off on a map at the bank an area within which property would not qualify as security for a loan.

"Banks and other lenders who blacklist city localities," Jane Jacobs wrote, "have done no more than take seriously the conventional lessons of city planning. Credit-blacklist maps are identical, both in conception and in most results, with municipal slum-clearance maps. And municipal slum-clearance maps are regarded as responsible devices, used for responsible purposes—among their purposes is, in fact, that of warning lenders not to invest here." Redlining had preceded slum-clearance maps (though perhaps not by much: Jacobs traces it back only to the Depression, when the banks exlcuded from the portfolios new loans in areas where the old loans showed high foreclosure rates); and it has become increasingly common in the years since the government gave up on "urban renewal" and began supporting "community development." The question of whether the lenders are causing or simply recognizing the hopelessness of an area can be extraordinarily complicated (as we shall see in the next chapter), and the FHA underwriters who refused to insure mortgages in declining neighborhoods were at least as important as the urban renewal planners in validating the banks' behavior. But the concepts of redlining and of land write-downs for redevelopment are precisely congruent.

The major support for city housing from the federal government in the years after World War II was Section 608 of the Housing Act, adapted from a wartime amendment that had made government mortgage insurance easy to get for developers of rental housing for war workers. The basic provisions of the wartime legislation were that the government would insure 100 percent of the "replacement cost" of anything a builder might put up, and that the judgment standard to be applied by the agency was simply that the project represented an "acceptable risk." This was on all counts considerably more generous than the original Section 207 of the Housing Act, which had permitted (and then only to charitable or limited-dividend groups or public agencies) insurance on apartment buildings only to a maximum of 80 percent of "value" after a finding that the project was "economically sound." Under the previous Section, moreover, FHA had controlled "rents or sales, charges, capital structure, rate of return, and methods of operation"; under the new section, private builders could organize as they wished and once nationwide rent-control had been ended they could charge whatever the market would bear.

At the war's end, the Truman Administration, advised by its economists that a hell of a depression was just ahead, fought desperately to persuade lenders to pump money into housing. At the same time, it was considered essential to keep interest rates low and to control both rents and the prices of building materials. To help returning veterans buy homes, the Veterans Administration was authorized to guarantee their mortgage loan. To help builders supply apartments for returning veterans, Section 608 was extended into peacetime. It quickly produced half a million badly needed apartments, but the price was a lot higher than the quality.

By 1948, the government was beginning to get some bad news about Section 608, and the insurance the FHA would write was reduced to 90 percent of replacement costs as of December 1947; but by then the genie had got out of the bottle. "Emboldened FHA officials," Charles Abrams wrote with deep distate, "had openly told builders they need have 'no risk capital or permanent capital investment' and that they could get back whatever money they invested 'before one spade of earth is turned.' FHA officials also told builders they could inflate their cost estimates, their land prices, their architectural and other fees. Builders did not hesitate to oblige. A whole web of deceptions was spun which made it possible for the knowledgeable to build projects with costs running into the millions without investing a dime. Some made dummy leases between themselves and wholly-owned subsidiary corporations at a spurious rent for ninety-nine years. FHA would then insure a mortgage on the leasehold and upon default would have to pay the fictitious rent for the duration of the lease . . . builders withdrew millions above what the projects cost."

A new verb entered the language: "to mortgage out," meaning to borrow more than the actual costs, permitting the builder to walk away with a profit whether he sold or rented his project or not. A Senate investigation in 1956 examined 543 mortgages and found that in 80 percent of them the builder had mortgaged out. Friendly estimators at the FHA had often approved loans that ran as much as 30 percent over the builder's costs, and land prices that were four and five times the builder's costs. "Failure to mortgage out," Louis Winnick wrote, "was due not so much to a sense of ethics as to a lack of building know-how or sheer bad luck." Adding insult to injury, the builders then took these profits as capital gains at a time when capital gains rates topped out at 25 percent and income tax rates went over 90 percent on the marginal dollar. Worst of all, construction standards were shoddy and design standards were stupid. Because the FHA estimated costs by room count, gave no credit for a large kitchen but did give half a room for an entrance foyer, the eat-in kitchen disappeared from American rental housing and the "dining alcove," a wretched nonroom, took its place. And all the benefits, of course, went to the builder; indeed, because the mortgage was inflated the rents had to be higher, and the tenant was the loser.

Section 608 died in a bloodbath of investigations and indictments before its benefits could be claimed by the developers of urban renewal sites. The write-down of land costs was an insufficient subsidy to convince builders that they could make money on rentals of high-rise apartments in dubious neighborhoods. Although part of the rationale for urban renewal had been the restoration of the cities' tax base, municipalities began granting real estate tax abatements to make the figures look better. The fish continued to lurk at the bottom of their pool, and in 1954 Congress passed Section 220 of the Housing Act, which set up a separate insurance fund for higher risk mortgages on properties in urban renewal areas and authorized loans at 90 percent of *costs* (including a builder's profit in the "costs," which meant 99 percent of out-of-pocket expenditures, except that a developer was separately required to put up a 3 percent equity before getting his loan).

Now there were entrepreneurs ready to build, but the lenders were wary; and Congress rewrote the charter of the Federal National Mortgage Association to permit what amounted to direct government purchase of mortgages insured under Section 220. About half the "private" money invested in urban renewal housing in the 1950s was in fact FNMA money duly appropriated by Congress. It turned out that the lenders were right to be wary: relatively few of the Section 220 mortgages were trouble-free (at one point *90 percent* of those held by FNMA were more than a month behind in payments), and roughly a fifth have had to be foreclosed.

The worst carnage came in the Section 221 program, created to help house people displaced from urban renewal sites, where nonprofit institutions were offered mortgage insurance for the full cost of an apartment house, at rates reduced to 4 percent (then 3 percent) by government subsidy. "You had a lot of developers with marginal land," says Philip Brownstein, Kennedy's Federal Housing Commissioner, "who would find a church or a charity with a heart of gold and a head of lead. By the time the developer got out, the sponsors couldn't remember how they had ever got into this thing in the first place, and it wound up on our doorstep.''

Some developers did very well by these programs. Congressman James Scheuer, for example, went from being just rich to being incredibly rich through urban renewal ventures in Washington, Cleveland, St. Louis, Marin County (California), Sacramento, San Juan, New Haven, and Brookline, Mass. (He was not a Congressman at the time, of course.) But the Mansion House experience was not unusual: *only* the packager, the man who knew how you navigated through the ''regs'' and the local politicos and the tax laws, showed a profit on the job. If the urban renewal housing was designed for the rich, there were only a few sites in what had been slums—where the surrounding neighborhood was likely still to be slummy—that might draw them. And if it was designed for middle-class occupancy there were cost factors, pre-tax and post-tax, that made it dubious competiton against the continuing and increasing lure of the suburban house. If it was planned for low-income people, an ever-increasing subsidy was needed to pay ever-increasing costs, in an economy where in fact the rents of privately owned existing units were rising much less rapidly than the costs of building new ones.

Three books—Jane Jacobs' *The Death and Life of Great American Cities* in 1961, Martin Anderson's *The Federal Bulldozer* in 1964, and Jay Forrester's *Urban Dynamics* in 1969—raised before the public the specter that government intervention in the economy of our cities has been not merely expensive for the good things accomplished (we could live with that) but has actually been harmful. Each of the three books is flawed by a central simplemindedness: Mrs. Jacobs', by the insistence that relatively self-supporting urban villages are ageless and vital, always and everywhere; Anderson's by the belief that ''free enterprise'' can solve all problems (Anderson was later Ronald Reagan's economic adviser in his run for the Republican nomination in 1976); Forrester's by what boils down to a two-factor analysis in which land for housing and land for industry are locked in mortal combat. But it is impossible to read any of these books without a feeling that its author has put his hand on something important.

Mrs. Jacobs was right in her denunciation of the Corbusier-inspired high-rise apartment development set in the greensward of a superblock without so much as a grocery store or an ice-cream parlor to serve the residents. Anderson was right in his claim that slum clearance under urban renewal spread more slums than it cleared and assisted the building of structures that would probably have been built without government help. Forrester was right in his projection that practices like those followed by Newark in the 1950s (when almost 60 percent of all housing starts in that city were sub-sidized), and by New York in the 1970s (when that city grabbed for more than one-third of all the urban renewal and subsidy money flowing out of Washington) would increase unemployment, de-moralization, and misery rather than reduce them.

What Congress saw, however, was all the vacant land—and the report of a Presidential Committee calling for the construction of 26 million housing units in the 1970s. The result was Section 236, which permitted HUD to subsidize apartment construction down to a 1 percent mortgage interest rate for low-to-moderate-income fami-lies—on a basis which once again permitted mortgaging out. It was as though the Section 608 experience had never been. Enormous tax breaks were made available to land developers for virtually no in-vestment of their own funds, fabulous quantities of urban housing got built, and the land cleared for urban renewal was finally oc-cupied again. In the four years 1969–73, the Nixon Administration oversaw the construction of three times as many apartment units for low-income people as had been built in the eight years of Kennedy and Johnson.

The results were even more scandalous than the study of Section 608 would have predicted. Section 608 housing went up at a time of grievous shortage so that everything was occupied (and, where loved, was put in shape and kept in shape), but Section 236 housing went up at a time when only a small minority of Americans really *needed* new homes, so that unsatisfactory units were quickly va-cated. The occupants of Section 608 were returning soldiers used

to a high order of social discipline; the occupants of Section 236
were veterans of the Sloppy Sixties. Costs of all apartment building skyrocketed with the jet power of government-induced demand
for new construction, and even with the subsidies the rents necessary to keep the ventures afloat rose beyond the means of the
moderate-income families for whom the program had been designed. Builders and lenders and unions had by the early 1970s
polished to a much higher sheen the art of screwing the government. Section 236 housing cost 20 percent more to build, per
unit, than comparable private apartment buildings in the same city
at the same time.

Of the four thousand apartment houses (with almost half a million
apartments) built before the Nixon Administration hurriedly shut the
window on Section 236, 1,600 had gone into some form of default
by 1975; and in 1977, after previously unimaginable "piggy-back"
rent subsidies had rescued from bankruptcy all that were possibly
salvageable, more than five hundred large apartment projects—a
billion-dollar investment—were still under water. A good fraction
of these buildings had turned into slums.

What urban renewal housing did achieve—and it was no mean
accomplishment—was the establishment of racially integrated
middle-income communities in the city. Southwest Washington
produced an integrated upper-income community. As early as June,
1963, Robert Weaver could point to forty-six "racially mixed"
urban renewal projects scattered through the country, and there have
been more since. Though some of these projects have tipped, more
are stable, and the existence theorem they offer has undoubtedly
been important in reducing the incidence of blockbusting and pan-
icky selling when the first black families move onto a block of
private houses.

But these things must be done very carefully if they are to work.
The shortage of middle-class housing for middle-income blacks has
persisted, and the opening of an avowedly integrated apartment
house will bring a disproportionate share of early applicants from
the black community; and if the occupancy looks predominantly
black, whites won't come. Moreover, except in a few places like the

metropolitan areas of Texas (the most integrated state—Fox & Jacobs has black salesmen selling to white homebuyers in Dallas), keeping a project biracial takes continuous effort. "The pressures toward resegregation of a racially mixed area," writes Morris Milgram, a pioneer and true believer in integrated housing, "do not allow a community to remain integrated for long without an active community program to maintain it." There is, of course, an open housing law, but most of the people who really care about integration shun it, because once the legal process is invoked—given the pious abstractions of the law itself and the inadequate information systems of the judicial process—housing managers lose all control over the selection of tenants, and the stable elements in the building (black and white) will flee.

"Lefrak City is remarkable in many ways," Samuel Lefrak told a meeting at the Harvard Business School in spring 1971, speaking of his development beside the Long Island Expressway in New York. "First, the facts. It consists of twenty eighteen-story apartment houses where 25,000 people live. On the forty-acre site are two office buildings, a post office, library, four schools, a 3,000-car underground garage plus a 1,000-car garage above grade, a shopping plaza, a 1,200-seat motion picture theater, swimming pools, tennis courts, squash courts, a health club, and much more. I built it all—the city did nothing. . . .

"Lefrak City is a prime example of the use of leverage. The property returns a bottom line profit of $6 million and has a value of $150 million. And those values, gentlemen, are not Wall Street paper values—they are land values, one-way values. Up! And up and up and up each year.

"In terms of meeting social needs, Lefrak City is a great success. It provided housing at an average rental of $50 to $45 a room. It offers employment right on the site. . . . Lefrak City has kept middle income families of all ethnic backgrounds within New York City. . . . Lefrak City pays New York City the sum of $5 million in real estate taxes annually. Before Lefrak City the piece of land on which it stands provided only about $16,000 each year in taxes."

Five and a half years later, Lefrak sat in his very large Manhattan office, old fowling pieces and a brass bas relief of Romulus and

Remus and the wolf behind him, heavy Spanish furniture, a rya rug
on the wall beside him, wing chairs in floral patterns for the guests.
He was wearing a vested green twill suit; a watch chain with a gold
shoe pendant rode high over his ample stomach. There was a red
Peter Max kerchief in the breast pocket of the suit, and a light brown
tie with dark brown horses' heads was neatly tied on his pink
checked shirt. Stout, balding, puffy-cheeked, vigorous. Only Max
Beerbohm could have drawn the spectacle tastefully or found an appropriate caption. Lefrak was smoking a cigar, and his blue eyes
concentrated on his visitor as he told the story of what happened
after a court—at the urging of black activist groups, the city's
Human Rights Commission, and the U.S. Department of Justice—
decided that he had been guilty of racial discrimination in accepting
tenants for Lefrak City, and would have to make up for it by taking,
no questions asked, all those applicants the activist groups would
now send his way.

"I am the victim," Lefrak said, "of a social experiment that
failed. OEO funded these groups to bus people here from all over
Harlem. It was the time of the agitation at Forest Hills, the city had
decided to make the public housing there a 75–25 ratio of white to
black. So what did they do with the surplus? Lefrak City got them.
Not only did I have to lower my occupancy standards, but I inherited the cancer. I lost my decent working people, they went to
Connecticut, Long Island. We lost in the last few years five million
dollars.

"Citibank called the lenders together; Lefrak City was divided up
between Prudential, Metropolitan Life, and the New York Bank for
Savings. Their greed had pushed me up on the interest, they hit me
for ten percent—instead of giving me aid and assistance were kick-
ing us—and now they were pushing on the amortization. But it was
a double-edged sword. Citibank said, 'Look, this guy's got a problem, and we've got a problem because he can walk away, he's losing money now.' The deal was to postpone amortization; they
wouldn't write down the interest rate. The New York Bank for Savings guy said, 'My board makes me get ten percent,' the son of a
bitch.

"Now I have a question. Do we get rid of the undesirables, or do we contaminate it further by following what the Justice Department wants us to do? The Housing Bill of 1974 passes with its Section 8, which said rent subsidy funds could be allocated to individuals, administered by the city housing agency, which gets a fee for it. I went down to see the Mayor; he said he'd asked for five thousand units. I said, 'Suppose I get you six thousand?' He said, 'Then I'll have a thousand units for finders, keepers.'

"I went to Washington and came back with twelve thousand units for New York City housing. So the Mayor decided he's got $150 million he can add to the general revenues of the City of New York, so they can vote themselves salary increases and lulus. He'll raise the rent in the city's housing from $200 a month to $400 a month, he'll keep the money. They wrote to Carla Hills [then Secretary of HUD] for an exception to the rules, for themselves; she turned them down in two days, and they realized they weren't dealing with just another landowner. Not that I used undue influence.

"I had to go to court anyway. I hired Roy Cohn to force both the city and the federal government to come through. Finally they did. We got rid of all the undesirables, the people who had come in through that Open Housing drive, and it isn't easy to evict seven hundred families in this town. They beat you out of four, five, six months rent, and the city had said we couldn't take more than one month's security. So I have seven hundred vacant units, and Section 8 money to subsidize new tenants. We are now [1976] relocating people en masse, primarily white, into Lefrak City; we got a special dispensation for it from Eleanor Holmes Norton [then chairman of the city's Human Rights Commission]. Our first rule is that we stay in hot pursuit, we jostle the bureaucracy every day.

"Fortunately, I can afford to hold seven hundred units vacant. We have the office buildings, and they're full—the Social Security Administration, ten thousand people. The federal government rents them. We brought that industry and those jobs to this city—they work twenty-four hours a day there, seven days a week. We pay the city seven million a year in taxes [they've gone up]; we run day-care centers, senior citizen citizen centers, a theater, even a camp, all the things the city does elsewhere with federal money, we have to do it ourselves. We also had to hire a private police force.

"From the housing end we hope to be self-sustaining. The office buildings and the retailing are profitable. It's all part of R&D. Anyway, what are you going to do?"

Anita Miller at the Ford Foundation (who doesn't much like Lefrak and sort of does like the people who pushed him to the wall) says the problem is really Lefrak's own poor property management, which shows up in everything from tenant selection to routine maintenance. An "intervention team" funded by Ford and HUD and the banks as well as by Lefrak himself has been busy at Lefrak City keeping the lid on, speeding the maintenance schedule and recruiting the necessary white tenants. Though most of the recruitable whites turn out to be elderly, and the mix of white elderly with black children and adolescents has not provided much fruitful "integration" elsewhere, Mrs. Miller thinks Lefrak City can be saved. An unofficial source at the Social Security Administration is less optimistic. "Our people, black and white, are still moving out of there," she said. "They're going out to Nassau County; they'd rather commute. Pity, too—it was so convenient for them. But they don't think it's safe anymore."

"The frequency with which one widely heralded federal program supplemented or replaced another," economist James Heilbrun wrote in a textbook in 1974, "finally gave rise to the suspicion that it was not merely the mechanics, but perhaps indeed the very premises of public policy that were mistaken. . . ."

Among the embarrassments of the urban renewal process had been the discovery—very surprising to the middle-class bureaucrats and academics who ran it—of the large numbers of well-kept, well-loved, low-cost homes owned by poor people that had to be torn down to make room for redevelopment in what the mapmakers had labelled slums. Frustrated in their plans to rebuild the cities and "wipe out the slums," the planners, bureaucrats, and legislatures turned their attention to what they began to call "neighborhoods."

Chapter 7

In the Neighborhoods

. . . as structures grow old and depreciate, and as the children of
resident families reach adulthood, there is a sharp reduction in the
neighborhood's vigor and in its ability to resist physical and social
encroachments. Combined with increasing housing maintenance
costs resulting from the structure's age, there is the threat of obsolescence as newer, more fashionable residential options become
available. As older residents gradually relinquish the constant
struggle to preserve the neighborhood intact, a process of owner
disinvestment may be set in motion. . . . The great bulk of central
city housing, especially in northeastern cities, was constructed in
the first part of this century. We are just now beginning to feel the
impact of a massive supply of outmoded housing, with more units
entering the obsolescence stage than at any previous time."

> —James W. Hughes and Kenneth D.
> Bleakly, Jr., urban studies scholars
> (1975)

Urban difficulties are not a matter of location so much as a phase in
the normal life cycle of occupied land.

> —Jay W. Forrester, computer mathe-
> matician turned social analyst (1969)

"You take an ordinary Baltimore brick and soak it in a pail of
water overnight, and you know what happens?" said Howard
Scaggs, president of the American National Building and Loan Society in Baltimore and chairman of the recently formed Neighborhood Housing Services unit there. "It dissolves. It's extraordi-

narily porous. You've never been able to forget about the outside of your house in this city: at the least, you've had to keep painting it.''

Baltimore is one of the nation's three great home-ownership cities (the others are Detroit and Philadelphia). Its working-class housing was not the tenement but the tiny brick-construction row house, twelve feet wide, perhaps forty feet deep, two or sometimes three stories high. An ornamented protruding cornice kept some of the rain off the façade. These were built by the tens of thousands, street after street, on the low-lying land embracing the harbor, south and east of downtown. Most of them were built before the turn of the century, without indoor plumbing, which was customarily added behind in a two-story wooden addition—kitchen on the ground floor, bathroom on the second floor. Often there wasn't room in the second-floor addition for a three-piece bathroom set; people do without a sink, and wash their hands in the small bathtub.

The one thing this housing had going for it coming into the post-World-War-II era was the fact that the façades had been maintained; they had to be, or the brick would simply wash away. But constant painting, as the owners of frame houses know so well, is an expensive and/or time-consuming proposition. Soon after the war, Scaggs says, ''the greatest salesman God ever put on earth came to Baltimore. Unfortunately, he was selling formstone.'' This is a word unknown to dictionaries, a kind of plaster aggregate that will adhere to brick and a wire mesh attached to brick. Properly sprayed and notched, it acquires the appearance of a series of slightly rusticated granite courses, all beige and pink. The stuff is immensely durable and all but impossible to deface, and it cleans with a hosing. On the individual house, it looks silly: visitors will finger it incredulously and shake their heads. On a whole street of houses, however (and such fully formstoned streets are fairly common in south Baltimore today), the effect has a perhaps deniable but credible charm. Put a granite flower pot to either side of the real stone (sometimes marble) stoop, which Neighborhood Housing Services has been doing, and you get an urban landscape of considerable originality, a monument to the indominability of someone or something. Here and there the brick façades remain, some with the long narrow windows that reveal an architectural desire to express the space on the surface.

The area where Neighborhood Housing Services is working in Baltimore is called North-of-the-Park, meaning Patterson Park, about a mile and a half from downtown through the great complex of Johns Hopkins Hospital and some high-rise inner-ring subsidized housing. Among its assets is Gordon's, one of the city's best restaurants. People work at the hospital, in the port, for the city, for Bethlehem Steel. There are only a handful of multifamily buildings among the four thousand structures in North-of-the Park, and all that handful are on street corners, with stores, some boarded up, on the ground floor. The streets are a continuous run of little twelve-foot-wide houses, more than 90 percent of them covered with form-stone, though nothing like 90 percent have in fact been renovated.

In the early 1970s, this neighborhood was badly on the skids. Most of the little houses were owned by landlords, who had come into possession of them originally, says NHS local director Tom Adams, "because they were people who had the trust of the community—the local undertaker, grocer, dentist, lawyer. The pattern was that someone died, and the survivors went to someone they trusted to buy the house. There was no public market. Money had dried up in the ethnic S&Ls. The presumption by most people was that nobody would want the house—this was not a nice place to live—people were going out to the suburbs. . . ."

The undertakers, grocers, dentists, and lawyers drifted away and pretty much neglected the houses. As George Sternlieb had found in Newark, the worst-kept facilities were not those owned and operated by the ruthless speculator, but those owned by an heir living some dozens of miles away, who never looked in on his property at all if he could avoid it; someone else picked up the rent and forwarded it. When the house became vacant, it had a tendency to remain vacant, while the vandals pulled it apart and maybe torched it.

One resident decided she wouldn't take it anymore: a brisk, small, solid woman in later middle age, Matilda Koval, wife of a television repairman; she is also an active member of the Coast Guard auxiliary, teaching prospective boaters to tie knots and the like. "There were too many absentee landlords and boarded up houses," she says, relaxing in slacks in her substantially remodeled living room, always the full width and now opened up to the full length of the house, with fluted pillars across the center of the room to hold up the second floor. The furniture is elegantly old-fashioned, with lace doilies for the stuffed armrests. ''We had two or three fires a week—the kids were setting the fires and then running. Trash accumulated in the alleys—nobody cleaned it away—and even in the streets. It really was getting to be a slum, and I wasn't going to have that around here.''

What brought the situation to a head was an announcement from one of the fire insurance companies serving the neighborhood that it was not going to renew people's policies. ''They said it was a target area with too many boarded-up homes. People came up to me on the street and asked me whether my insurance had been dropped, too; they were all complaining. We formed a block club and got together with SECO.'' The initials stand for Southeast Community Organization, a holdover from a vanished federal program. With help from SECO and the parish priest, Mrs. Koval sent well-publicized invitations to the mayor, the city council, the health department, and assorted other magnificos in Baltimore to come on a walking tour of the Patterson Park area and look at what was happening. ''You name them, they were all here,'' Mrs. Koval says with grim satisfaction. ''We didn't take them just on the streets; we took them in the alleys. We didn't care if they got their feet dirty.''

Among the institutions involved with SECO was Neighborhood Housing Services, an odd hybrid organizations sponsored by the Federal Home Loan Bank Board and the Ford Foundation. The Home Loan Bank had been sensitized by Congress to the fact that the Savings & Loans they supervised were withdrawing from the neighborhoods where the poverty program was most heavily funded. With loans unavailable for resale, maintenance expenditures shrank and the housing stock ran down. In a society where most personal wealth is in the form of equity in a house, the people who lived in these areas were becoming steadily poorer, despite the infusion of federal grants.

It is easier to denounce ''redlining'' than to suggest alternatives, for we live in a statistical age, and place is one of the natural categories to use when processing information. ''The truth of the matter is,'' says Lou Winnick of Ford, ''that the lending experience in these older neighborhoods has been dim. If you're a missionary you can work case by case—but in the real world we live by averages. Insurance companies are not irrational if they refuse to insure seventy-two-year-olds and do insure thirty-two-year-olds, even though some seventy-two-year-olds will live longer than some thirty-two-year-olds. The seventy-two-year-old says he's been redlined: he's from a long-lived family and he still has all thirty-two teeth in his head.''

No bank or savings association could as a matter of business judgment risk an uninsured loan on an apartment house in *most* urban neighborhoods today. An internal memo at one of New York's largest savings banks reported to top management in early 1976 that ''in an increasing number of instances, property values after an extended period of amortization payments reflect a higher loan-to-value ratio than when originally funded [that is, even in an inflationary time the market value of the insured building has dropped by more than the amortization of the loan]. In . . . good middle-income properties where no significant change has occurred in the neighborhood . . . gross rental income has increased by less than one third of the rise in expenses.'' That's New York, with rent control and the third highest real estate tax in the country; but roughly comparable statements could be found in the files of banks in Chicago and Cincinnati, Pittsburgh and Milwaukee. And the rational fear of lending to a neighborhood's apartment houses easily becomes what may be an irrational distaste for lending to the single-family homes in the same area.

Once the banks and savings associations stop lending money into a neighborhood, its deterioration is guaranteed: the prophecy fulfills itself. Owners can see no reason to invest their own money, even if they have it, because in the absence of mortgage financing on resale there is no way they could get back their investments. Single-family and multifamily alike deteriorate, and if density of occupancy is diminishing in the city the redlined neighborhoods become candidates for abandonment. Housing stock and human confidence are lost, and the maintenance of both is important to the future of the city.

''How do you move away from the confrontation?'' asks Lou

1974, and we worked out the basis for the corporation. By August
we had a corporation in being, an IRS certificate so we could receive tax-deductible contributions, and a computer set up for fund-
raising. In January 1975 we went into business.''

Ford put up $50,000 to get the Baltimore NHS started, with
$50,000 more to come after Scaggs and his friends had raised
$200,000 themselves, and had secured commitments from the Bal-
timore S&Ls to make every bankable loan NHS would bring in.
Among the major decisions of the planning period, taken against
Ford's advice, was to limit the use of the city code inspection force,
relying instead on volunteer inspection by local people. Director
Tom Adams, a rotund but dieting level-headed ex-seminarian recruited early on from the poverty program in Waverly, Rhode Island, was not averse to using inspection as a threat. "It's a tool," he
says mildly, "to help investor-owners decide if they want to sell''—
but he wanted to be sure NHS decided whom to threaten, to avoid
terrifying and evicting the elderly. (Mayor Warren Widener of Berkeley, California, has gone even further to keep the peace: "We tell
people we're doing a survey, not a code inspection.") City code
inspectors could not legally distinguish between a violation in a
home owned by an absent landlord and a violation in a home owned
by an occupant struggling along on social security.

Scaggs insisted that there be no publicity at the beginning. "Until
you know something is going to work," he says, "publicity is just
another frustration for people. This was a project of three groups—
neighborhood people, city people, and S&L people—who are traditionally at each other's throats. We agreed that nobody would speak
'representing the community' or 'representing the city' or 'representing the banking industry.' Everybody there was to represent
Neighborhood Housing Services. We put our office in a corner
building that had been boarded up for ten years, and it didn't cost us
one penny: all voluntary labor and donated material." In what must
have been an esthetic statement, albeit a wistful and mistaken one,
an incongruous cedar shake was applied to the upper stories of the
NHS office in lieu of formstone. No question it stands out in the
neighborhood.

"The city," Scaggs continues, "made a commitment to concentrate services. Sanitation was the first and biggest thing. Some of the alleys around here were just impassable with debris—old refrigerators, mattresses, you couldn't get a truck through. They cleaned it out: the idea was, once you get it in shape you can keep it in shape.'' The city also cleaned up Patterson Park itself, built tennis courts around the periphery to draw people from downtown and supply bright lights at night, restored what had been in the nineteenth century a popular bandshell, guarded the lavatories, and planted grass and shrubs as needed. New street lighting was installed in North-of-the-Park, and $15 million was spent on a storm sewer to end flooding in this part of Baltimore. The state had already given the neighborhood a full complement of new elementary schools: no old building was still in use. The parish Catholic Church invested by completely remodeling its basement as a community center, the meeting place for all community groups; and a new parish priest became active in the committees. The local S&L office was modernized top to bottom. Meanwhile, NHS went to work on the housing, block by block rather than by individual building.

Houses for rehabilitation came from three sources: the city, which had title to sixty-three vacant properties seized for nonpayment of taxes; investor-owners who were more than happy to get some money out; and residents, who planned to resume occupancy when the work was done.

The city sold twenty-three vandalized and burned-out hulks for one dollar each to ''urban homesteaders,'' who would do much of the rehabilitation work themselves (five were taken, and done proud, by Korean immigrant kitchen help at Gordon's). Another fourteen were rehabbed by the city itself for sale to homeowners, and seventeen were put in shape for rental.

To help investor-owners sell, and to guide purchasers who rely on its services, NHS maintains a roster of appraisers who also estimate the cost of the necessary repairs: ''It's no-stop shopping for a potential buyer,'' director Tom Adams says cheerfully. In general, Adams reports, absentee owners have been getting sums about midway between what another investor might pay and what they could hope to sell for, after a long wait and before payment of commission to real estate brokers, if an unsponsored homeowner bought the property. Prices have been rising. In the early months, most houses sold for as little as $3,000. Between July 1976 and February 1977, homes sold by investors to tenants averaged a little over $6,000, and homes sold to outsiders averaged about $6,500; sales in process in early 1977 averaged $7,644 to sitting tenants, $8,416 to outsiders. The three-story houses completely reconditioned and equipped by the city sold for up to $20,000, greatly encouraging rehabilitation work on privately held property.

The area is geographically divided into three community associations, each of which has elected or appointed a block leader for every block in its jurisdiction. NHS hears from the block leaders about houses that become vacant, gets in touch with the owner (Adams knows who owns every parcel) and notifies its list of people who want to buy. Keeping on good terms with the investor-owners is important, first because they often own more than one house and may make others available if one deal goes well; second, because they are the prime sources of information about tenants who have put in applications to buy and renovate and must now be vetted for loan purposes. NHS arranges waivers of downpayments for people with perfect rent records, and especially for tenants who have had deals with their landlords—not uncommon in areas like North-of-the-Park—by which they got rent reductions in return for keeping the place in good condition. And NHS has its own "high-risk" loan fund, using money from Ford and assorted donors (including the Morris Goldseker Foundation, named for a man a congressional committee once labeled the most unscrupulous urban real estate dealer in Baltimore). After two years, NHS had processed for the neighborhood 283 loans totaling $2.3 million, of which $1.4 million was from the banks and savings associations, $550,000 from the city, and $373,000 from NHS itself. There had been one default, of $600.

Most of the work has been done by contractors from within the neighborhood. "It's one of the advantages of being in a working-class area," Adams says. "At first they were reluctant to become involved. The city has a reputation for being slow pay, and the red tape was bad. We pay within a week." Much of the work, even apart from the "homesteaders," is done by the purchasers themselves. This observer was taken to visit a formerly burned out shell being put in shape by a lean and muscular baker in his fifties and his wife. He was mounting a railing on the stairs, and she was laying vinyl tile on a bathroom floor. The washing machine was due to be delivered that afternoon. He had bought the structure for $1,700, NHS had arranged a loan of $10,300 for him—"and," he said, "I had some money of my own. I figure $15,000. Where can I get a house for $15,000?"

Adams knows of twenty-two cases where no loan at all was required to pay for extensive rehabilitation, and cannot even guess at the number of people who simply went to the bank where they had their checking or savings accounts and without any intervention by NHS borrowed the necessary money. "As soon as a few houses started to look better," he says, "the whole thing changed." He can tell anecdotes by the dozen, about people who had planned to move, looked at house prices in the suburbs, looked at what they could buy if they stayed home and put some money in the old house, and decided the old neighborhood could make it after all. "If you'd come into this neighborhood the way it was," says Howard Scaggs, "you'd have got the impression nothing but poor people lived here, but that wasn't true; there were lots of people with moderate means, working hard."

Ford's Winnick says that all successful rehabilitation programs come out that way: "The building permits are always much greater than can be explained by the loans. They're not great borrowers, they don't like paying interest; the money comes out of the mattress. It isn't just in America. We've had cases in India where you couldn't explain the building permits by the loans. I've talked to economists about it in other countries, and they all say you never know about the informal flow of savings, but it's enormous." Bill Whiteside says, "Buildings are a little like the human body. They regenerate themselves." It's never easy to measure how much is invested this way, because the people making the investments have a felt need to dodge the tax assessor.

North-of-the-Park is safely in the recovery room. The neighborhood has even drawn its first permanent bachelor, always a good sign. (Matilda Koval went over to snoop at his work in progress: "It really is pretty, with a patio in back and a redwood deck. On the second floor the whole back room is a bathroom, with a sunken tub. And that house was all boarded up.'') The danger is that prices may go too high, drawing speculative purchasers and tax assessments that punish renovators who need all their money to pay their newly incurred debts. As always in the modern city, too, the police protection must hold up. The community groups have developed a system with the police department, by which each resident gets a number to call and a number to give, changing every month. ''The kids have police radios,'' Mrs. Koval explains, ''and when you give your name and address the kids know where it is and who you are, so a lot of people were afraid to call. Now, with the number system, the police know where to go and the kids don't know they're coming, so the police come around the corner and catch them.''

The streets and alleys have been kept clean, and Matilda Koval's monthly meetings at the church are well attended, usually with politicians and officials of the sanitation and police departments in the audience, available for questioning. Asked how long he thought North-of-the-Park would need such special attention, Scaggs said seriously, ''As long as it's in existence. It's well worth all the affection it can be given.''

Rehabilitation programs can be run by public agencies, too, and also by private builders with an instinct for picking their way through thickets of bank procedures and public regulation. Perhaps the outstanding publicly run program is in Tampa, Florida, where A. William Benitez directs a Community Improvement Office in the Metropolitan Development Administration. Benitez grew up in construction because his father was a builder, and came to public office in the mid-1960s when work was slow in his father's shop. It is quietly important to his job that he supervises the housing code inspection unit as well as the fancier grant and loan operations. Benitez is the author of *Housing Rehabilitation: A Guidebook for Municipal Programs,* published by the National Association of Housing and Redevelopment Officials. A relatively young man— not yet forty—he comes to work in a small rear office one flight up, wearing a leisure suit and cowboy boots. Like most people with

Spanish surnames in Tampa, he speaks with a rich Southern accent, the way people with Italian surnames in New York speak Brooklynese.

Benitez has money from the city government for a high-risk loan fund ("that's the only kind of loan I make"), from FACE (Federally Assisted Code Enforcement), and from the federal Section 312 program, which makes what are not supposed to be risky loans at 3 percent, for rehabilitation of one- to four-family houses. One of the strangest phenomena of the declining months of the Ford administration was the attempt to dump Section 312. "I watched them trying to destroy a program that returns a million and a half a month," Benitez says disgustedly, "and promoting subsidy programs from which nothing ever comes back." Fortunately, Congress supervened. There is also a federal *grant* program, Section 115, for the same purpose, but the Tampa city government is afraid of it, knowing no criterion by which one family should receive and another be denied such a gift. Where the family can't afford to pay anything toward the costs of its rehab job, Tampa makes available from the sale of a bond issue a no-interest mortgage loan, to be recovered when the house is sold or passed on at death.

There are limits to the uses of rehabilitation. The Boston Redevelopment Authority insists from its experience that "A housing restoration program must distinguish between problems which can be resolved by a single injection of public resources as contrasted with problems that will require more sustained public support. In areas where household income was adequate to guarantee routine upkeep, one-shot fix-up funds like federal 312-115 loans and grants, coupled with area-wide code enforcement, were effective in returning areas to good condition. But in some lower income areas, these tactics only temporarily improved housing because inadequate cash flow is again leading to deferred maintenance." Benitez might go a little farther than that, but not much: he wants to save neighborhoods in "early decline," and in his book he cautions that "under no circumstances should the percentage of dwelling units of the 'dilapidated' and 'terminal' categories exceed 10 percent of the total number of dwelling units in the selected area."

"Rehab is not a substitute for demolition," Benitez says. "When you start a program it's imperative you start where you're guaranteed success. If you put all your eggs in the wrong basket and you blow it, you're finished. This is a difficult sell, and believe me it's a sell—you have to be able to drive the politicians through, invite the TV cameras in, after the first year.'' His first area in 1968, then, was a Hispanic neighborhood east of the city, developed originally in the 1920s in the first Florida retirement boom, with houses on plots fifty by a hundred or larger facing paved streets, some of them divided roadways with planting (now overgrown) in the middle. There were 1,500 units, of which 1,100 had been tagged by his inspectors as ''deteriorating but not dilapidated.'' It was an area in ''early decline.''

The tactic is to tell people that their neighborhood has been chosen as a ''target area'' for improvement, and wonderful opportunities are opening up for them to better their housing. Neighborhood meetings are called to explain the program, leaflets are slipped in mailboxes, and only then are the code inspectors sent around. ''We do a lot of PR first,'' Benitez says, ''but when you get the letter it's written into the law. There are code violations in your house that *must* be corrected.'' These were all single-family homes, almost three-quarters of them owner-occupied, virtually entirely white.

For jobs costing less than $15,000 on his inspectors' estimates, Benitez developed a list of contractors interested in doing this work (there were about two dozen of them in 1977); for jobs costing over $15,000, the law required him to advertise. (''The response to the ads is zilch,'' Benitez reports. ''Small contractors don't read the papers, and large contractors aren't interested in rehab.'') The homeowner himself chooses names from the list: ''As a homeowner,'' Benitez says, ''I don't want to be lumped in with five other homeowners. I'm going into hock for twenty years—I want it to be *my* job.''

The city caught a fair amount of flak for its first choice of a neighborhood where the average income ran about $10,000 per family, but Benitez fought back. ''We don't go into the worst areas of town,'' he says, ''but all our programs work.'' Another problem was the federal government's desire to have rehabilitation conducted within areas Washington had designated for urban renewal, a desire backed with a one-third subsidy to the city for work done there. "They want you to rehab in areas where the drainage is bad," Benitez says. "How can you expect people to get excited about their new house if every time it rains they can't get out of it without a damned boat?" The first area Benitez chose required only thirty-five loans from the city itself; all the rest were deemed "bankable" by the bankers. "We touched only thirty-five houses directly," Benitez says, "and seven hundred that were below standard were brought up to standard."

Later jobs—the most recent in an all-black neighborhood beyond a disastrous housing project to the southeast, in an area where the city has never paved many of the streets—have been harder to make work, especially because Benitez insists that the standards on his rehabs must be as high as the FHA standards for new construction. "We want a thirty-five-year remaining economic life, and that almost always means a new roof. We go in and we find the roof was put up on two by four rafters, twenty-four inches on center, and there's no way we're going to put new sheathing on that; we'll go up and put on roof trusses." Particularly when dealing with wartime and immediate postwar construction, the rehab contractor must sometimes rebuild the house on its piers. ("You find termite damage," contractor Andy Barrios said, brushing sawdust from a luxuriant mustache and looking around at a crew cutting millwork to size in a shotgun house. "Poor people can't afford to take care of their houses the way rich people do.") Benitez admits that often his owners have to spend as much on rehab as it would cost to build a new house; but the availability of Section 312 money at 3 percent means that the owner pays considerably less per month on his mortgage (and pays it off quicker, because Section 312 loans are written for twenty years) than he would pay for new construction.

Also, he says, the processing burden is less: "The government has created such a monster when you subsidize new construction. That contract is two hundred pages long." It is hard to believe (but must be true) that getting approval for new construction can be more paper-bound than the procedures Benitez outlines in his book, more than two-thirds of which by bulk is devoted to sample forms. He lists twenty-eight separate steps in the restoration area process, and no fewer than sixty-seven forms, every one of which must be filled out.

(This is of course one of the great advantages of the private NHS: "Some people came with a professor," says Baltimore's Tom Adams. "He said, 'Tell me about the regs [regulations]—who do you apply to?' I said, 'Forget it—there aren't any regs.' "

The cities themselves can make these problems: a Catholic priest working to rehab apartments in New York's East Harlem reports that only a hundred or so units in slums are in work in that city because the buildings department will not issue a certificate of occupancy until all the work on all the units is completed, and if a rehabbed unit is not occupied immediately vandals destroy it overnight. "The rules of the game," he told a conference in June 1977, "are at a point where there's no game any more—there's just the rules.")

Because he as well as the owners must operate under a code compulsion, Benitez tends to make loans more to the property than to the person. Still, the specific of what should be done to remedy violations can be matched to the owner's resources. "If it's an elderly person on a pension, hey, he can't afford to maintain that house. So we'll put on aluminum siding and eaves, so he won't have to paint at all. But I have no qualms about making a twenty-year-loan to people of any age. The oldest owner I've made a loan to is ninety-three. That was two and a half years ago, she's in fine health and she loves her house. And she has a fine house—the people who take it over will be glad to maintain it."

What Benitez has going for him, of course, is the immense vitality of the Tampa Bay metropolitan area, which every year ranks in the first three in the nation for population growth. There are some rubble-strewn lots in Tampa where once there was housing—the Spanish district in Ybor City is the most visible—but on investigation it always turns out to have been the government's doing: an urban renewal project was started, taken through the demolition phase, and then for some reason halted. Benitez can run into furious objections to his program from homeowners who are perfectly content with their house as it is and have no desire to take on debt to fix it up, but by and large the people who spend money on Tampa housing can do so with perfect confidence that the housing market will return their investment to them—plus a profit—when the time comes to sell. This vitality also, of course, secures the banker's loan.

In St. Louis, almost nobody who lives within the city limits can feel that confidence. The entire middle ring of the city is a zone of devastation that must be experienced to be believed—great blocks of dirt (grass won't grow because of all the debris), with occasional surviving houses still poking up onto the horizon. From 1950 to 1975, the population of the city of St. Louis dropped by 40 percent, and what remained was economically feeble: in the 1970 census, one-eighth of the white population consisted of women over sixty-five. The homes where all these families had lived were demolished: what else could be done with them? One of the strangest fallacies in contemporary discussion of urban problems is the idea that people leave the cities because the housing is abandoned; obviously, the housing is abandoned because the people have gone.

North of the expressway that bisects this city, the Germans, the Irish, and the Jews have all fled to the western and southwestern suburbs; and now the blacks they fled from are fleeing themselves in the other direction, to the northern suburbs. South of the expressway on The (famous) Hill (where Joe Garagiola's brother still waits on tables in the restaurant), the Italians have held on, and between The Hill and the Mississippi on the south side an older middle-class neighborhood clings to superbly built homes that can be bought for about half of what such homes would cost elsewhere in America.

Residentially, this was the best built city in America. The Baltimore brick has to be covered somehow or the house leaks; the St. Louis brick is impervious to wind, weather, and age. Tearing down the houses of St. Louis, the contractors carefully stack hundred-year-old bricks on pallets to be taken to the railroad and shipped off to cities where luxury builders want antiqued first-quality brick for new construction. Moreover, the residential *layout* of the town was extraordinarily satisfying, with many private roads serving self-contained middle-class areas of twenty to fifty homes on wooded lots. Even the row housing, much of it, had a sturdiness and a size most unusual in the late nineteenth century, when most of it was built. Gone now, a lot of it, forever.

It's hard to say what went so desperately wrong in St. Louis. Urban historian Brian Berry argues that in those handsome houses, "behind a façade of evident structural soundness, lay antiquated plumbing and heating, crumbling plaster, ancient wiring, and rotting floors." School desegregation may have had a more severe impact on residential choice in this city, just north of the Mason-Dixon line, than on others. St. Louis had less chance to keep those who wished lower density living than almost any other city in America, because its municipal borders had been restricted to a tight arc around the river by a referendum in 1876, when city dwellers were still trying to avoid the costs of supplying services to those settled on the outskirts. (An effort to reverse the results of this referendum almost a century later failed in both the city and the suburbs.) Government policies consistently made things worse, reaching a peak of shell-shocked inanity in June 1973 with an elaborately printed Development Program which after two years of rapidly accelerating abandonment simply reprinted a 1971 Plan Commission recommendation for the construction of 45,000 new dwelling units within the city limits. Bungled urban renewal dispersed the least civilized element of the community into neighborhoods that were on the edge anyway. The last nail in the coffin was the federal Section 235 program, which by encouraging poor people to purchase older homes they couldn't possibly maintain helped transfer sound houses to owners who could do nothing with them, in the end, but run away.

"This city is abandoned and isolated to an extent that you couldn't believe modern, civilized men could have ignored it," says Leon Strauss, whose Pantheon Corporation is the biggest redeveloper in town. He is a chunky engineer with long graying hair and a thick sandy mustache, whose office is a reconditioned private house in an endangered neighborhood near Forest Park, where the great Fair honoring the Louisiana Purchase was staged almost three-quarters of a century ago. A New Yorker originally, brought up on the Lower East Side, where his father was an officer of the International Ladies Garment Workers Union, he came to St. Louis after two years in Israel, building homes to empty the DP camps. One of his first jobs was on the Pruitt-Igoe housing project so publicly destroyed a few years ago: "It's quite an experience to see something you built destroyed in your lifetime." He had risen to a vice-presidency in the giant Millstone Construction Company when he quit "because I couldn't stand by and see a city I love go down the tube with nothing being done about it. I don't blame anybody for running, or for wanting a house out in the country. I do blame people for turning their backs." He started on his own in 1972 with only $20,000, but, he says, "the banks knew about me."

What makes Strauss's operation feasible is that St. Louis has written off its losses. No bank is still carrying these vacant lots in its mortgage portfolio as though they held income-producing apartments, and the city knows that no real estate tax revenues can be derived from that land in its present condition. Thus the city can compete against its suburbs with a considerable cost advantage. "I can buy land less than two miles from the center of St. Louis," Strauss says, "for a cent and a half a square foot, and if there's a building on it, they throw that in. So I can afford to give people something nice for their money."

Strauss is working mostly in four areas, one in the relatively stable south side (where he pays considerably more than a cent and a half a foot), one around the medical schools where he has some help from the universities, another on a 100-acre rectangle of once middle-income small apartment houses and detached homes on the other side of the park. On this one, he has a $3 million initial commitment from the Mercantile Bank for a project that will eventually recondition or build 2,500 residential units—and the loan is at the bank's prime interest rate, "though the only security they have is a first deed of trust on abandoned property." What is probably closest to Strauss's heart, however, and furthest advanced, is a development district called Jeff-Vander-Loo, which abuts on the eerily vacant fifty acres that once held Pruitt-Igoe. His "JVL Housing Corp." has already rehabilitated about 1,400 units and continues to own and operate about three hundred. "I could go broke there," Strauss says, not without concern.

Jeff-Vander-Loo is a mix of three-story row houses with the main entrance on the parlor floor, reached by high stone steps; small detached houses; new low-rise apartments—and vacant lots. Strauss has been converting the row houses to three-flats and building the low-rises; he avoids single-family jobs. ("Poor people," he says flatly, "can't afford to own homes.") Rents run $102 to $150 a month on federally subsidized mortgages. The area is virtually all black. Strauss speaks of four major factors in the rehabilitation:

First, active community organization headed by an "inspiring" black leader, Maeler Shepherd.

Then, financial help from the National Corporation for Housing Partnerships, a federally chartered housing invesment fund of $100 million raised by selling shares to large corporations that wish to do something socially useful and are willing to take their payoff in tax deductions (getting NCHP in the package was a triumph for Strauss: he reports that George Brady, the president of NCHP, "told me, '*No* scattered-site rehab. I know Jeff-Vander-Loo, and I'm not going to have that shit in our portfolio.' I said, 'Come look at it; these are special people.' He sent eighteen people here, and they were *tremendously* impressed.")

Third, the Mennonite Church, which wanted to do something in the city and established a new church in Jeff-Vander-Loo. "They've been the rock," Strauss says. "Lots of tradesmen are Mennonites, which helps us get workers. They're saints; they're real—like the kibbutzim in Israel in the 1940s."

Finally, and from Leon Strauss's point of view most important, Don Vinson, his own manager in the area. Vinson, a dark, handsome man with a brush mustache and gray-flecked curly hair, had been a bus driver who liked to throw darts at a bar where Strauss used to go for drinks when he worked for Millstone. "This was 1968," Vinson says, "and we had these, you know, social problems. Leon had these attitudes . . . we used to come by his house, at two in the morning and wake him up, just to talk." Millstone adopted an affirmative action program, and Strauss recruited Vinson from the bus company for a training program. When he left Millstone he took Vinson with him.

Vinson's office, neat as a surgery, takes up the parlor floor of one of the converted three-story row houses. The walls are hung with framed black-and-white photographs that have special meaning for him. One, for example, shows a uniformed football team of black high-school kids in 1948. Vinson points out some faces: *"He*'s a

movie star, on Kojak; *he*'s with a construction company; *he*'s a lawyer, *he*'s with Monsanto—we had some good people in those days,
too.''

Vinson's most significant task is accepting tenants (there is no
need to recruit them: when J-V-L Housing Corp. announces the
opening of a new section, the fire department complains about the
crowds). ''The whole key is perspectives,'' Vinson says. ''Leon
has a perspective as a developer. You as a writer have a perspective.
Me—as a black, born in this community—I have my perspective.

''The black community would work,'' Vinson continues, ''if
people had more control over who came in, and could write their
own rules. A politician owes someone a favor. His kids may not be
disciplined, we don't want them. The politician says if he got a better home his kids would shape up—but that's not true. Giving someone a house doesn't change his sense of personal and social responsibility. You have to make judgments on your experience. You do
employment verification, credit check; find out where he's renting
now; and someone from the neighborhood drives by the house. People can be ever so poor and still keep up the appearance of the cur-
tains. If people are on the borderline, you go look *into* their house.
How's the hygiene? Are the kids running all over the place, can't
make 'em sit down?

''You have to use your mother wit. School—you check the kid's
age with his grade. If he's a year or two behind in school, nine times
out of ten that kid is going to be a problem. Single women—it's impossible to have a housing complex in this earth without single ladies. I can't keep somebody out for that: my mother raised five by
herself, and she didn't do too bad. You have to compare these
women with your mother, your wife, your sister, your female associates.

''Unconsciously, you judge people by a style—the way they
dress, the rhetoric, habits of lying. . . . But you have to get behind
that when you're deciding who to rent to. I want an economic
mix—I don't want one hundred percent rent supplement families.
Otherwise you might as well put Pruitt-Igoe up again. We want people who work to live here. Maybe the federal government will prove
us wrong, but we want working people.''

Vinson has a number of targets he knows can't be hit. ''If the city, the state, and the federal government want to do something to enhance the values of Jeff-Vander-Loo,'' he says, ''the best thing would be to set up a private school system here, one for boys and one for girls. We're not talking about what's against the law, we're talking about what people need. That would guarantee stability and demand and growth in the community. When they talk about community stability, of course, they're always talking about the black community. The white community moves and nobody worries. But we've never shown that kind of tranquillity and growth.

''We need a community where people can take pride in the schools—and in their streets. If people come into a community and don't do right, we shouldn't need four, five, six months to get rid of them. It's not being dictatorial—it has to do with judges believing in the city they're serving. These knuckleheads have to get it into their heads that if you want to save the city you're going to have to have some order.

''Get me a sweeper truck that sweeps the streets three times a week, a sewer truck that flushes the sewer every week. Then people will start cleaning up. Sure, stuff will wash out of the vacant lots, but if you assign a crew to the area and just *do* it, we can take care of the rest. I lived eighteen months—*achtzehn monat*—in Germany, on the German economy, when I was in the service. The kids dropped candy wrappings in the street there, too—but the city cleaned them up. In 1948 around here—when they took that picture of our football team—we were poor as church rats; but our streets were clean. The city flushed those streets. Take this neighborhood—we have boundaries—and see that here the streets are swept very methodically. It would change the whole complex of this community.

''And if you can't do that, don't holler that the city is falling down. What we need in this city is a good black community. That would force other blacks to say, If you're going to Vinson's neighborhood, you'd better get your act together. That's what people say when they're going to Ladue.''

Night had fallen by now, and Vinson took his visitor around the neighborhood to show off what had been accomplished, what remained to be done. On one of the street corners, even on a fairly cold night, there was a crowd of young men. "That's a place," said Don Vinson near despair, "that sells liquor on credit, which is against the law. You think the police don't know about it? This city has the personnel; they need the supervisors."

In its 1971 study of housing abandonment, the Urban League drew its examples from seven cities: New York, Chicago, Detroit, St. Louis, Cleveland, Atlanta—and Hoboken. The last obviously has no place in this list: squeezed in a strip between the Palisades and the Hudson River opposite New York, it is home to only 50,000 people, less than half the number that could be housed just in the dwellings abandoned every year in the big city across the river. The purpose of including Hoboken seemed to be that this place was *really* doomed, a hopeless combination of overcrowding and abandonment, its residential neighborhoods soon to be reduced to the universal rot of its abandoned passenger docks. Less than six years later, the Wall Street Journal carried a front-page article about how Hoboken was the rehab wave of the future, a town where a high fraction of an antique housing stock had been brought up to snuff, the working class was well housed, and the middle class was returning. A "Dear Citizen" letter from Mayor Steve Cappiello, offering the owner-residents of the city's five-to-ten-family tenements free architectural assistance and twenty-year rehab loans at 6 percent, claimed that as of early 1976 "new," "newly rehabilitated," and "improved" housing units totaled more than 15 percent of the city. "Property values in Hoboken," the mayor proclaimed, "are rising steadily and will continue to rise."

Hoboken in fact has substantial assets: the combination of the Erie-Lackawanna railhead and the river makes it convenient for heavy industry; Stevens Institute is a thriving science school; and the garment district of New York is more easily accessible from Hoboken (by express bus through the adjacent Lincoln Tunnel or by Port Authority subway train) than from most parts of Manhattan itself. It had never been hated, as slums are; even the Urban League admitted that despite growing abandonment, "There is considerable pride of ownership and a considerable amount of home improvement activity." What was wrong was the sheer age of the city, most of its housing built in the nineteenth century, and the absence of mortgage money for purchase or rehabilitation. There was an urban renewal plan that called for tearing down much of the multifamily housing in the city (and eventually no small proportion of the brownstones too), but there was no reason to believe it would help.

"We decided to work in the areas of transiton," says Joseph Cicala, the city's athletic young housing director. "If we didn't stabilize what was worth stabilizing, we were going to become Newark or Jersey City. Those were the people who were moving out, programmed by the media to move out. They wouldn't invest money: you offered them a home improvement loan, they'd say, 'Why? All the walk-up apartments are being milked—look at the tenancy. And the garbage falls in our backyard.' The only thing that was holding them was that financially they were trapped; at the prices for Hoboken housing six years ago, they couldn't get anything out if they sold.

"What we had to do was take care of the multifamily. This was contrary to Hoboken's master plan: the planners wanted to tear down all these eighty-year-old buildings. We went to Washington, got ourselves designated a Project Rehab city, got a special pot of money for relocation similar to urban renewal. Then we took what everybody called the Tootsie Roll flats, because Tootsie Roll had used it for worker housing, to import cheap labor from the islands [i.e., Puerto Rico], and we put that in shape. When we began doing some others, the people in the brownstones, who had sat back for a year and a half, started coming in. We had some innovative programs—like, we took the first 20 percent of the fire insurance risk, to make insurance available cheap in Hoboken. To get the first private home improvement loan in Hoboken we had to go to Jersey City—no bank here would take it. Between 1971 and 1974, more urban home improvement loans were made in Hoboken than in all the rest of New Jersey put together. We have zero defaults."

The star of Cicala's multifamily show is an avuncular figure in an Eisenhower jacket and a cloth cap, Walter Barry, who came to Hoboken in 1972 to found what he calls Applied Housing Associates. He had been a union organizer, international representative for the United Electrical Workers in the New York-New Jersey area, before he retired in the mid-1960s. His son Joseph was emerging from law school at about this time, not particularly eager to practice law, and the two of them went looking for something socially contributory they could do together. They found housing, in riot-battered Newark, raised money, and rehabilitated eight hundred units in three-story and six-story wood-frame buildings in the western part of town. Luckily, they were just finishing these labors when half a dozen of the major insurance companies decided they had to do something about the cities and formed North American Development Corp. as a consortium for investment in low-income urban residential realty. The Barrys were able to sell out to North American, which wanted to get a fast start, and which presently went broke in Newark as elsewhere; and the West End of Newark became what the Urban League calls a "crisis ghetto." "We learned from that," says Walter Barry, "that you must be selective in your housing stock—it's very difficult to rehibilitate wood frame."

Other lessons were learned, too. "A rehab project," Barry says, "should abut a stable community, so you can use it as a beachhead to move deeper. If you start in the center, you get inundated by blight. To attract decent tenants, they have to believe the neighborhood is coming back. Then, the project must be large enough to make an impact, and to make sure the economics are there, to pay for a maintenance staff and an activities program, make the housing into a community. Finally, without tenant selection everything will go to pot. You have to get stable families, which doesn't necessarily mean two-parent families. A family with a strong mother and children may be better than a family where the husband and wife are always beating each other up. Having attracted stable families as a base, it's easier to handle some marginal families: peer pressure is more important than anything a landlord can do. We've never had graffiti on our walls, because the kids who might do that know they would be—literally—killed."

The Barrys first big Hoboken project was a pair of block-long seventy-year-old brick-exterior railroad flat tenements that backed onto each other, with the usual clotheslines and filth in between. Forming an investment syndicate and borrowing through the federal Section 236 program (under which mortgage interest rates are reduced to 1 percent), Barry completely gutted the insides of the two buildings, leaving nothing but the bearing walls, and installed elevators and hallways, laundry machines, modern equipment. The old steam radiators were torn out and a hot-water heating system substituted. Between the two buildings he bricked over the pavement, tore out the laundry poles, supplied a leisure area for older people, a playground for toddlers. The result, for a little more than $8 million, was an almost self-contained 316-unit four-story apartment complex, home for nearly 1,200 people, 60 percent of whom pay their own way without further subsidy at rents that range from $145 a month for an ''efficiency'' to $215 a month for an apartment with four small bedrooms.

Applied Housing now has about nine hundred apartments rehabilitated and under management, with six full-time maintenance men (who double as investigators of applicants for apartments: one of them visits each prospective tenant, at home). There are also twenty-six part-time ''supers,'' elderly residents who serve a quasi-concierge function, keeping an eye on things, checking up on maintenance requests, etc. One of the maintenance men works a three-to-eleven shift, so working people can call about problems. Costs are watched carefully. The temperature in the boilers, for example, is reset every hour, to avoid wasting heating water when the outside temperature doesn't require it. After four years without a rent increase (tenants pay their own gas and electricity), not only was the mortgage current in spring 1977 and the replacement reserve full to the brim, there were some extras available—money to start a teenage club in the basement, for example, and to install a sauna near the maintenance men's locker room. (''If you're working on a boiler,'' says maintenance chief Eugene Fernandez in a Spanish accent, ''a tub never gets all the dirt out; but a sauna does.'') The secret is 99+ percent rent payments on a budget written with a 5 percent vacancy factor.

On the units the Barrys were rehabilitating in 1977, some of them a few miles away in North Bergen, the costs run about $30,000 each, 20 percent above the costs in Hoboken five years before— over a period when construction costs generally are up nearly 50 percent. ''There are economies built into a situation where you have a team working together over a number of jobs,'' says the Barrys' architect, Joseph Vitullo. ''The architect learns from the carpenter on the job.'' Walter Barry breaks in: ''And both the architect and the builder learn from the maintenance man.'' And Fernandez adds, seriously, ''We're closer to the tenants; that's how we get the information.''

Joseph Barry has been the theoretician of this business, and has done some speaking around the country on the prospects for urban rehabilitation. ''What you need,'' he says, ''are housing entrepreneurs, people who have a long-term perspective. To make these things work, you have to create a whole class of urban housing operators.''

Rehabilitation is not a new idea. Title I of Roosevelt's first housing Act in 1934 had offered loans for ''renovation and modernization'' (showing that redundant language is part of the genetic code of federal housing programs). Though slum clearance was a central target of urban renewal, it became clear before the 1950s were far advanced that the academics and the government were tarring with their slum brush a great deal of housing that was both sound and functional in its neighborhood. Kennedy's housing administrator Robert Weaver admitted ''a failure to differentiate between an area of low-rent housing and a slum. Thus, until recently, little attention has been paid to the incidence of home ownership.''

As early as 1954, Congress ordered the FHA to consider ''the feasibility of achieving slum clearance objectives through rehabilitation of existing dwelling units'' before approving urban renewal programs. Kennedy's first Housing Act in 1961 laid considerable stress on home improvement loans, and the Administration promised to preserve 128,000 of the 235,000 surviving dwellings in the proclaimed urban renewal areas. Moreover, while earlier loan programs for rehab purposes had been in the three to five year range, under the 1961 Act home improvement loans could be insured for twenty years. The costs were thereby immensely reduced on the borrower's budget. A four-year self-amortizing loan for $5,000 at 5 percent will require a monthly payment of $115; a twenty-year self-amortizing loan for $ 5,000 at 6 percent (a higher interest rate, to compensate for the longer term) will require a monthly payment of less than $36. Weaver in 1963 acclaimed "FHA's revision of standards to reflect a realistic attitude" and "its insurance of mortgages for the purchase and rehabilitation of existing structures"; and announced that "the newer approaches are expected to accelerate results at a much more rapid rate."

Nevertheless, not much happened. The banks were reluctant to lend into deteriorating neighborhoods even with federal insurance, and the homeowners, believing the situation to be hopeless, saw no reason to incur debts for the purpose of throwing good money after bad. In Boston, where an aggressive Redevelopment Authority was led by a man with a genius for rubbing people the wrong way, pressure came on the banks from all quarters to open their pocketbooks in the declining neighborhoods, and a positive response was found—BBURG, Boston Banks Urban Renewal Group. This consortium agreed to accept virtually all loan applications for owner-occupied housing that the members had individually turned down; but the demand was relatively small, and the loss rate high.

In New York, the city itself ran and funded a Municipal Loan Program to finance experiments in rehabilitation. In one spectacular venture—with a grandstand erected across the street for the crowds, a band playing, Mayor John Lindsay in attendance—two gutted tenements on the Lower East Side were renovated by lowering prefabricated rooms through the roof, all ready for occupancy. Admittedly, this process cost more than building new houses from scratch (the federal government put $250,000 into the spectacular), but the costs would come down, wouldn't they? with experience. They didn't; no further houses were rehabilitated by this process; and a decade after the fanfare event, the reborn tenement was torn down. In Harlem, a block of less antique tenements—thirty-seven buildings with 458 apartments—was made into a demonstration project, usually cited as a triumph (by Morris Milgram as late as 1977), but in fact now almost indistinguishable from the other examples of decay around them.

In a study for the city's Housing and Development Administration in 1967, Rhonda Radisch found that rehabs did offer "decent, safe, and sanitary" housing, but "at standards considerably below those of new [subsidized] construction . . . in proportion to the costs involved.'' And the subsidy was very large, for the city not only excused the rehabilitated properties from taxation on the increased value but actually gave a complete abatement of the taxes previously assessed against the building, for a period of nine years, in effect repaying the builder for his investment. (Under the city's new so-called J-51 program, with real estate taxes considerably higher, this abatement runs for twelve years, and in most cases is worth more than $5,000 per unit.)

Over the years New York's Municipal Loan Program made 211 loans involving 3,500 apartments; by 1976, 95 percent of the loans were in default. One wonders who got the money, and how. This was a $100 million program. Repayments on the loans were supposed to run about $750,000 a month in 1976; the city's actual receipts were under $200,000 a month. It should be noted that the original purpose of this program was to convince the banks that they had been unreasonable in redlining the areas of the city in which the city itself was now investing. Applying for money under the federal Community Development Program in 1975, New York claimed that ''the City has demonstrated its capacity to rehabilitate structures of all kinds.''

The 1968 Housing Act put the federal government into this business with both feet, creating a ''special risk insurance fund . . . not intended to be actuarially sound'' to support the Section 235 program under which moderate-income families could purchase new or rehabilitated single-family homes with down payments of as little as $200 and mortgages that could be written down to a 1 percent interest rate. (In fact, they were not: 3 percent was as low as HUD went.) As of spring 1977, this insurance fund required a congressional authorization for a one billion dollar deficit; and more than a hundred thousand homes around the country had gone into forclosure. The damage done to the cities was incalculable. ''Most major cities in the United States,'' Mayor Richard Daley told a Senate hearing in 1975, ''have been left with thousands of abandoned and vandalized structures in what had been desirable neighborhoods.''

''We had two thousand foreclosures in Oakland,'' says Les Coplan of the San Francisco Home Loan Bank. ''Some of them were

Section 235 loans; the rest resulted from the breakdowns that followed 235.'' Barbara Brown of Melbourne, Florida, was working
for a small mortgage loan company when the Section 235 program
was booming: "I've never seen the abuses of how they qualified
people for those loans—people on welfare, waitresses, people making less than $ 400 a month, and it wasn't a stable $400.'' After all,
the government was insuring the loan.

Some sound homes came out of the construction phase of Section
235, in tracts like the Irvine Ranch and the Shimbergs' Town and
Country development outside Tampa, where builders simply made
some of their new houses available for purchase under the program.
In these two sections, and doubtless in others, a visitor cannot tell
which house is owned by a man paying market rate interest on his
mortgage and which house is owned by a subsidized purchaser. (If
the neighborhood knows, however, there may be trouble: people
who are stretched to pay for their homes may be extremely unkind
to neighbors who are getting *their* homes for less, especially if the
subsidized family turns up with a new car or starts digging for a
swimming pool.) Far more common has been the situation where
the difference between the Section 235 house and the conventionally financed house is immediately noticeable. "The FHA set
the price at $25,000,'' says Chicago community organizer Gale
Cincotta, "so then the trick was to build it for $12,000.'' Bill Beni-
tez adds, "Nobody objects. People don't see why a guy who is paying $75 a month should have housing as nice as people who are pay-
ing for it themselves.''

In the rehab section of the program, the problems were much
more severe. Nobody in Washington seems to have realized how
neatly Section 235 rehab, by deliberately and consciously lowering
the average income level of the occupants of the houses, fitted into
the established patterns of blockbusting in the urban centers. For
years, unscrupulous brokers had been buying up houses in deteriorating neighborhoods for extremely low prices, frightening their
owners with the threat of a wave of black occupants (sometimes
supplying the first of the wave themselves, with help from some
friends in the welfare department). The houses for sale were not
available to black purchasers directly, because the banks wouldn't

give them mortgages; but through the intermediation of the white brokers, who could get mortgages, black families would buy on "contract," paying perhaps three times as much as the broker had paid. When they fell behind in their payments, the broker would repossess the house, seizing whatever equity the black occupant had built up, and sell it to somebody else. Now under Section 235 such brokers could get a much better price, and all at once, doing a little cosmetic rehabilitation and arranging an insured FHA Section 235 mortgage for the purchaser. The broker could keep that extra profit, too, because the old-time residents from whom he bought the property were not up on all the federal programs, as he was. A fair number of these swindles, especially in Detroit, were sales to people virtually swept off the streets, whose capacity for maintaining a home or the payments on it was visibly nil; the broker supplied the $200 down payment himself, sacrificed a small portion of his profits in gifts to somebody at the savings association and somebody at the FHA, and got out on the golf course early every afternoon. Profits could then be heightened still further by repurchasing the mortgage from the government at a discount, in the certain knowledge that it could presently be foreclosed, and the insurance claimed, at list price.

Even this ill-designed program could be made to work on a small scale, however, given careful enough supervision. In San Francisco, a private development fund with Ford Foundation help set up a Buyers' Agent Program, and HUD awarded four hundred Section 235 authorizing certificates to the fund rather than to the prospective owner or the building contractor. By selecting low-income people who seemed likely to be able to handle home ownership—and by requiring a six-month course in the subject from every purchaser ("we don't want people buying into these houses who don't have the right to move in," said Mrs. Elizabeth Eudy, who ran the program)—the fund held mortgage defaults in its part of Section 235 to less than 2 percent.

The experience in multifamily rehabilitation under Section 236—"Project Rehab," as HUD Secretary George Romney called it—was even worse than that under 235. "We had it here," says Bruce Rozet of Los Angeles, a whizkid developer from the Stanford

Research Institute who has been playing the angles of subsidized housing for the past ten years. "HUD decided that if you could buy at $4,000–$5,000 a unit and rehab for $3,000 [these figures are as of 1969], you would have a viable model. So they called some developers in and told them about it. You got scatter-site redevelopment, with nobody caring how they were going to manage the apartments after they were finished; and it was consummate disaster."

Here, too, an ill-conceived program could be rescued by the right participants. In New York, the Upper Park Avenue Community Association preserved risky rented rehab housing, sponsored mostly by the Bowery Savings Bank, by giving tenants a detailed course before they came (you don't wash wood floors; all garbage must go in the incinerator; the exterminator must be called *immediately* if you see any beasties), and by *not* giving them a lease for the first year of occupancy. The energetic neighborhood ladies who ran UPACA inspected every apartment every month; and anyone who flunked the inspection had a choice of taking the course again or getting out. Both Leon Strauss and the Barrys are Section 236 rehabilitators, whose work could scarcely have been done without such government help. But they are cheerful exceptions to a grim rule: across the country *more than half* the Project Rehab multifamily projects became what was politely called "financially distressed," which from a tenant's point of view means no maintenance.

Rehabilitation is probably a useful tool for only a limited selection of buildings in a limited selection of neighborhoods. It "deserves a try," Charles Abrams wrote in 1965, "only when projected net income compensates for the projected trouble and risk. Costs are a guess when one breaks plaster. Fixed estimates are hard to obtain, and when the job is finished, the contract price will usually be swelled by extras." George Schechter of United Housing Foundation says, "I think it's a horror to invest public money in rehabbing tenements—these buildings can't survive with three kids in an eight by nine room. The processing takes a minimum of eighteen months—do you know what happens in that building in eighteen months?" The true cost of a housing unit, moreover, must be found by amortizing the initial expense over the years of its occupancy, and there are relatively few rehabilitated apartments that have as many years ahead of them as the new apartments put up even by mediocre builders.

Nor is the building the only problem: "A venture in rehabilitation," Abrams writes, "means gambling on the neighborhood more often than the structure. . . . The section may be so bad that no face lifting will enhance its economic complexion." In summer 1977 a New York City Planning Commission that was agitating for an anti-redlining law in the state legislature complained bitterly (and publicly) about the allocation of federal Section 8 subsidy money to a pair of tenements on a hopelessly decayed street in the South Bronx.

If the neighborhood is coming up, of course, owners and builders need little federal incentive to make the homes better. Kenneth Patton of the Real Estate Board in New York, commissioner of real estate in the Lindsay administration, dislikes the word "rehabilitation" and the expensive work it usually implies. (Especially in New York, where the unions announce agreements with the city to cut labor costs on rehab projects but then never actually sign the contracts that would carry out the publicized agreements.) "What you ought to do," Patton says, "is strip the walls, install new wiring and plumbing, and then turn the property over to the housewife for Home Improvement."

Boston has been employing procedures not far off Patton's prescription, subsidizing the value of the work rather than the specifics. Someone who wishes to rehab a house he owns can apply to the Redevelopment Authority for an appraisal; and then call the same agency for a reappraisal on completion of the job. The Authority will refund 20 percent of the value an appraiser says was added, by whatever means, up to a certain ceiling. "That taps an immense potential for do-it-yourself repairs," says George Peterson of the Urban Institute, "which previous programs have not done." As of summer 1977, the Boston program had made no fewer than 12,000 grants.

In all cities, a high fraction of the most important rehab work is being done "illegally," without notice to the buildings department or (especially) the tax assessor. There are plenty of streets where the façades are very misleading. How one designs housing policy to en-

courage illegalities, I don't know; but the man who could find a way
would be a great public servant.

Barring some such joyous burst of constructive illegality, rehabilitation is likely to remain a minor if significant weapon in the war to
save the cities, benefitting hundreds of thousands but not millions.
What has misled people is partly a false syllogism, partly a false
hope, and partly some false bookkeeping.

The syllogism is that because people in good neighborhoods
modernize, repair, and maintain their homes, a program that pays
for modernization and repair will make good neighborhoods. Unfortunately, the world is more complicated than that. Unless the factors
that have produced the decline in the neighborhoods are remediable
by improvements in the housing stock alone, which will be uncommon, the tendency is for the rehabilitated housing to decline relatively quickly (within five years) to slum status. "We have five
preservation neighborhoods," says Morris Sweet of New York's
City Planning Commission; "the harder we work on them, the
worse they get."

The false hope was stated flat out by Leon Wein and William J.
Quirk in 1969 in a *Cornell Law Review* article entitled "Homeownership for the Poor." "The underlying theory of homeownership," they wrote, is "that the change of a building's tenure
from rental to ownership will result in a stable building regardless of
the neighborhood." This is all but impossible even for rich people,
as the abandonment of magnificent mansions in all our cities demonstrated. (There are occasional exceptions: I remember a visit to
Gardner Cowles in Minneapolis, in a splendid home maintained in a
neighborhood which had otherwise turned almost entirely to institutional and low-income use.) Far more likely is the result suggested
by an official of the Newark Housing Development and Rehabilitation Corporation: Homesteaders, he said, "are being lured by the
apparent low cost of homeownership into spending most or all of
their savings on the purchase of a parcel which they cannot afford to
adequately rehabilitate or maintain. In a sense, the city is trapping
many of these people into a worthless and hopeless investment."

The false bookkeeping occurs in many ledgers—in an underes-
timate of the costs of rehabilitation itself, a failure to add up the

many investments the municipality must make to change the image of a declining area, an overly sanguine estimate of the economic life remaining to a rehabilitated older structure, and especially an inadequate projection of the maintenance that a rehabilitated home will require. For some time, the owner of a rehabilitated house in a neighborhood struggling for renewal will have a negative equity in his home—that is, the sale value on the market will be less than the money invested in the property. Individuals who get in trouble will thus be drawn toward walking away from the house rather than fighting it out, and the losses from such abandonments mount rapidly, both for the lenders to the program and for the neighborhood as a whole.

In the case of rental properties, the drain of even a few vacancies is so severe that it quickly triggers the process described by a New York State Moreland Act Commission as "a race and gamble. . . . [T]he owner can cut costs only so far and still provide his tenants with any service at all. Once the rent revenues, minus debt service payments, are no longer covering his minimum operating costs, his only sensible course is abandonment. The whole baleful sequence . . . immediately imperils the neighborhood. Every other owner close by must decide quickly how much time he has left to recoup the balance of his investment. . . ."

There is a theoretical difficulty with even a successful rehabilitation, in terms of costs and benefits, because rehabilitation seeks stability in a dynamic world. New housing, even if it is not intended as "replacement housing," inevitably opens up existing units for occupancy by others. This filtering process, as noted earlier, is traditionally the means by which lower income families have improved the quality of their residence. But rehabilitation blocks filtering, which means that the benefits go mostly if not only to the occupant of the one unit.

Some neighborhoods rehabilitate themselves through natural vitality—one thinks of New York's Brooklyn Heights and Yorkville; Chicago's Lincoln Park and Old Town; Washington's Georgetown, Capitol Hill, Adams-Morgan; Kansas City's Westport; San Francisco's Cow Hollow and Pacific Heights; Boston's Beacon Hill; Calhoun-Lakes in Minneapolis; the Garden District in New Orleans;

are about the only spectacles that rival in joy the sight of people at work improving their home. Thus every rehab project gets a great press, especially a project involving photogenic young people who will shout "Venceremos!" at a television camera or mutter great truths like "A journey of a thousand miles begins with a single step." Extrapolations from the experience of the shanty villages of Latin America, whose residents have no place to go, will not help us much in deciding what to do for urban neighborhoods in developed economies. In deteriorated areas, the rule that people make their own amenities becomes significantly misleading. What is important about the rehabilitation movement is that it can direct outside resources to the nonhousing conditions of neighborhoods that can make good use of these resources but cannot command them without help.

In areas where there is forceful leadership, public money available for investment in municipal infrastructure, and preponderant reliance on earned incomes rather than transfer payments, emphasis on housing rehabilitation rather than new construction can be economical and effective. This does not, however, begin to take care of the problems of the people who simply cannot pay the major part of their own housing costs. "What do you do," said Howard Scaggs, thinking about Baltimore, "with the people who absolutely can't afford it? You get back to subsidies, and I hate those damned things. I've been watching the government pour billions into housing and I keep looking at boarded-up homes." But as a realistic matter we are not now talking "subsidies"; we are talking about some form of public housing.

Chapter 8

Public Housing

Nobody ever writes about public housing from the point of view of the people who say, "Things will be better when we get into the project."

—Tim Sullivan, New York City
Housing Authority

This year started very badly. As part of the Target Project Program at Hunters Point, we're setting up tenant management offices in different areas, and the first one to become operational was headed by a very promising young man, twenty-three years old. Unfortunately, he got murdered New Year's Day—not necessarily in the line of duty, but it was a very shocking thing.

—John A. Crowder, San Francisco
Housing Authority (1977)

Though the lines of the flats were clean and straight—their main asset—those blocks which were half a dozen years old already seemed to reflect a certain weariness of the spirit. Perhaps it was that skimpy detail spoke of the accountancy of remote public bodies; the entrances and stairways had already the neglected air of a place not loved; the whole impression was of too much public anonymity, of a space-saving set of buildings for those lucky enough to get in, but which proclaimed no satisfying new way of community life.

—T. R. Fyvel, on council housing in
London, 1961

On a hillside southwest of Winston-Salem stand two hundred or so not very unusual small houses, brick-faced, four to an acre, well

173

kept on curving roads. No garages, and not much ornamental shrub-bery—but no other possible hint that this is public housing. It is, though—part of the Turnkey III program that permitted local hous-ing authorities to contract with private builders for the construction of detached homes that could be made available for purchase by low-income households. Between 1971 and 1974, before the Nixon Administration closed the pipeline, Winston-Salem built 967 of them on seven sites, at costs ranging from $19,000 to $22,000 per house, with a community center at each site thrown in. Even in Winston-Salem, it would have cost more to build these homes as apartments in high-rise buildings.

The sales literature for Winston-Salem's Turnkey III reads:

"Now don't you just tell yourself that you never could own a home. There's a new program in Winston-Salem that builds new homes for working folks and sells the homes at prices and monthly payments that fit the workingman's pocketbook.

"This is no trick offer that is going to take advantage of you. It's a government program designed to give working people a chance to own a good home in a good neighborhood.

"This program is called Turnkey III and it's administered by the Housing Authority of Winston-Salem. THERE IS NO DOWN PAYMENT!"

The arrangements are not like those of the private market. The housing authority retains title to the houses, and the "Homebuyer" takes possession as a month-to-month tenant on a "lease-purchase" contract. In effect, the homes have been financed by a twenty-five-year mortgage in the form of a tax-exempt bond issued by the au-thority at interest rates of about 3.75 percent. The federal govern-ment through an Annual Contributions Contract pays both principal and interest on these mortgages. If the Homebuyer remains in place for twenty-five years—and stays poor enough to qualify for the full Annual Contributions Contract treatment, which as of early 1977 meant his income was no more than $7,200 a year plus $300 for each child living home—the house will become his property without further ado, though if he sells it within two years of taking title the housing authority can recapture its subsidy from the sales price. At any time in those twenty-five years, a Homebuyer who comes into some money or rises on the salary scale can purchase the home by arranging private financing for the unamortized part of the mortgage. If he leaves without buying, however, his house is simply put up for "resale" by the authority, again on lease-purchase terms, but at a price which must be the higher of two independent appraisals. The original Homebuyer then gets back only $9 a month on what he paid in his "lease-purchase" contract, provided he returns the property in apple-pie order. The hedge against inflation—the appreciation of the house—benefits the housing authority unless the family hangs on.

The monthly payments are roughly 20 percent of the family's income, less an allowance ($30 a month in 1977) for heat and utilities. In fall 1976, occupants of Winston-Salem's Turnkey III paid the authority an average of about $93 a month. From these payments, about $20 is taken for the operating expenses of the authority, about $15 is set aside for a Nonroutine Maintenance Reserve, and $9 is credited to the Homebuyer's Ownership Reserve for later downpayment on the house or refund on departure. Some of the rest goes to the city in lieu of real estate taxes, and what's left over is returned to the federal government to diminish the Annual Contribution. In fall 1976, 235 families were paying more than $125 a month, 301 were paying less than $75 a month.

To occupy these houses, reports David Thompkins, deputy executive director of the housing authority, "people must have a job and must have had it for six months. And they must keep it. These residents go to work every day." The average income in fall 1976 was $7,300 per family. Regardless of income, Turnkey III residents are expected to pick up considerable burdens. The $30 allowance did not cover the actual costs of fuel and utilities at 1977 prices. The lawn is to be seeded and fertilized twice a year at the occupant's expense, the furnace motor is to be oiled, broken screens and windows replaced, damaged floor tile reset, etc.—by the Homebuyer himself. With help from North Carolina State University and Forsyth County Technical Institute, the housing authority offers Turnkey III customers courses in everything from home decorating and household repairs to money management. For most of the residents, housing takes close to 30 percent of income: "To afford a home,"

Thompkins says, "is a hell of a cost." Even so, turnover in Winston-Salem's Turnkey III is almost exactly at the national average for people moving out of the homes they own—8.4 percent.

There have been a few problems relating to construction defects, especially furnace motors and windows, where the builders used cheap goods that have had to be replaced ("I recommended that HUD allocate some more money," Thompkins says; "get some good inspectors out there and see we get our money's worth.") On the whole, though, Turnkey III has been good housing. In effect, what it does is give the less well paid black worker (Turnkey III in Winston-Salem is all black) a life system comparable to that of the unionized white worker. And demonstrate (with a little help from government) that like the white worker he can handle the problems.

It is scarcely surprising, of course, that working people willing to pay 30 percent of their income for housing can make a go of public housing at a density of four units per acre. And perhaps one should not be surprised that the total subsidy required in Winston-Salem's Turnkey III (tax exemption on the authority bonds, real estate tax reduction by the city, and direct help from the government) amounts to about $100 a month, less than half the subsidy required in the average public housing project of this age in the United States. (On public housing built in 1977, the average required subsidy per unit runs more than $300 a month.) Winston-Salem's Turnkey III can be taken simply as an illustration that suburban living is cheaper and more satisfying for most poor people (and easier on the governments that assist them), just as it is for most middle-class people.

Turnkey III also illustrates, however, the overall success of public housing in Winston-Salem. Homes for purchase represent only about 30 percent of the Winston-Salem public housing effort, which in sum supplies 3,200 homes for about 11,000 people—roughly 8 percent of the population of the city. By northern standards, the proportion of these homes that deliver decent, safe, and sanitary housing seems extraordinarily high.

Part of the reason is historical: "In the South," says Henry Schechter of AFL–CIO, who used to be the housing maven at the Congressional Research Service, "they always put the public housing on the main street, because it was the nicest buildings they had.'' And some of it is probably sociological: ''Southerners,'' says Thompkins, ''are more disciplined than people from other parts of the country.'' A former FHA official remembers ''all those meetings of NAHRO [National Association of Housing and Redevelopment Officials] where the southern cities sent delegations of little old white ladies who looked into people's apartments every month to see they were doing right. Well, you can't get away with that in the north.''

Disciplined or not, the people who wind up in conventional public housing in Winston-Salem come from the same beat-up groups that populate public housing elsewhere—apart from the elderly, the residents are almost exclusively black, 95 percent of them are female-headed families (mostly with lots of kids), and half are on welfare. In one project, there are about 360 adults and 1,100 children in about 340 units. Yet there is no vandalism, and little more feeling of danger from criminals than can be found in private housing areas of the city. Though there are some signs that drug and robbery problems are moving south, James Haley, the executive director of the housing authority, still finds himself worrying less about murders in the office than about worn grass and mud on the grounds—''There's some heavy wear and tear because of children, and, you know, the trees grow, and there are trees under which grass won't grow.'' Some of the credit obviously goes to policy, management, and design.

The policy is succinctly stated by Thompkins, an amiable but earnest, rotund black engineer, native of Winston-Salem, who went to Greensboro A & T and came back in 1957 to work for the housing authority. ''Public housing,'' he says, ''is an excellent program, but people have to participate in paying for it.'' Residents paint their own apartments—the authority provides paint, brushes, ladders and dropcloths, but not labor. They cut the grass in their area, and tend their yard: the authority provides the use of a power mower. ''I've had a few residents say, 'That's your responsibility,' '' Thompkins reports. ''I sit them down: 'You're only paying $24, $25 a month, and that includes heat and utilities. . . .' Public housing is not a giveaway program. After the government builds that housing for him, the tenant has a great responsibility. We give a resident a unit in A-1 condition, floors waxed, windows clean—people are more likely to take good care of it—and people have to mind to keep it in A-1 condition. We do an internal inspection every six months; we ask, 'What have you done with our walls and floors?' You're not abusing the tenant when you do that; you're looking after the government's property.''

Winston-Salem's housing authority also collects all the rent: delinquencies and vacancies between them run about 2 percent of the monthly budget, and a lot of that gets cleaned up later. If the bill is not paid by the eleventh of the month, a community service representative pays a visit to find out why; if the representative can't find the tenant, a notice of eviction is mailed immediately. ''This housing authority *never* pays a tenant's rent,'' Thompkins says. ''But we know the sources, the places where our social service department can go to get a tenant the rent if there's no way he can find it himself. Our church-sponsored Section 236 housing has a 20 percent delinquency rate; they've all gone broke and they want us to take it over. They got into this because they thought they were doing a good thing; they never realized it took management.''

In terms of management, Winston-Salem runs lean and courteous; there is both a civilization and an efficiency to the operation quite unexpected by any Northern visitor. Headquarters is a two-story building in an urban renewal area, in the shadow of the authority's high-rise apartment house for the elderly; the rear windows look out on the slope where the residents of that building plant and tend an authority-sponsored garden. The entrance area is carpeted and brightly lit, couches and upholstered chairs for waiting visitors, on the table the magazines the residents like—*Photoplay, Modern Screen, Ladies Home Journal, Family Circle, Glamour.*

Cooperation with other city agencies is close: the city maintains a park adjacent to each of the projects, and staffs some of the community service systems (which are pretty ambitious: ''The new philosophy of Public Housing,'' says the authority's official booklet on *Resident Services,* ''is to solve the Low-Income Residents' problems as well as to furnish them with safe and sanitary shelter''). Resident councils hold meetings under Roberts Rules of Order, and their presidents attend the meetings of the housing authority board.

Disruptive tenants are more likely to be booted by the resident councils than by the housing authority—"they're more severe than we would have been," says executive director Haley. "We have liquor houses—apartments where people sell liquor by the shot—but we don't have to do anything about them. The people who live there turn them over to the police."

Project managers and senior assistants rise from the ranks: Thompkins, who is in charge of operations, says he knows everybody who works for the housing authority, and walking the projects with him is indeed one social call after another. "I have a staff of only twelve people for this whole place," he said, looking around the 500-unit, fifty-three-acre Happy Hill Gardens (the city's oldest, most of it completed in 1952). "You can tell I'm getting work out of them."

Except in the high-rise for the elderly, there isn't a hallway in the Winston-Salem housing projects. The most densely built of them, with about 15 units to the acre, is a string of two-story barracks, but each barrack is divided into sections that are separate homes with their own back yards and front porches and doors to the outside world. Most of the other projects are free-standing fourplex units, some of them one-story, twelve units to the acre. The apartments themselves are minimal in size, as public housing was supposed to be in the 1950s, built without closet doors, bathtubs without showers; but the walls are sturdy lath-and-plaster, and renovation money from Washington is now installing new kitchen equipment and bathroom fixtures (including showers), sliding doors for the closets, and aluminum windows with thermopanes. At all four of the projects built in the 1950s, there is a three-year waiting list.

In choosing new families for these projects, incidentally, Winston-Salem now gives preference to employed parents. "We need the mix," Thompkins says. "Welfare families never go on trips. Working families go off, say to Disney World, come back, show people pictures; and the welfare kids get a contact with the outside, develop ambitions of what they're going to do some day."

Congress first interested itself in housing in 1892, when the Commissioner of Labor was granted $20,000 for an investigation of the slums in the nation's fifteen largest cities. The investigators found New York, Boston, and St. Louis just awful, Chicago, Baltimore, and Philadelphia considerably less poisonous, but couldn't think of much the federal government could do about the problem other than restrict immigration. In 1908, Theodore Roosevelt appointed a temporary Housing Commission which recommended that the federal government condemn, purchase, and rehabilitate much of the nation's slum housing, but the proposal was never seriously considered. In fact, direct ownership of housing by national governments has not been an appealing solution to slum problems anywhere. Though national governments arrange the financing, heavily subsidized housing is owned by local governments in England and the East Bloc, by church-related groups in the Netherlands, by autonomous housing associations in France and Denmark. The United States Government found itself briefly in the homes business in 1918, when 16,000 detached houses and apartment units were rapidly built for war workers by a U. S. Housing Corporation, but these properties were hurriedly sold off when the war ended.

The Depression put Washington back into the housing business. Hoover held a conference on Home Building and Home Ownership in 1931, and in 1932 the Emergency Relief and Construction Act authorized federal help for housing producers who would agree to limit their profits. One very large housing project was in fact built by this law—Knickerbocker Village in downtown New York. Roosevelt's National Industrial Recovery Act went a step further, permitting help for either private limited-dividend companies or state and municipal agencies. Unfortunately, there were no state and municipal agencies ready to build housing, and the limited-dividend builders took advantage of provisions in the new law to dump onto the government the failures of the preceding three years, establishing a pattern that has endured to our day.

Roosevelt and Harry Hopkins and Harold Ickes as Secretary of the Interior then put the new Works Progress Administration to the task of building housing that the federal government itself would own and operate, and sent Rex Tugwell off to build the Greenbelt towns. The financial arrangements were that the government wrote

off 45 percent of the costs and budgeted to recapture the rest from
tenants' rents over sixty years, at 3 percent interest.

"This solution to the low-rent housing problem," Daniel Y.
Sachs wrote cautiously in the *Handbook* of the National Association
of Housing and Redevelopment Officials, "satisfied no one. Local
governments . . . wanted to retain control." When a federal court
in Louisville in 1935 denied the government the power to take land
by eminent domain for housing purposes, Roosevelt instructed the
Solicitor General not to appeal the case (mostly, no doubt, in fear of
what the "Nine Old Men" of the Supreme Court might do to the
whole WPA program if given a chance); and the New Deal went
looking for other procedures.

It is worth noting, however, that the housing projects that cele-
brated their fortieth anniversary in 1977—most notably, Techwood
Homes near Georgia Tech in Atlanta and Harlem River Houses in
New York—were built directly by the federal government with
WPA labor, and were built exceedingly well, including for decorative values some of those marvelous squared-off muscular soap-
stone workers of the 1930s to be the household gods. "The oldtimers tell me," says Tim Sullivan of the New York housing authority, who was about a dozen years shy of being born when Harlem
River Houses was completed, "that these houses were built so well
because the workers were taking a kind of revenge for the rules
which required that the buildings have such poor facilities."

Techwood still looks fine, too, four-story brick buildings on a
gentle slope, embracing grassy areas; its manager Edward Riley
says it has about fifteen more good years in it, though in Atlanta the
exposed hot-water pipes are a cruelty in the summer. The transition
from white to black at Techwood, still in process, has had painful
moments, but there is evidence around that the place is cared for: a
stern warning notice on the tenants' association bulletin board
begins, "Some people have been keeping filthy apartments . . ."

The durability of these early structures, with their thickly plastered walls and ceramic tile hallways, gave what turned out to be
misleading signals about how long later housing projects could be
expected to give service. "They will stand for a century," Charles

Abrams wrote as late as 1965; but the odds today are that despite heavy expenditures for rehabilitation many of the projects built in the 1950s will not survive for the forty-plus years necessary to retire the bonds that built them.

The substitute for direct federal construction and ownership was wheeled into place with the Housing Act of 1937, still in essence the charter for the 3,200+ local housing authorities in the United States. By the terms of this act, a new Housing Authority in the Department of the Interior was authorized to make loans to local agencies to construct low-rent housing, and to promise sixty-years of "annual contributions contracts" that would pay back the loans. The law fixed a maximum construction cost per room and set limits on tenants' income. Local governments were expected to provide 10 percent of the costs of building the projects, and were required to guarantee that they would demolish one existing slum housing unit for every unit built with federal help. Private contractors would bid for the work. Municipalities and school districts could impose real estate taxes on the value of the new projects, to be paid by the tenants in their rent. Counting everything together, the projects were supposed to be break-even ventures for local government. Significantly, the first purpose announced in the preamble was not to create housing stock, but "to alleviate present and recurring unemployment." By 1942, 175,000 units, mostly in two- to four-story apartment houses (elevators were an impossible luxury in the early days of public housing), had been built in 290 communities. Another 195,000 units of war-related "permanent" housing were added to this stock by 1946 in areas where war industry or military bases had created new demand for housing.

The conservative Congresses of the immediate postwar period were not disposed to expand a program of this sort, but when Harry Truman was elected on the platform of the Fair Deal, public housing was thrown into a higher gear. The Housing Act of 1949 opened with the rhetorical promise of a target so high it could never be hit: "a decent home and a suitable living environment for every American family." As part of the pursuit of this purpose, the national Housing Authority was authorized to support the construction of 810,000 units of public housing in six years. (Inflation and the

Korean War supervened, cutting annual production down to 35,000 units.) And some changes were made in procedure.

To get the cost of public housing off the federal deficit and the national debt, local housing authorities were told to raise their own money in the private bond market, the bonds to be both federally guaranteed and tax exempt. Part of the subsidy was thereby moved out of the budget and into the loss of tax revenues from the holders of tax-exempt bonds. Federal guarantees were offered for the entire cost of building the project, with the annual contributions contract to pick up the full debt service on the bonds. The duration of the bonds was reduced to forty years. Local governments were forbidden to assess real estate taxes against the projects, and were told to substitute a tax of 10 percent of the "shelter rent" (i.e., costs of debt service and maintenance, but no utilities) as a payment to the city in lieu of taxes.

Maximum construction costs per room were still set in Washington and were set low. Public housing apartments, Martin Meyerson and Edward Banfield wrote in 1955, "were built to be inexpensive and to look so as well. Rooms in the projects were small. There was often no storage space for suitcases, tools and large toys; ceilings were not plastered; floors were of trowelled concrete; woodwork was waxed rather than painted; bathtubs were not provided with showers; most closet doors were eliminated; heating and other pipes were exposed; and in some projects there were no dividing walls between living room, kitchen and dining space."

Moreover, the projects were to be deliberately set apart from the city—to be better than the city, the leaders of the movement thought. Chicago's housing director Elizabeth Wood, in a call for "bold planning," specified "an unalterable determination to relocate all transit and traffic in such a way that the blighted areas are subdivided into superblocks averaging, let us say, about eighty acres, through which no streetcar or other public transit passes, through which no traffic goes. . . . Little plans will require a redoing of the total job within three generations." An earlier theoretician, James Ford, had written in 1936 that a large project had an "increased chance of maintaining its distinctive character because its very size helps it to dominate the neighborhood and discourages regression." Small projects, by contrast, "may slip back and show evidences of being overpowered by the surrounding conditions instead of acting as an improving influence."

It is impossible to recapture these days any sense of how much just plain *good* was going to be done through the construction of large-scale public housing of this kind. Myres McDougal and Addison Mueller of the Yale Law School wrote in 1942 of "the well-documented facts that slum clearance and the provison of sanitary low-rent housing decrease danger of epidemics, raise general public health, reduce crime, cut juvenile delinquency, reduce immorality, lower economic waste by reducing health, police and fire protection costs, make better citizens, eliminate fire hazards, increase general land values in the vicinity, cut the accident rate, and prevent the cancerous spread of slums to uninfected areas." As John Heimann says sorrowfully, "New buildings were going to wipe away all the shame and degradation: America the blackboard."

And so we got the fifty-acre superblock of Pruitt-Igoe, now demolished, Boston's Columbia Point, mostly now boarded up; Philadelphia's vacated Rosen Apartments; Newark's desolate Columbus Houses; Chicago's menacing Cabrini-Green; the Fort Greene Houses in New York that so horrified Harrison Salisbury: the "shoddy shiftlessness, the broken windows, the missing light bulbs, the plaster cracking from the walls, the pilfered hardware, the cold, draughty corridors, the doors on sagging hinges, the acid smell of sweat and cabbage, the ragged children, the plaintive women, the playgrounds that are seas of muddy clay, the bruised and battered trees, the ragged clumps of grass, the planned absence of art, beauty or taste, the gigantic masses of brick, of concrete, of asphalt, the inhuman genius with which our know-how has been perverted to create human cesspools worse than those of yesterday."

Andrew Greeley, observing the Chicago scene, wrote, "fortunately for the ethnics, they stopped being poor before the reformers could set up high-rise public housing and dependency-producing welfare legislation to 'undisorganize' them."

About seventy-five of the original tenants of Harlem River Houses were still living there forty years later, and four of them sat around the other day in the converted apartment that serves as a manager's office, talking about the old times. Three were women in their sixties and seventies, two black and one white, all at one time or another officers of the tenants association; one was a man in his early fifties, black, brought up in this project and returned after military service and marriage, now the project's delegate to the city-wide Residents Advisory Council. The New York City Housing Authority by itself would be one of the fifteen largest cities in the United States, with 550,000 people living in its properties.

"I moved in on October 6, 1937," said Mrs. Chloe Hamilton from under a flowered hat. "There were 574 tenants. If you were three in a family, you got two rooms [she doesn't count the kitchen], four in a family got three rooms. Tenants that lived four in a family couldn't make more than $2,200 a year. We paid weekly rent: on the ground floor and the top floor the rent was $5.45 a week; on the second and third floor, it was $5.35. Utilities were $5.10 a month.

"We formed the tenants organization to keep the project clean. We were people with federal government jobs, railroad jobs, moving jobs. This place was built by the WPA, but a WPA worker didn't make enough money to move here. You had to make five times the rent, and even a man in the post office didn't make that. Then so many people were forced to move because their incomes were too high, because you couldn't make too much, either."

Among the attractions was (and is) a nursery school. "It was nine to twelve," said Mrs. Loretta Ferguson, the white member of the group. "We paid fifty cents a month. Then there was some twelve-to-three. You didn't get a free lunch like you do now. We'd bring out the children's equipment and in the afternoon the teenagers would put it away."

"There were pits there with full sand," said Harry Holmes, "and a sprinkler system in the center; that's where we played. Back in 1939, at the 32nd precinct, they had a PAL—one of the first organizations in public housing—to develop relations with youth. That's long gone."

Mrs. Pearl Carpenter remembered that there had been housing guards in those days too. "Did they carry guns?" asked manager Patrick Coleman, a large black man probably more casual with his

tenants than usual, because he was being rotated out of this still soft
berth into one of the problem projects. "No," said Mrs. Hamilton,
"sticks—they didn't need guns."

"Oh!" Mrs. Ferguson burst out, "our children were so good!"

"Women didn't work," said Mrs. Hamilton.

"We stayed home and took care of the children," said Mrs.
Ferguson. "We brought them out clean in the morning—"

"There wasn't a dirty child in this project," said Mrs. Hamilton—

"And we took them home, fed them lunch, and brought them out
clean again."

"We had a meeting," Mrs. Hamilton said. "Judge Watson was
here. A woman asked, 'When should you start training a child to do
what's right?' Judge Watson said, 'Before he is born.' "

"You lose control when the child goes to public school," said
Harry Holmes.

"My son went to Resurrection," said Mrs. Ferguson. "If he
wasn't at school, they called to find out why. My son wanted to play
pool, and my husband found a place where they'd call if he wasn't
there on time."

Harry Holmes said seriously, "There was a . . . *concern* might
be a good word. Children here knew that if they got rambunctious
somebody was going to come out of the woodwork. I do it still—I'll
chase kids all the way over to Brathurst Avenue somewhere if I find
them doing something wrong. The generation brought up here has
moved a great step forward—from the responsibility the authority in
those days put on parents to bring up their children."

Every June—in 1977 they moved the date to September, to mark
the actual anniversary—Harlem River Houses has an alumni re-
union at the sand pit for the men and women who were the children
brought up here. "They come from *Chicago,*" said Mrs. Ferguson
proudly, referring mostly, it should be noted, to a black community.
"Doctors, lawyers, detectives . . . Cecil Holmes, the president of
Motown, came from California. There's Audrey Small, who's the
fashion director of *Ebony.* . . ."

Things have been going wrong at Harlem River Houses too. The
basement recreation rooms are no longer used, manager Coleman

reports, because they can't be made safe. The sand is gone from the sand pit. Iron bars were put on the first-floor windows after a wave of robberies, and now the tenants want them on the second-floor windows, too. They look bad, because the housing authority somehow didn't use rust-proof paint on them. The little public library branch that had been in the project from the start is now open only two mornings a week. The Great Society youth, art, and counseling programs are all gone. The exterminators who used to come once a quarter now come only once a year—and the problems they deal with have grown much worse since the Environmental Protection Administration forced the project to substitute a compactor for its former incinerator. "The garbage is out there from Sunday to Tuesday," says Mrs. Hamilton. "In hot weather. We won't die from smoke but from *germs*. They have something the porters should spray in the compactor room, but they don't do it.

Still and all, the people who live in Harlem River Houses consider themselves fortunate. It's a walk-up, and the rooms are small, but nothing is inadequate. "This," says Chloe Hamilton realistically, "was for peple who couldn't afford to live in nicer housing"—and for the money it is still an admirable buy. The New York housing authority likes to show that crime rates within its projects are lower than those in the surrouding neighborhood (a public-relations ploy that tends to backfire, as the people in the surrounding neighborhood blame the residents of the project for their crime problem; and the city does spend $35 million a year on a special housing authority police force, peace officers who carry guns). But Harlem River Houses undoubtedly is safer than the high-rise project a few blocks north, where the Polo Grounds used to be. "The high-rises," says Harry Holmes, *"capture* these people."

Still, everyone wants to talk of the past, and express regrets about its disappearance. There were indignities: Mrs. Hamilton remembers that their first night in the project they didn't have the use of their furniture because everything had been taken away to a fumigating plant for sterilization, then a routine procedure. (Pioneer housing manager Abraham Goldfeld of New York's Lavanburg Houses noted approvingly in 1937 that in Manchester, England, the local authorities "have recently experimented with the object of finding a way of extracting the cyanide gas from the bedding.'') But for the hazards and indignities of those days, there was at least a reason. Now it's all brainless, other people's theories applied to the lives of the poor, the great complex of ''rights'' the poor might indeed ''enjoy,'' except they can't afford them.

Harlem River Houses was one of very few public housing projects to be built with storefronts on the street, and Mamie Lee has run Mamie's Reweaving and Mending in one of the shops for more than thirty years. ''It was *beautiful* when I came, '' she said. ''I think it *changed hands,* this building. If they screened tenants the way they used to, and fixed the sidewalks . . . You used to get a ticket if you didn't clean the sidewalk.''

Manager Coleman said, ''We can't screen tenants any more, Mrs. Lee. The courts won't let us.''

''They used to get them out of here quick if they were bad,'' she said thoughtfully. ''They'd even put their furniture out on the sidewalk. You know, there's some people who just don't care. And it's *Miss* Lee, Mr. Coleman. I'm not married yet. . . .''

Charles Abrams was a warrior for public housing: ''It was,'' he wrote in 1955, ''the first major effort to provide decent housing at rents families of underprivileged minorities and immigrants could afford. It was the first program to establish by actual practice that Negroes and white could be integrated into housing and communities without friction. . . .'' By 1965, he had to note that ''In Philadelphia, where 80 percent of the dislocated families qualified for public housing and 67 percent were actually referred for occupancy, less than 15 percent moved in. In New York City's West Side, only 16 percent of the 68 percent found eligible said they would live in a project. Only 3 percent of the site dwellers in one Detroit renewal area entered public housing—most preferring the slums. In a large renewal area in Los Angeles, less than 1 percent of the inhabitants were willing to occupy public housing.''

If you ask the people who live in public housing what went wrong, the all but unanimous answer will be, as Mrs. Hamilton put it, ''they let in all those *welfare* families.'' This answer will also be given by those who are on welfare themselves.

Among people who could not imagine living in public housing projects, it is widely believed that they were built for the purpose of providing homes for welfare families. In his book about an imbroglio in the Forest Hills area, New York mayoral candidate Mario Cuomo offered a syllogism: "The housing program is supposed to assist the most needy. Welfare are the most needy." As early as 1958, Salisbury was blaming the conditions in the projects on the "fact" that "Charity, welfare, and relief cases get first choice . . . indiscriminate application of this means test populates Fort Greene or Red Hook almost exclusively with that segment of the population which is least capable of caring for itself. . . . By screening the applicants to eliminate those with even modest wages the project community is systematically deprived of the normal quota of human talents needed for self-organization, self-discipline, and self-improvement. It becomes a catch basin for the dregs of society."

In fact, public housing under the intial legislation was designed for the working poor. The local authorites were not permitted to charge rents below those necessary to pay operating expenses (in 1949, the high-water mark, public housing was almost a break-even proposition for the federal government: the local authorities actually refunded 84 percent of that year's Annual Contribution, because rents collected produced a surplus over operating expenses). "Relief," as it then was, simply did not give enough for its recipients to pay those rents. "Obviously," Senator Robert F. Wagner said in a speech supporting the Housing Act of 1937, "this bill cannot provide housing for those who cannot pay the rent.' 'By 1958, many of the jigsaw pieces of subsidy by which America has created a dependent lumpenproletariat of urban blacks had been put in place, and the requirement that first priorities in public housing go to people displaced by urban renewal and highway construction had begun to lever larger numbers of the dependent families into the projects. But in New York City, by far the largest housing authority, it was still easier for a camel to go through the eye of a needle than for an umarried, unemployed mother to get into an established housing project with her kids. (Even in 1977, less than 30 percent of the residents of New York housing projects were on welfare.) One found concentrations of welfare families only in those projects where, because of location or reputation (somehow, these places get a smell from the start), nobody else wanted to go. The original criticism of the St. Louis Housing Authority in connection with Pruitt-Igoe was that, confronted with the need to fill 2,400 units very fast, it simply loaded up the buildings with the sweepings of the streets: in the mid-1950s, before the plague of female-headed families had struck the black population, 55 percent of the households in Pruitt-Igoe fell into that category.

It was the Housing Act of 1965, establishing the rent-supplement program and freeing the local housing authorities from the obligation to pay their operating expenses out of the rent collections, that fully opened up the housing projects to welfare recipients. "In the intervening seven years," Oscar Newman of N.Y.U. wrote in 1972, "the high-rise buildings to which they were admitted have been undergoing systematic decimation." The deterioration has been accelerating: in 1975, 79 percent of the residents in the Boston Housing Authority's conventional rental projects had no wage earner in the family. At Boston's Columbia Point, now mostly evacuated and boarded up, 62 percent of the households were receiving Aid to Dependent Children. A horrified Congress in 1975 changed the law to extend the grounds on which local housing authorities can refuse to accept—or can later evict—tenants who seem likely to harm either the property or their neighbors. But nobody seems to have told the judges.

Sociologist Rober Gutman of Rutgers offered a general theory in 1970: "The inhabitants of the projects in the 1930s and the inhabitants of the projects today are samples drawn from two very different populations. During the earlier period, some people with middle-class ideals and values were forced to live in slums because of reverses suffered in the Great Depression. But the economic circumstances in which this group of slum dwellers found themselves did not diminish their commitment to their house as an object according to whose properties they ranked themselves and in terms of which they anticipated others would rank them. When they were relocated to newer, more handsome, better equipped housing projects, these temporary victims of the Depression saw themselves as acquiring a possession consistent with their middle-class ideals and aspirations. . . . The tenants of public housing today are drawn from levels of the class structure which are less likely to regard the house as a significant possession influencing social ranking, perhaps because their life history leads them to invest their loyalties on objects which are more easily lovable, such as automobiles.''

. . . And yet. There remain those housing projects in Winston-Salem, with the same welfare population—house-proud, a lot of it. Oscar Newman found in New York that the difference between high-rise and low-rise projects was much more significant as an explanation for crime rates than the ratio of welfare families. In New York, Newman argued (with a scatter graph to prove his point), the number of robberies in a housing project was directly proportional to its height. Newman put much of the blame on the standard design of the high-rise apartment house, with its ''double-loaded corridor'' (i.e., apartments on both sides of the hallway, and stairwells in which bad guys can hide, through which they can escape). Nobody can police those stairs and corridors: the resident has too many neighbors he or she doesn't know. ''A building which is open to entry by anyone, and in which it is very difficult to distinguish resident from intruder,'' Newman wrote, ''is one which is criminally vulnerable.'' In St. Louis, a Ford Foundation-sponsored study of ''tenant management'' programs reported ''that from December 1971 through March 1974 the crime rate at the low-rise Peabody housing project was between 50 and 80 percent lower than that at Darst, a high-rise housing project directly across the street.''

The English social psychologist Pearl Jephcott has argued that people who adjust well to high-rise living ''are people who are self-sufficient and socially rather 'above average.' The wider their experience of men and affairs the more they can cope with sharing services and undertaking mutual obligations with a large number of other people.'' It seems reasonable enough to believe that situations in which people ''live on top of one another'' require a greater social discipline than situations where they are separated by greensward; plausible, if not quite so reasonable, that the human animal like others has a certain need for ''territoriality,'' as Desmond Morris has put it, a need that can be frustrated by culture—but only at a price that people with limited resources from that culture may not be willing (or able) to pay.

The revulsion against the high-rise as a place to bring up families is a world-wide phenomenon. (Illustrating how unfashionable the high-rise has become in West Germany is an official report that "air turbulence carries into upper floors the origins of illnesses affecting the respiratory organs." This recent finding is especially impressive in the light of the historical evidence that the one health benefit un-questionably achieved by moving people out of slums and into high-rises was the decline in the incidence of infectious respiratory disea-ses.) Having been brought up myself on the fourteenth floor of two apartment houses—and having raised my own children in a twelve-story building where the fact that we were on the third rather than some higher floor was totally irrelevant—I do not automatically un-derstand the problem. The freedom represented by the chance to command from an early age the use of an urban infrastructure (par-ticularly public transportation) seems to me to outweigh by far the discomforts and social demands of crowded living. But you can't study this subject without recognizing that most parents do not see the world that way—and they've got to pay for the housing. And, of course, a high-rise with a doorman and super who comes to fix the toilet may be a quantum jump from the high-rise where you let your-self (and who else?) in with a key, and submit maintenance requests through a bureaucracy.

High-rise housing for the elderly works fine, once they get used to it, for the same reasons that mobile home parks for the elderly work fine: the crowding leads to a sense of mutal support. ("I can't tell you," says Florence Berman of the Clarksburg, West Va., housing authority, "how many elderly came into our projects with-out a friend in the world and when they left—usually feet first—everybody loved them.") Older people take up less space in the world, fear destruction, cherish what they own, and tend to have more important worries than the color of their neighbor's skin. Extra services, from medical help to a communal kitchen, a bus to church, an advocate with Social Security, a bingo game, are far more easily provided if their recipients are all housed on one small piece of land. Contrary to previous theory, most old people today would rather live with other old people, and in poor communities they do indeed have reason—it is *the* shame of our cities—to fear the adolescent young. A study of six high-rise projects for the elderly showed a range of 58 to 83 percent in the proportion who did not want to live in a building with teenagers.

Oscar Newman says a certain amount can be done by designing even high-rise apartments for safety. He admires Davis & Brody's Riverbend in New York, where people reach their duplexes via broad exterior walkways, and each apartment has a partially bricked-in private entranceway from what used to be called "the sidewalk in the sky." Breaking an individual building into separate sectors each served by its own elevator and small hallway will yield much better security than the usual elevator bank and long hallway on each floor. (But Pruitt-Igoe was thirty-three relatively small buildings averaging fewer than eight apartments to a floor.) Floor-to-ceiling windows on the fire stairs, depriving criminals of their privacy in escape, may also make a difference. The most important change that can be made in an existing project, however (this is still Newman), is probably external: fencing off some of that meaningless greensward that was so popular through the 1950s and 1960s to relate each piece of it to a specific building, increasing the comfort and security of the residents of that building in using it and decreasing the apparent danger in space that "nobody" owns.

At best, one finds relatively few successful high-rise houses for low-income families. Problems in public housing are most likely to be manageable in designs like Harlem River Houses, where the four-story structures line the perimeter of the site and the common area inside is approached through entrance tunnels, and where the small number of apartments per entry allows residents to spot both neighbors and strangers on the stairs—or, indeed, on their approach to the entryway.

Low-rise alone, however, guarantees nothing. In Charlotte, North Carolina, a new public housing development of two- and three-story apartments that looks most attractive when seen from the road to the airport turns out on closer inspection to have suffered boarded-up and burned-out units, and terrible maintenance around the doorways, before achieving its third birthday. In Winston-Salem itself, the church-operated subsidized housing completed in the 1970s, all of it two and three stories high, shows outside doors to hallways ripped off their hinges, broken windows patched with boards, graffiti on the walls. (Graffiti, says Bruce Rozet of Los Angeles, who specializes in recapitalizing busted church-operated housing, is the mark of Cain: "Once they start defacing the property, you're dead.") Perhaps the most miserable public housing project in America is the set of 1,100 units in two-story and three-story narrow structures built in 1953 on the hillside above the old navy yard in San Francisco—Hunters Point, in many ways a luxury site, where the views are spectacular and the winds are such that the sun shines on days when downtown is fogged in. Planning its revival with $16 million of HUD Target Project money, the San Francisco housing authority commissioned a survey by a black public opinion polling firm, and learned that "respondents felt they were most unsafe walking from the car to the House," and that only 29 percent thought it would be safe to ask a neighbor to keep an eye on the place if they went away for a few days. Almost three-quarters wanted no part of a tenant-staffed security force: they wanted uniformed police. (More than half, incidentally, owned their own washing machine.) But the San Francisco authorities believe that their high-rise "Pink Palace" in the Western Addition urban renewal area is an even worse sewer than Hunters Point.

Asked for their own view of the prime cause of the deterioration of public housing, a considerable fraction of local administrators will blame the "Brooke Amendment" of 1969, by which local housing authorities were forbidden to take in rent more than 25 percent of a tenant's income (after it is reduced by a formula deduction plus $300 per child). In theory, the federal government would make up the difference between that less-than 25 percent and the authority's normal break-even rent; in fact, Congress did not appropriate enough money to fund the program fully, and Nixon impounded some of what was appropriated. Even after the money began flowing through, the delay between the submission of evidence by the local authority and the actual payment of the bill by Washington ate up the entire operating reserves of the projects (and then some). The Amendment exempted welfare families whose rent was separately paid by the welfare department, but the Nixon Administration was committed to the notion that the poor must learn to manage their own resources; and the Department of Health, Education & Welfare issued orders forbidding local welfare departments to separate out rent payments from the federally financed share of welfare grants. In project after project through the country, maintenance work ground to a halt.

"If you're running a well-managed program," says Ray Minor of the Kansas City housing authority, "you're visiting every year, you're making sure the residents are taking care of their units, and then your income is reduced by a Brooke Amendment. . . . Suddenly, your maintenance has to go on a deferred basis, and your buildings deteriorate. You begin to have a marketing problem, increasing vacancies. On top of that, in those days, the only way you could keep out someone was if he'd been a tenant before and had left owing you money. And one vandal can tear down more in thirty minutes than my maintenance crew can accomplish in a week."

Adding enormously to the damage in the early 1970s was the rapid increase in energy costs, coupled with a design error that had been policy for almost twenty years. Most public housing built after the early 1950s was "master-metered": that is, the local power-and-light company delivered electricity (and sometimes gas) to the site, and the authority took care of distributing it from the entry point. The rationale was that the elimination of individual metering saved construction costs, and the bulk purchase of power meant lower rates. Energy costs were thus built into the rents which, by the Brooke Amendment, had to be less than 25 percent of tenants' income. The planners had underestimated both the capacity of the human animal to waste what comes for nothing and the maintenance costs of aging gas and electrical conduits on the property. There was no place from which the payments for these costs could come, other than building maintenance.

As the maintenance goes bad, so do people's attitudes toward their housing. In the early 1970s rent strikes proliferated, depriving the urban authorities of what little money remained for maintenance or security. One of the running sores of the public housing community from 1937 to 1970 was the restriction on tenant incomes, which meant that people who did well, the potential leadership group of the project, were forced out. By the time this rule was dropped under the Housing Act of 1974—and people were permitted to stay if they were willing to pay rent increases commensurate with their increased income—the maintenance in most projects had declined to the point where nobody who could get out was willing to remain. Half the nation's big-city public housing authorities were trapped in a vicious cycle of low-maintenance/vacancies/further-deferred maintenance/abandonment.

In 1974, Congress began appropriating large sums—more than these buildings had cost to begin with—for the restoration of "target projects." Housing authorities were again permitted to inquire—as any private landlord would—whether an applicant was likely to be a good neighbor to the existing tenants. (Arthur Solomon of the Harvard-MIT urban studies center denounces the New York City Housing Authority for continuing "to apply an admissions policy favoring 'better families' and the maintenance of a 'good family environment.' " How viciously racist it is, really, to demand that poor black Americans alone in the world must be denied protection against the presence of criminal vandals, psychotics, and drug-dealers next door!) By then, unfortunately, the monomaniacal lawyers of the poverty programs and bullheaded judges had combined to deprive local housing authorities of the flow of information from police departments and school systems necessary to make an intelligent selection. For many projects, and possibly for some urban housing authorities, the bell has tolled. HUD is pretending desperately that there are no more Pruitt-Igoes—permission to tear down other projects has been repeatedly requested and refused—but, of course, there are.

When an authority can no longer impose upon its tenants the idea that their project must be maintained because it is government property, the last line of defense is the argument that they should take care of it because it is *their* property. Though the problems of public housing have not in truth been the result of "monolithic landlordism," as the cant line of the 1960s had it, the central bureaucracy unquestionably can be unresponsive to tenant concerns, real as well as imaginary. (There are, of course, a fair number of imaginary concerns in this community. James Cox of the Los Angeles Housing Authority tells a charming story about a lady who called to insist

that painters had to come immediately to her apartment because
" 'the Virgin Mary is going to be here at six o'clock this evening
and I want the Virgin Mary to be impressed.' . . . I didn't want to
tell her that I had my doubts that the Virgin Mary was going to be
able to keep the appointment. So finally I said, 'Well, you know it's
about 4:25 now; we close at 4:30. There's no way in the world for
me to get somebody there and do all that in the next hour and a half.
I don't know what I'm going to do. Do you think you can talk to the
Virgin Mary?' She said, 'I think I can.' And she was just as
nice. . . ." But there is nothing imaginary about the roaches and
the rats, the stench, the broken windows, the nonfunctioning elevators, the neighbor dealing drugs, the rapist on the stairwell, the
broken pavement in the playground, the transmission trouble with
the car that ate up the rent money—all of which may well be treated
on a don't-call-us-we'll-call-you basis by an overworked bored
clerk downtown.) Computerization, by confusing the quantity with
the quality of information available at headquarters, may worsen
both centralization and unresponsiveness. "It's a hoax," says housing lawyer Richard Baron in St. Louis, "to pretend that the computer gives you a substitute for someone getting off his ass and
walking up to the top floor to see if the maintenance man really
swept the landing."

Baron, a lean young man with thinning blond hair, is the father of
"tenant management" in St. Louis and its theoretician nationwide.
From Detroit originally, he was graduated from the University of
Michigan Law School in the mid-1960s and came down to Legal
Aid in St. Louis on assignment from the Office of Economic Opportunity in Washington. He found himself in the middle of one of the
first big rent strikes, provoked (this was before the Brooke Amendment) by a major rent increase imposed by the housing authority on
tenants most of whom were trying to live on the very low welfare
payments of the State of Missouri—$124 a month in 1969 for a
woman with three children.

In essence, the housing authority was seeking to restore the original principles of public housing. St. Louis was full of vacant properties. Maintenance and security had fallen through the floor in the
housing projects—especially the enormous Pruitt-Igoe project, de-

signed by Skidmore Owings & Merrill with the three- and four-bedroom apartments on the top stories and only a single elevator per building. (This design won an American Institute of Architects prize, a splendid illustration of how much architects care about livability.) With rents very low in a city where the vacancy rate in private housing was 10 percent, families with any sort of income at all were getting out of the dangerous and stinking projects. The housing authority hoped with its new rent schedule to find enough money to puts its buildings in shape, and undoubtedly (though this was never admitted) to drive away some fraction of the dependent households whose inability to cope was poisoning the life of the projects. Because there was so much vacant housing in the city, the economic expellees would not in fact be without a roof of some kind over their heads, and public housing would be saved.

The resulting rent strike lasted for nine months, and drove the housing authority to the edge of bankruptcy. (More than a hundred employees had to be laid off.) With help from Baron and his colleagues at Legal Aid, the tenants delayed action in the courts on the authority's hundreds of eviction notices, and were able to bring about a HUD investigation. The HUD team recommended a solution based in large part on the tenants' demands—including a tenant representative on the authority's governing board, maximum rents of 25 percent of each tenant's income, continuing consultation with a Tenant Affairs Board, and mediation of other issues by Harold Gibbons of Teamsters Local 688, a large, ruminative man who as International Vice President had kept the Teamsters respectable on social issues through Jimmy Hoffa's presidency of the union. Gibbons froze the housing authority out of the negotiations, worked up a deal with the Mayor and HUD and the rent strike leaders, and organized a community-wide Civic Alliance for Housing to supervise a restructuring of a demoralized and destitute housing authority. Part of the deal, signed in October 1969, required the creation of a "tenant management program," a term nobody could define.

The definitions were found mostly by Baron, Gibbons, the new housing authority executive director Thomas Costello, and rent strike leader Jean King, a large, gallant, and highly articulate welfare mother with a big Afro and an irresistible grin. The final struc-

ture has the housing authority as essentially a holding company and
service center, supplying skilled major maintenance and repair
crews and computerized rent-collecting, payroll, bill-paying, and
accounting services to a group of separate corporations (some
tenant-controlled, some controlled by churches or social agencies)
which run the projects on a day-to-day basis. Each corporation
screens and accepts its own tenants, and hires people for office work
and routine maintenance. In the four Tenant Management Corporations (TMCs), each responsible for a project of roughly six hundred
units, the executive staff and most of the work force are tenants
themselves.

But Baron's grand design involves a good deal more than housing
management. "There are countless efforts of the federal bureaucracy to coordinate funding for this community," he says. "The
Act of 1974 set up federal regional councils around the country
under the Office of Management and Budget. But when you wade
through all the paper and all the levels of bureaucracy, the people
who are charged with integrating programs don't know what that
means in programmatic terms.

"Our position is that people born and brought up in public housing are far more capable of handling the social and human problems
of the families that live in these projects. We have taken HUD
funds, Title 20 HEW social service funds, Labor Department training funds, Law Enforcement Assistance Agency funds, education
funds, and we've linked them all together. *We* are the community
development agency. I really see the TM strategy as a microcosm.
We are doing nation-building, taking the customs and traditions that
have evolved in low-income communities and using them to stabilize these communities. We've trained mothers to run day-care
centers to release other mothers to go cook for the elderly, and we
have Title 20 funds for both.

"The Office of Management and Budget publishing an A-95 set
of regs [A-95 is the coordination category] is like a Small Business
Administration making a loan to a neighborhood group to start a
grocery store, when the A&P is going out of business in these
neighborhoods. Governments keep programming failures and then
say, 'See, there is nothing to be done.' So our TMC has a joint ven-

ture with Interstate United, a food conglomerate. They get a fee. They do training and budgeting for our elderly kitchen program. We made the marriage on pure economic terms—everybody agreed it made sense as a business venture. We have jobs, good food, a happy corporation, and a happy community group that's participating in the profits. We have eleven employees, pay them three dollars an hour so they'll have enough money to get off welfare, count on nine dollars of sales per employee hour. I see a daily cash flow report on a weekly form, we know what everything costs—$1.14 per meal in the day care center, $2.14 per meal for the elderly. The cash flow is positive.

"It's not that the country doesn't have enough money in urban programs," says Baron, expressing the ultimate heresy. "We don't know how to spend the bucks we got. If we could get eighty percent efficiency on the dollars we've already got, we could save eighty percent of these inner-city neighborhoods. If federal agency planners and local officials really knew how to orchestrate federal programs, intervention strategies would be developed all over the country that would stabilize these distressed communities. But we have very few technicians. . . ."

There is, perhaps, a little more theory than practice here. The two-year evaluation of TMC done in late 1975 by George Wendel and the Center for Urban Programs at St. Louis University—on commission from the Ford Foundation, which has put considerable money into this venture—reported that "TMCs had not played leadership roles in the coordination of social service delivery systems." But the Center gave Baron's people high marks for performing the "hard management" functions; and Baron's own brief for the nationwide expansion of TMCs conceded that "program efforts in the area of soft management directly depend on . . . success in collecting rents, decreasing vacancies, and providing prompt maintenance and custodial service. Tenant management is, above all, a real estate management program. . . ."

To a visitor's eye, moreover, it looks pretty good. At the Darst project, where Jean King is now the resident manager, a long wall by a walkway is charmingly painted with a proletarian picture of work in process and children at play. The wall was painted two years ago by the tenants themselves under guidance from art students at Washington University of St. Louis; what counts is that it remains unmarred in a project where people used to shoot out the windows with shotguns. Among the ideas for better security being tried out at Darst is a card-in-a-slot and key-number system, like the systems that operate the new cash machines for the banks, that substitutes for keys in opening the hardened glass doors to the elevator lobby in one of the buildings. Darst also has a creche, run by an experienced English administrator, for babies six months to two years of age—the only such funded by HEW in the country.

Rents have been raised, and rent collections are at 99 percent. The TMC office is a delight, elegantly Scandinavian, a superbly intelligent use of Ford funds, with a carpeted and wood-paneled board room for the elected representatives of the tenants and carpeted private rooms in which the staff can meet with tenants to discuss problems. Because the housing authority in its holding company capacity collects and posts the rent, the office receives little of that routine traffic that wears down most project offices and requires special security precautions.

One of Darst's seven buildings is still a burned-out hulk, and the entrance lobbies of several of the others are still much the worse for wear, but the atmosphere is one of recovery. Current HUD regulations allocate extra money to housing authorities with increased rent collections and decreasing vacancies, and St. Louis under its holding company system has qualified. Miss King has Community Development money to put apartments in shape and Target Project money to help rent them (she has furnished a model apartment, just like a private builder, to persuade shoppers). It's not cheap to do: "Most of our vacancies," she says, "have been vacant for five to thirteen years, which means that if nobody else vandalized them the maintenance men did." But it appears to be coming along.

Much less can be claimed for the Boston version of Tenant Management at the Bromley-Heath project in that city, also supported by foundations (but not by Ford) and more widely publicized. The annual report for 1976 from this venture proclaims that "the TMC has maintained its steady growth in maturing as an outstanding innovative management model, still the first and only example in the nation of recognizing the capability and resourcefulness of a comprehensive tenant-controlled public housing community. This distinctive prestige is illustrated by the TMC being selected as an officially designated community achievement by the federal American Revolution Bicentennial Administration. . . . Specific accomplishments are many and each involves and deserves the interest and attention of professional and lay persons. . . . The TMC assumed the direct administration of the Community Patrol—a security force directed by tenants. The Patrol is the sole example of this model of a security organization. . . . There is a long waiting list of applicants for Bromley-Heath and the quality of management services available. . . . Yes, Bromley-Heath is a community of people and things. Flower and vegetable gardens, bountiful and beautiful, blossom and grow in Bromley Heath as a reflection of the growth in spirit and confidence of the residents. The TMC is proud of its record. . . .''

Virtually none of this is true. Bromley-Heath is a disaster area, clearly a project en route to abandonment, and the annual report is simply an example of what people think they can get away with when they have Cambridge, the foundation world, and the Washington bureaucracy on their side. A quarter of the units are vacant, and almost 40 percent of the rest are not paying rent—even though TMC has persuaded the welfare department to pay directly, deducting from the monthly check, for a quarter of the residents. Whole wings and (worse) many individual units of the high-rises are burned out, and on the day of my visit not a single elevator was working in any of the buildings. There were graffiti on the walls. For the many apartments available at Bromley-Heath there is, obviously, no waiting list at all; the only people applying to get in are mothers with very large families, who list with the Housing Authority (which does the TMC intake in Boston) for every project. There were lots of young people loafing about, snapping their fingers.

"They steal the sinks, the radiators, the pipes," said Mrs. Lois Hamilton, assistant director. "Now they even steal the plywood from the windows when we board them up. The kids get into the vacant apartments, turn on the water, and damage all the apartments in the line. The elevator problem is hopeless—repairing them is money thrown away. The kids have learned to get into the elevator housing on the roof. We talk to the parents, tell them to keep their kids out of the elevators. They're not built to take the loads, or to have a box propping the door; that blows a fuse. These projects were built for people trying to get on their feet, not for people who will be living here generation after generation. . . ."

The formal structure of the Boston TMC is easy to fault—despite the assertion of *tenant* management, the fact is that neither the executive director nor her assistant lives at Bromley-Heath; and the contract with the city housing authority turns over to the subsidiary not the tenant selection process it can do well but the data processing and accounting jobs it can scarcely do at all. Still, the problems in Boston public housing—the fifth largest authority in the country—may be beyond resolution. There is a viciousness at large in this city that—for example—burns down a $400,000 community center in the Archdale project before the doors can open. Boston may be the only housing authority that runs segregated white projects, not because it wishes to do so but because race relations are so bad that no black family could be protected if it moved into one of the South Boston projects.

Egged on by the footling communes of young lawyers that dot the Boston landscape, both the federal and the local courts have interfered destructively in the management of the projects, at one point forbidding evictions for nine long months during which the authority sank to the edge of bankruptcy from nonpayment of rent. A black girl in the office of one of the projects was near tears as she told that day's story, of a white judge who had refused to evict a woman with eight-month-old twins, who had deliberately not paid her rent for four months: "We've got other women with babies who pay their rent," the girl said, "—or used to pay. Once the word gets around that you don't have to pay the rent, it's like cancer. And then where do we get the money to keep the place going? By the time you get a judge educated to the fact that when one tenant doesn't pay the rent it cuts the services for the others, he gets promoted to a better court. . . ." Indeed, as a serious matter of self-praise—apart from the gush in its publicity—the Bromley-Heath TMC points out that it has successfully evicted several disruptive tenants, while the housing authority finds it all but impossible to get rid of anybody.

Similar claims of effectiveness in the policeman's role are made in St. Louis, on behalf of "disciplined, trained residents who, without compromise, are prepared to enforce community-developed standards against those who persist with unjustified rent delinquencies or antisocial behavior. Difficult management decisions which have traditionally been avoided are now being made and accepted by public housing residents who have found new strength in their own communities." But it is individual managers, not anonymous "tenants," who really make such decisions, and one cannot be quite certain that this process should be deprofessionalized. In the Neighborhood Housing Services committees in Anacostia, the bankers have been shocked by the remedies residents suggest when borrowers become delinquent on their loans. In tenant management, the structure will be dangerously vulnerable to the choice of manager.

Jean King told the St. Louis University researchers, "My Board of Directors hires me, not the tenants. I led the rent strike in my building. They still look to me for leadership. . . ." Asked what would happen if she left, she replied, "The place would fall apart. . . There is no one who wants the job. They think the idea is a good one [tenant management] but managing is a lot of work." Until the original leadership, which is first-class , has been successfully replaced, skepticism about tenant management will be inescapable.

Beyond that, one has an itchy feeling that turning the projects over to their tenants is a variant on Michael Harrington's rule that capitalist democracies nationalize their failing industries to give socialism a bad name. Richard Baron's advocacy of tenant management has a bitter edge to it. "These people," he says, "are never going to penetrate the mixed economy of the United States. It does well for eighty percent of the people, but they are the other twenty percent. There's no point subsidizing an operation that's going to hell—it's like the bombs in Vietnam, just gone. You and I aren't going to solve the problems of these communities. Do you step away and let them go to hell, or do you establish the preconditions to let them take care of themselves?'' If tenant management's triumph is that the flies will conquer the flypaper, the solution is much less satisfying than what we had hoped we would buy with public housing.

Chapter 9

Income Mixing

I have just been in a country from which we have nothing to learn.
> —comment by The Netherlands Minister of Housing to a reporter as he returned from a trip to the United States in spring, 1976

Over the years, public housing failures have most often been blamed on the location of the project—it was in the middle of the ghetto, and thus automatically infected by the surrounding blight, or it was too far away, in places where the residents lacked cultural support. In Chicago in 1969, Judge Richard B. Austin ordered the housing authority to build 75 percent of all new public housing ''in predominantly white communities and at least one mile from the edge of the nearest black ghetto . . . and to avoid clearing the selection of project sites with the aldermen of the wards involved.'' (This decision put a stop to all public housing construction in Chicago.)

On the other hand, Len Downie of The Washington Post denounces Atlanta's Lake Meadows, ''in a remote corner of Atlanta that lies in suburban DeKalb County'' where ''everything was a distant automobile or bus ride away''—which would be the norm in the places where Judge Austin wants the public housing. In Boston, M.I.T.'s Arthur Solomon is bitter about Columbia Point— ''desolate sites, entirely cut off from the rest of the city''—but this

location, fifteen minutes from downtown Boston by bus, out in the
harbor with light and good air and water on three sides, was considered hugely desirable when it was predominantly white "veterans
housing" right after the war; and in summer 1977 there was a plan
afoot to recycle it as middle-income faculty housing for the nearby
Boston campus of the University of Massachusetts.

The emphasis on "mixed-income" neighborhoods grew increasingly strong through the 1960s, as neighborhoods themselves grew
increasingly segregated by income. It was recognized but not accepted that income segregation might be a way—perhaps the only
way—to achieve racial integration in residency for the black middle
class. Evidence piled up that wealthier communities—safe in the
certainty that their land values prohibited the construction of public
housing—were more supportive of social goals in housing than
lower-middle-income communities. As early as 1949, a constitutional referendum in California, to require local voter approval before federally sponsored housing projects could be authorized in a
municipality, was voted down in the richest suburbs and carried by
an outpouring of ayes in the cities. Congress had imposed on cities
wishing urban renewal or public housing assistance the local creation of a "workable program" to assure that federal money would
not be wasted. In the 1960s it was discovered that this requirement
had provided an easy out for municipalities that did not wish such
help; and now the pressure goes the other way, as suburbs are
warned that aid for projects everybody does want, like sewers and
parks (and down the pike maybe even some school aid), will be
withheld unless the community accepts its "fair share of the burden" of subsidized housing. The upper-income suburbs do their
token thing—*noblesse oblige*—but as of 1977, the lower-middle-income suburbs will have none of it. On the medium time-horizon,
they control the levers of political authority in America; and what's
more, they are right: poor people get out of the ghetto really only
when they make some money.

The high point of efforts to force integration of income levels on a
piece of land came in Lyndon Johnson's abortive "New Town in
Town" program, which was to build much new multifamily housing on property owned but no longer needed by the federal govern-

ment. Martha Derthick has described rather horrifyingly how this program came into being: "One morning in August 1967, as he was sitting in his bedroom at the White House and talking to Special Assistant Joseph A. Califano, it occurred to the President that federally owned land in the cities could be used for housing. Within hours, his staff had assembled a working group from the executive department to figure out how this could be done. . . ." Within three days, this task force had produced a "plan" to build four to five thousand housing units—70 percent of them subsidized and 1,500 for the very poor—at Washington's Fort Lincoln, site of an abandoned juvenile training center. Three architects—Paul Rudolf, Moshe Safdie, and Harry Weese—were put in competition to design the first four hundred units (250 low-income, 150 moderate-income). In January, Edward Logue, master builder of redevelopments in New Haven and Boston (actually a lawyer), was brought down at a fee of $295,000 to start putting the thing in the ground. In February, the President announced to the Congress that "Here in the nation's capital, on surplus land once owned by the Government, a new community is springing up."

By then the project had already run into a buzz-saw of opposition from the black, middle-class homeowners on the surrounding streets; and their unremitting antagonism eventually killed it. Derthick explains: "The President's distance from local politics made it difficult for him to analyze the housing from the perspective of local officials, and to calculate the advantages and disadvantages of housing developments as they would. 'He did not understand,' one of his aides said later, 'what a mixed blessing low-income housing was for the cities.' " Ten years later, the first buildings of an all middle-income project were opening in Fort Lincoln.

Yet there was nothing new about any of this. Subsidized elderly housing, which does not carry the threat of crime, vandalism, or deterioration of the local schools, can be built almost anywhere without much protest; but any sizable project for poor families has always been extremely hard to plant in a neighborhood where people are paying their own way. Detailing the struggle over the placement of the first postwar projects in Chicago, Meyerson and Banfield quoted a truck driver who led community resistance to public housing in the southern part of the city: "A social worker came up to me one day and said that I should feel sorry for these people and that I should take some responsibility for them. Now, I'm not a rich man but I worked hard to raise a family. I went to work when I was sixteen and I supported three younger sisters. When they grew up and got married, then I got married and now I'm raising a second family. I told her this, and she didn't have any answer for me. How can she tell me that I owe these people something?"

Since the middle 1960s, then, the federal government's main effort has been to provide housing for the poor without building projects. Various "scatter-site" Turnkey programs were developed by which the authorities could contract not only with private builders to design and build housing for the poor, but also with private management companies to operate the resulting buildings for a fee. Nothing about the properties when completed was to say they were "public housing." Shannon & Luchs, a Washington, D. C., firm that had one of the experimental contracts to operate two middling-sized mostly low-rise developments (155 and 246 units, respectively, the latter being half elderly) argued that "the institutional label of public housing . . . contributed greatly to the stigma now felt by many tenants of public housing. . . . [Our] tenants . . . report that now that they are living in housing 'managed by Shannon & Luchs' rather than NCHA [National Capital Housing Authority], they are regarded differently by their neighbors. In turn, their pride in living in private housing seems to affect, at least in part, their willingness to put forth greater effort to improve the entire property."

More ambitious was the "Section 23" program of "leased housing," first authorized in 1965, expanded in 1970 when local housing authorities were directed to devote to this purpose at least 30 percent of their receipts under their annual contributions contract with the federal government, and exempted from the Nixon freeze on new federal housing commitments in 1973. Under this program, housing authorities were ordered to find existing structures—which could be single-family homes as well as apartments—that could be rented for the benefit of their clientele. The authority itself became the tenant for the private landlord, and sublet to the resident. With costs of new housing rapidly increasing, Section 23 looked economically right. "Building more units is far most costly than using the existing housing stock," Arthur Solomon of M.I.T. wrote, "irrespective of whether the costs are calculated from the viewpoint of the federal government—direct HUD expenditures and foregone tax dollars— or in terms of total resource costs. Our calculations indicate that *twice* as many poor families can move from substandard to standard housing when the government leases existing units."

The basic requirements of Section 23 were that the rented facilities meet FHA standards and that the rents be no greater than the "fair market rent" for comparable accomodations, a number that would be drawn from the tea leaves by the HUD area offices. The critical and largely arbitrary nature of that number seems not to have been appreciated either by Congress or the HUD bureaucracy: set too low, it prevents the authority from finding accommodations for its people; set too high, it gives windfalls to landlords, not infrequently some rather unsavory landlords, and may push up the rents paid by all poor people in the community.

Section 23 offered significant noneconomic advantages. Leased housing took the "welfare" stigma off the occupant—and, perhaps even more important, compelled the local housing authority to choose occupants likely to comport themselves as reasonable tenants of a privately owned housing unit. Destructive Section 23 tenants would get the housing authority in bad trouble with private landlords, impose financial penalties on the authority, and jeopardize future leasing from other landlords. Presumably, the same restrictions on selection procedures applied to Section 23 housing as to public housing—the courts would not, say, permit a housing authority to reject an applicant just because her adolescent son, living home, had a record of torturing prepubescent neighborhood girls with cigarette butts. (After all, boys will be boys; the mothers of monsters have civil rights, too; this family has to live *somewhere;* and they're all blacks anyway.) But there was always so large an applicant group for Section 23 that the authority could make sensible selections without triggering judicial intervention, provided everybody was careful not to write memos.

Together with its companion "rent supplement" program— which paid some of the rent for special categories of low-income people who had found their own housing (elderly, handicapped, displaced by government action or natural disaster)—Section 23 meshed well with federal programs to promote the construction of moderate-rental housing for moderate-income families, which had begun to come on stream in the early 1960s with Section 221. And in 1968 there came Section 236, designed to be the prime vehicle carrying the production of subsidized housing to a targeted 600,000 units a year. Under Section 236, in theory, all tenants would pay at least a "basic" rent representing the costs of operating the building on a 1 percent mortgage. In reality, the government had placed the first foot on a slope that has turned out to be extraordinarily slippery and extraordinarily steep.

The proprietors of projects funded under Sections 221 and 236 naturally made their apartments available to the local housing authority under rent supplement and Section 23 programs: for the first time, the government had a program that seemed to assure a mix of low-income and moderate-income people under the same roof. Once Section 236 was in gear, other possibilities opened up, for HUD had authority to give an interest subsidy under 236 to only *part* of an apartment project. Now it became possible to mix three income levels—the poor under Section 23, the lower middle class under Section 236, and the middle class at market rents. How well this worked is a matter of some dispute, because the Section 236 program became a disaster of such dimensions that no one factor standing alone can have caused more than one small part of it.

One of the problems in Section 236 was that many projects simply could not rent their apartments even after the gigantic subsidy of a reduction to 1 percent in mortage interest rates. (And it is a gigantic subsidy: on a thirty-year, $30,000 mortgage, the difference between a 9 percent and a 1 percent mortgage is almost $150 a month.) One of the reasons undoubtedly was that so much of the Section 236 housing was badly built: "We're going around now trying to plug the holes," says Jerome Steinbaum of Los Angeles, whose firm manages these things for HUD all around the country, "They used aluminum wiring without proper safeguards, they put on faulty roofs; and the materials inside, like plumbing fixtures, were poor." (His associate Sam Parnas puts the situation more vividly: "They threw this shit at us like you'd throw water out the window. Where were the inspectors?") Another reason some apartments remained unoccupied in some projects was that shoppers didn't like what they saw in the halls and lobbies.

Income mixing in any event can work only if nobody talks about it too much: indeed, it works best if the image is one that denies the facts, preferably with a light touch—as at Chloethiel Smith's Laclede Village in St. Louis (of all places), where manager Jerry Berger rented a Rolls Royce the first month to carry the first kids in the project to school. The number of people who wish it known that they live in an apartment house with the poor is relatively small. "We have thirteen midwestern projects," says George Brady of the National Corporation for Housing Partnerships. "Twelve of them are yielding a total of $239,000. The one in Chicago is losing $250,000. Why? It's one-third a 236 project, out in a suburb: the developer printed the fanciest brochures you ever saw. He has tennis courts, saunas, real corn architecture. But there was a series of do-gooder articles in the paper about the poor living with the rich. He was 92 percent rented; he went right down to 60 percent. Then he got under financial pressure, and he panicked: 'Better get tenants in, I need the dollars. . . .' "

What killed Section 236, however, was not programmatic failure (always a marginal matter to the federal government, anyway: the constituencies a program develops count for more than its accomplishments), but the costs. HUD had estimated in testimony to Congress in 1968 that the costs of providing new housing would rise by 2.75 percent a year from 1969 to 1978; the real increase has been about 12.8 percent a year. And in one of those bits of cleverness that were so common in the 1960s, the real costs of Section 236 were initially concealed by switching the subsidy system from what were in effect government loans to annual government grants to reduce borrowers' costs on private loans. The result was that the federal government, which can borrow at the lowest rates, began paying private market mortgage rates to finance subsidized housing.

"The apparent motivation of the Act," William J. Quirk and Leon E. Wein wrote shortly after its passage, "is to avoid at all costs the appearance of large expenditures in the federal budget." They quote testimony by Senator Robert Kennedy, describing the

Section 236 subsidy procedure as "an unsound and needlessly expensive procedure for financing the housing units which it seeks to provide . . . the Federal Government will have to pay 75 percent more interest each year on every dollar of mortgage issued than it would if the mortgage were financed in accordance with the present practice . . . the same subsidy dollars provide only 57 percent of the housing that they would if mortgages were issued under current procedures." Kennedy's figures were off—the gap between government borrowing costs and mortgage rates narrowed for reasons we shall examine in Chapter 16—but he was right as rain on the issue. Year by year, as the 236 program grew, appropriations to add identical numbers of units required *geometrically* larger amounts of money.

And the 1968 Act programs took off almost immediately. Subsidized housing starts had never been above 100,000 a year before 1968; in 1970, they crossed the 400,000-unit mark. Each new unit built under Section 236 (or its companion Section 235) committed the government to annual payments looking out thirty to forty years. When the Nixon Administration put a moratorium on new commitments in January 1973, HUD Secretary George Romney told the Congress that he thought the costs to the government of continuing these programs through their ten-year authorization would climb above $100 billion. Unlike some others in the Nixon collection, this impoundment stuck: Congress grumbled but did not act to revive the programs.

Enoch H. Williams of the church-affiliated Housing Development Corporation in New York commented in 1977 that "we were just getting the momentum when all of it was stopped. It's surprising there was no howl from around the country, considering all the lobbying that had gone into getting the act passed." In fact, though they thought Romney's figures were high (they were really low), even the professional housers were frightened by the developing fiasco. "We'd never looked at the long-range implications of those programs," Congressman Thomas L. Ashley admitted recently, "except in the dark of night."

Eventually, Section 235 was reprieved on a small scale, and the execution of Section 236 was accomplished slowly: projects that had won approval before January 1973 were still opening their doors in 1977. After mid-1973, however, the only significant continuing government subsidies for the construction of housing were the guarantees to builders of housing for the Section 23 public housing leasing program (which became, as we shall see, the chassis for an entirely new federal approach to these problems)—and the Farmers Home Administration in the Department of Agriculture. As the almost-only game in town, FmHA housing programs grew like weeds: from $790 million of new loans made in 1970 to $2.5 billion in 1976. This adds up to something like 125,000 units of low-cost housing a year, and no discussion of social housing should omit it.

Briefly, then: Farmers Home is that rare federal bird, a direct lender. (To keep FmHA figures off the budget, its loans are then sold to the public through the Federal Finance Bank, and the agency professes to call them "insured," but in fact it's as direct a lender as any bank.) There are 1,770 Agriculture Department county agents around the country, and the applications for mortgages are made to them. The borrower need not be a farmer, and the location need not be rural: Congress has authorized loans in towns up to 20,000 population, and there are lots of areas twenty miles from fair-sized cities which Farmers Home may consider "rural in character." The income cutoff is relatively high: in 1977, a family with three children could borrow under the "low-income" category with an income up to $11,473, and under the "moderate-income" category with an income up to $17,368.

The house itself must by law be "modest in size, design, and cost." In 1976, when the median single-family house nationally cost $45,000 to build, the average Farmers Home house cost $23,267, and its average size was 1,057 square feet (as against the national average of about 1,600 square feet). While the cost of new housing has been rising nationally at a rate of almost 13 percent a year, the cost of FmHA homes has been going up only about 6–8 percent a year. Loans run thirty-three years, at interest rates that may be as high as 8 percent, but the agency is empowered to reduce rates to as low as 1 percent to keep housing expenses under 25 percent of the family income. As of 1976, about 60 percent of Farmers Home loans were receiving "interest credits," and the average rate on the outstanding loans was 2.9 percent. The family must submit to "income verification" every two years, and if the income has risen the interest rate will rise with it. Over the cutoff point, the family must refinance privately, at whatever the private mortgage rate may be. I suggested to L. D. Elwell, who administers the housing programs for FmHA, that it must be all hell to get people to make that change. "Oh, yes," he said; "it is."

"Farmers Home is the star of housing programs for Congress," says Henry Schechter of AFL-CIO, whose members are heavily represented among the purchasers. "It gets a lot of housing built, cheap." How good that housing is and how carefully FmHA makes its loans are matters of some dispute in the business. Elwell's modest office is hung with pictures of attractive homes, some of them imaginatively designed in the California manner, but builders tend to be scornful, and an academic consultant to tract developers remembers "walking down a street with FHA housing on one side and FmHA housing on the other. One of them had two-car garages and air conditioning, a flagstone walk to the door, quality aluminum windows, an overhanging roof with an interesting texture—because the builder had to sell it. The other was just a bunch of wooden boxes. All we could do was laugh."

A builder in West Virginia says of Farmers Home, "You have to build a batch of them quick and get to hell out, or the deterioration of the first houses keeps you from selling the new ones." Though there is unquestionably a lot of utility-grade lumber and bottom-of-the-line equipment in FmHA housing (the cost, after all, is literally half of what FHA-insured housing sells for), much of this comment may be snobbery. Elwell, a lean, serious man in steel-rimmed glasses, who rose through the ranks from a county office in rural Illinois, says flatly that his office applies the FHA Minimum Property Standards, and his people inspect to see they are met: "There is *no* difference."

In recent years, loan delinquencies have been very heavy. FmHA has a confusing mission from Congress: it is not permitted to lend to anyone who could get a mortgage elsewhere, yet its mortgages are not supposed to be risky. The theory when the agency was started in the 1930s (under another name); was that rural areas were light on lenders—in the 1930s, especially light on solvent lenders—so that perfectly good potential borrowers had nowhere to go. Today the distinction is that private lenders require a downpayment, and FmHA does not. "We're not looking for deadbeats," Elwell says; but with the rapid expansion of the agency's lending in the 1970s (which has involved a good deal of lending through builders, who get a "conditional commitment" that a loan will be made to any "eligible buyer" they produce), FmHA seems to have acquired a number of customers who could bear that brand. In June 1976, 21 percent of all FmHA individual housing borrowers—153,000 homeowners—were more than a month behind in their payments, up from only 3 to 4 percent in the early years of the decade.

On the other hand, while FHA loses *85 percent* of its insurance guarantee on the average foreclosure, FmHA rarely takes a significant loss, because rather than go through a court proceeding (during which the householder in possession tends to let the house deteriorate mightily), FmHA agents try to find new purchasers for the home. At those prices, after all, there are lots of customers—especially when no downpayment is required The incentive for the original owner is that by selling to the purchaser the agent brings around, he can usually walk away with some money in his pocket.

As direct lenders, moreover, the county agents keep in touch with their borrowers in ways that mere insurers never can. Elwell remembers from his days in the field "a fellow who had a Shell station. His accounts receivable piled up, he lost the station, and he was in a state of shock. I came knocking, because he hadn't made his payments: 'Anything wrong?' He told me. I said, 'Can you get a job?' He said, 'How do you do that?' I sent him to the Illinois Employment Office, he found something, and began paying the mortgage again. We can give moratoriums, to people who are in trouble."

In 1977, FmHA had some 800,000 loans outstanding (a small fraction of them for rural multifamily housing), and almost as many people were living in Farmers Home houses as in public housing. America having become a country where the average level of civilization is higher in rural than in urban areas (and of education, too, as measured by the National Assessment), there is considerable reason to agree with the Congress that the FmHA program is a success, though the delinquency figures do argue that it cannot be expanded at the rate Congress seems to want.

Between early 1973 and fall 1974, a brand new system for subsidized housing was developed by the Department of Housing and Urban Development and sold to the Congress as a new "Section 8" of the Housing Act. Like many other Nixon programs—which await the revisionist historians of a decade from now—it was a mix of proposals from the libertarian right and the academic left, without much leavening of common sense or input from people actually working on the problems. The result is hardly credible for sheer wrongheadedness: monstrously expensive, strikingly ineffective, and deficient in both horizontal and vertical equity. Carried out to its planned conclusion (which will not happen), it would do for the costs and efficiency of housing in America what Medicare and Medicaid have done for the costs and efficiency of health care.

Section 8 of the Housing Act—the significance of the number, which is also used for psychiatric discharges from the armed services, has been noted by many—is a step toward a housing allowance system, probably best defined as income-maintenance-for-a-purpose. As such, it fits in with a worldwide trend toward what Raymond Barre (who later became premier of France) called "aid to the person, not to the bricks," and with the ingenious attempt in Holland to get rid of the incubus of rent control without punishing the poor.

The initial and correct assumption is that the problem with poor people's housing is not the housing itself but the poverty: in Nathan Glazer's words, "higher incomes in the poverty group would effectively solve their housing problems even without new housing policies." (We retain for present purposes the perhaps pious hope that higher incomes will also improve people's behavior vis-a-vis property.) The segregation of low-income people in low-income neighborhoods results from their inability to pay the housing costs of life in better neighborhoods. Thus the government's role should be to supplement what would be a normal housing allocation from the family's income (between 15 and 25 percent, depending on family composition), bringing that number up to a figure that will enable the family to pay the "fair market rent" for decent housing of the requisite size in the given area of the country. The HUD area office determines what constitutes "fair market rent."

The recipient does not *have* to spend all this money for housing— as the HUD sales literature stresses, "the program places the choice of housing in the hands of the consumer and rewards him for being a 'smart shopper.' . . . With appropriate training and information, eligible families can gain sufficient understanding . . . so that they may evaluate the quality of dwelling units as they shop. . . ." The reward for smart shopping does not go entirely to the Section 8 beneficiary, however; he has to split with the government any savings he accomplishes by finding an apartment below the "fair market rent."

The reliance, then, is on the market, a basic libertarian principle—but strongly supported here by the academic left. "Additional purchasing power provided by government-assisted leasing arrangements," Arthur Solomon wrote while Section 8 was wending its way through Congress, "promises to counteract existing physical and environmental deterioration by helping to stabilize demand and thereby create an investment climate in which owners will be more disposed to conserve or even enlarge the supply of decent housing at moderate prices. This is *central* to any policy that is to avoid the same disappointments encountered by community development reforms of the 1950s and 1960s."

Many poor people might shop to find cheap-for-its-own sake rather than good-at-the-price, so the market cannot be allowed to go unfettered. The beneficiary of the Section 8 allowance will thus not be permitted to shop the entire housing market—he will be restricted to accommodations that meet the FHA Minimum Property Standards, perhaps as modified by some input from the local housing authority, which administers the program. Moreover, for both policy and public relations reasons, the recipient will not be permitted to spend his Section 8 grants on luxury: nobody who occupies housing that costs more than the dictated "fair market rent" can receive a Section 8 allowance.

By its nature, Section 8 is an "existing housing" program, but it was clear to those designing it that the construction unions and the home builders—the heart of the "housing lobby"—could block any bill which did not at least make a gesture toward production subsidies. In any event, as the HUD *Fact Sheet* admits, "For the program to work, there must be an adequate supply of modest, decent housing at or below the Fair Market Rent." To stimulate production, then, the law permits HUD to enter into contracts with builders guaranteeing them subsidies for the low-income tenants they can recruit to new apartments that charge the "fair market rent."

Construction costs having risen so rapidly in recent years, the "fair market rent" on old housing will not be enough to pay the bills for new housing, so the Secretary of HUD is empowered to guarantee rent payments 20 percent above what has been determined by the local office to be the "fair market rent" for this area. The HUD contract can be taken to the bank and pledged as security for a mortgage. For builders with exclusively private financing, the maximum length of the Section 8 contract is twenty years (longer than the typical multifamily mortgage, which is usually written to be refinanced at some point in the building's life). For builders drawing their funds or mortgage insurance from a state housing program, the Section 8 contract can run for forty years.

Recipients are divided into two classes: "Very low income" (less than 50 percent of the median for the area), and "moderate income" (between 50 and 80 percent of the median). Income mixing is assured in new Section 8 housing, in theory, by requiring the builder to rent 30 percent of his apartments to "very low income" tenants and some individually negotiated fraction to tenants who can afford to pay the "fair market rent" without subsidy. Among the theoretical advantages of Section 8 is that the subsidy *could* be "shallow"—recipients' income is to be "verified" annually, and should incomes rise faster than the operating costs of housing, the government's obligation could diminish. On the other hand, if "fair market rents" rise more rapidly than the incomes of Section 8 recipients, which is what has been happening, the government's obligations would increase. In housing built or rehabilitated under Section 8 guaranteed rents, incidentally, the annual income verification is to be done by the landlord, which opens an exciting field for profitable interaction between landlord and tenant.

An immediate and practical advantage to HUD from the institution of Section 8 was that it provided FHA with a tool to rescue a good fraction of the bankrupt Section 236 projects by steering Section 8 beneficiaries to them at increased rents. As much as half the initial allocation of funds to Section 8 may have gone to these bailout operations. Beyond that, it is hard to see why anyone who took the time to think about it could have supported this program, either at HUD or in Congress.

The analysis of what is wrong with Section 8 breaks into several discrete categories, depending on the location of the analyst. Any of these objections standing alone should have been sufficient to sink it at birth.

(1) From the government's point of view: the new construction section is unbearably—and, in New York, unbelievably—expensive, with quick profits built into the scheme for anyone and everyone.

When America started subsidized housing in 1935, the federal government built the units itself, using federal employees in the Works Progress Administration, financing the project through the sale of government bonds.

Then the money was channeled through the local housing authority, which acted as planner, architect, and general contractor, but hired private subcontractors; and the financing was done through municipal bonds, which benefited purchasers through the tax exemption on the interest income.

Thence we went to "Turnkey," with fees to private architects and profits to contractors—justified in the result, to be fair, by the greater efficiency of private enterprise: where Turnkey and housing authority projects were built at the same time, Turnkey came in quicker at a lower price per unit.

Next, Sections 235 and 236, where profits were supposed to accrue to the banks as lenders, and to investors on a limited-dividend (plus tax shelter) basis as well as to everybody in the construction business.

Now, with Section 8, the entire panoply of profit possible in the private market is to be loaded onto the subsidy, because the beneficiary simply rents from the private owner. With all those hands in the pot—landowner, lawyer, developer, architect, lawyer, contractor, lawyer, investor, lawyer, bank, title company, lawyer—at least thirty-five cents of every subsidy dollar is siphoned off into excessive fees and windfall profits.

The program has come on stream at precisely the worst time, when the rents needed to support new multifamily construction in any part of the country are way above what anybody but the rich now pays. Thus the "fair market rent" system must be corrupted— no lesser word will do—if any housing is to be built under Section 8.

In 1976, bailing New York City out of its municipal mortgage on Manhattan Plaza, a 1,700-unit disaster on West 42nd Street, HUD agreed to a subsidy of $1,800 per room per year—$600 a month on a two-bedroom apartment—to run for forty years, at a contracted total cost of $460 million (and it could go higher). In early 1977, HUD signed a contract with National-Kinney for a 490-unit Section 8 apartment house in New York, to receive guaranteed annual rentals of at least $4.8 million for forty years. To reach these figures, the area office had to assert that the "fair market rent" for a two-bedroom apartment in New York is $737 a month (roughly two-thirds of the median family income in the city)—and then take 120 percent of that! It will clearly be impossible to rent any of these apartments on an unsubsidized basis, so the income-mixing feature goes out the window. This project also benefits by financing with state tax-exempt bonds, and by reductions in New York City real estate tax assessments. The total subsidy from the three levels of government runs about $1,200 a month for a two-bedroom apartment. Of course, that includes utilities.

It might be noted in passing that five families can be provided with new houses of their own in the Farmers Home program for less than the total subsidy needed to put one family in a new apartment in New York under Section 8.

In April 1977, the National Leased Housing Association held a pair of one-day seminars in San Diego and New York to present news of Section 8 developments. The topics were as follows:
- The 1976 Tax Reform Act and Section 8
- Tax Treatment of Limited Partnerships with Non-Profit Partner
- Outlook for Carter Administration Tax Reform

- Structuring ownership entity for syndication
- Role of the syndicator—broker dealer
- Optimum timing of syndication
- Role of accountant and lawyer in the syndication process
- Securities Act concerns in the syndication process
- Special problems of syndicating state agency or conventionally financed projects
- Developers guarantee of operating deficits
- Pay out of syndication proceeds
- How much is your project worth?

A ticket to the one-day seminar (including lunch) was $135.

Oh, boy.

(2) From the businessman's point of view, despite all these profit opportunities, the risks are greater than the returns. Consistent with free-market principles, HUD made the first allocation of contracts for new Section 8 housing the subject of an auction; and since the auction occurred in the middle of the 1975 housing collapse, a number of builders put in bids. Within a year, most of them had backed out, forfeiting some part of their deposits. Partly, their retreat was the result of having tried too hard at the start, making optimistic assumptions about costs that fell far short of what the estimates turned out to be once the drawings came off the architects' tables. But much of the disenchantment with Section 8 new construction reflected the results of a closer look at the scheme by the investors and lenders whose commitments the builder required before he could proceed.

In general, the rules require the initial developer of a Section 8 project to retain ownership for a number of years. And, obviously, Section 8 is a rent-controlled situation. So was Section 236. The investors' experience with Section 236 was that when they needed a rent increase from HUD they couldn't get one—always not for months, often not for years. And that was under a Republican administration. Though "fair market rents" under Section 8 are to be adjusted every year, the fact is that the government and the owners of this property will have directly adverse interests on the occasion of each review—and HUD will have the General Accounting Office looking over its shoulder, ready to pounce on every "unwarranted" increase but not in the least concerned about situations where the delays or denials of rent increase applications put the landlord out of business. The high rents on the original contract will be really necessary, too, once the building is in operation, because the builder and his colleagues in the National Leased Housing Association will have creamed the profits, mortgaging out on big fees and inflated land and construction costs, before the first occupant arrives.

In theory, the operator of a Section 8 project has the same power to select among applicants that any private landlord has. In fact, HUD will pay 80 percent of the rent on vacant units for sixty days, provided the landlord does not reject any "eligible" applicant. The standards of eligibility are those of the local housing authority, quite possibly enforced by a court. And mistakes will be extremely difficult to correct, because the housing authority must approve all evictions. Thus the owner could find himself with his own little Pruitt-Igoe very quickly, especially if the slightest slackness appears in the rental housing market. Every tenant is free to take his Section 8 subsidy anywhere; if new tenants look troublesome to existing tenants, they can easily move out—much more easily than before, because their power to bid for alternatives has been enhanced by their subsidy. Finally, because the "fair market rents" being established for new construction under Section 8 are in fact much higher than those the apartment could command on the market, the owner will be unable to rent them at all if Congress ever cuts back the supply of beneficiaries—and there are few legislative predictions more certain than the coming disenchantment of the Congress with Section 8.

Three years after the passage of the Act, considerably fewer than 100,000 starts had been accomplished under Section 8, even including the starts that were really transfers of previous applications under the Section 23 leased-housing program, administratively reclassified to make the new program look better. Some number of those starts, moreover, are fakes—efforts by builders to get a hole in the ground and get the first payments by the banks released so they can recover their origination costs before declaring the project bankrupt. Toward the end of 1977, construction began to increase, because Washington put the heat on the regional offices to produce numbers whether the projects look sound or not. Even so, in fall 1977 HUD began to float trial balloons for a new Section 248, which would combine the best (or maybe worst) features of Section 236 and Section 8 to raise the level of subsidized housing starts.

Because so much of the early Section 8 appropriation was used to fill the vacant apartments in Section 236 projects (or to handle special disasters like the Lefrak City story in Chapter 6), it is hard to know how well the "existing housing" segment has worked. Here the general complaint by the local housing authorities is that HUD has set the "fair market rent" too low, which probably means that owners (not unreasonably, given the massive paperwork requirements and the special eviction restrictions) want a premium to accept Section 8 tenants. A fair number of Section 8 recipients, apparently, have used the money to stay where they are, reducing what had been an intolerable rent burden, or giving the landlord a rent increase, presumably in return for an upgrading of the apartment. These are worthy accomplishments (Anita Miller of the Ford Foundation insists that this element of Section 8 is crucial to urban rehabilitation); they may or may not be worth their cost.

(3) From the social point of view: Section 8 benefits too few people at the expense of too many. More than 40 percent of the population has an "adjusted" income that falls below 80 percent of the unadjusted median. Assuming everyone eligible for Section 8 subsidy got it, and the average subsidy was $150 a month, the annual cost would be more than $55 billion. That's clearly too much money. If we restrict the recipients to 10 or 15 percent of the eligibles, however, which is about the most the budget will stand, we create an irrational distinction among beneficiaries of social policy. And we still probably pump in enough new rent money to raise the rents of everybody in the country—especially the rents of the 85 to 90 percent of the Section 8 eligibles who *don't* get the subsidy.

Statements that will be found true always and everywhere are hard to discover in any social science—but the rule that increased demand will raise prices unless there is increased supply looks about as solid as the Second Law of Thermodynamics. In rental housing today, additional supply can be arranged only at higher prices, enforcing the predictability of the results if the government pumps up the demand curve. Thus the poor who are *not* in the program will wind up carrying its burden. The higher the "fair market rent" is set to stimulate supply increases, the greater the burden on the poor who don't receive the subsidy. In the towns, individual owners of small rental buildings may let the chance slide—RAND reports that in Green Bay, Wis., where an average $73 monthly allowance takes care of the housing needs of families with annual incomes of $3,500, rents have not risen. It wouldn't be true in the larger cities.

To secure income mixing—"spatial deconcentration of the poor" is the phrase preferred at HUD—Section 8 fair market rents may be set at different levels for different sections of a metropolitan area. In Chicago in 1976, for example, the rent that could be paid for a two-bedroom apartment by an applicant for assistance under Section 8 ranged from $500 a month in one of the better suburbs to $175 a month in one of the old parts of town. Meanwhile, the average grant for housing in the Aid for Dependent Children and Old Age Assistance programs works out to about $100 a month per household. Some welfare families get five times as much as others for their housing expense; at the top of the scale, some poor families through the political equivalent of an Act of God receive a gift of housing far superior to what can be bought by the middle-class families who also carry the burden of taxation to pay the subsidy bill. Somewhere there may be an involuted concept of justice or equity that approves such results, but if there is, it's beyond me. A faintly sinister cast is given to this stupidity by the fact that the underlying policy would have virtually no public support if the supposedly sovereign people were ever consulted.

Nobody has solved effectively the problems with which the HUD theoreticians wrestled on their way to Section 8. "In most countries," J. W. G. Floor of the Netherlands Housing Ministry says, "the situation as regards rents and subsidies is highly disorderly, if not chaotic." But nowhere else is public policy so *foolish*. The great strength of the economic marketplace as a source of decisions in America has historically enabled governments to display eccentricity without doing too much damage. This freedom for rhetoric, for appeal to constituencies rather than reality, must diminish as the size and pervasiveness of the government grows. In housing, we have reached the point where our present and predictable future difficulties really are the government's fault. "Today," as a report from HUD put it in 1974, "there is not a significant aspect of the vast, diverse, and complex housing market that is not affected by government action in one form or another." The costs of even the best-intentioned mistakes have grown too large.

There are some sensible things our various governments can do to improve social housing programs in the United States, once it is recognized that the problem is one of allocating housing supplies that are always and everywhere below the demand for housing at the artifically low prices a social housing program must set. We shall go looking for potentially viable policies in the concluding chapter. Before we can get much of a grip on that, however, we need to know more about the realities of this peculiar world—about builders and buildings, salesmen and managers, money and banks.

Part III

REAL PROPERTY

Chapter 10

Builders and Buildings

"THIS IS GOD! All you people down there are to clear out by the end of the month. I have a client who is interested in the property."

> —poster appropriately adorned with fierce bearded head and pointing finger, seen on the wall of a secretary's cubicle at the offices of Fox & Jacobs, Dallas

It's fun to see these things come out of the mud.

> —Preston Martin, former president, Federal Home Loan Bank Board

It certainly doesn't look like a factory that turns out more than $100 million worth of product a year, this open shed fifteen miles from downtown Dallas. There are only about two dozen men working in it, moving fast, pushing two by fours past a circular saw, tossing the wood onto tables to be assembled into frames (a pneumatic hammer pumping blunt nails to avoid separating the wood fibers), stapling eight-foot lengths of thin, dark Masonite boards onto the frames with the help of a giant press that slaps in forty or so staples at a shot. Eight in the morning till midnight, sixteen hours a day, five days a week, all year long. In 1977, Fox & Jacobs was running at a rate of almost five thousand homes a year, twenty-five per working day, more than four-fifths of them for the Dallas market (the rest for Houston). Housing is the nation's most deconcentrated business, but not in Dallas. In some years, Fox & Jacobs has built 35 percent of all the new homes in that city, 65 percent of all the new homes at the lower and lower-middle end of the price spectrum. Charles Rutenberg of the expansionist U.S. Homes says he closed down his Dallas operation "because we ran into the power of Fox & Jacobs."

The shed is pretty much open to the elements on both sides. On one side are the railroad tracks and the freight cars disgorging great structures of stacked lumber, from Weyerhauser, Georgia Pacific, anyone else who can supply guaranteed quantities of #2 wood for delivery on long-term contract. On the other side are Fox & Jacobs' own trucks, great monster road haulers, made more capacious by a dip in the center of the chassis, a rack over the cab, an extended tail. Each of these trucks will carry the panels, studs and plates, trusses and sheathings, necessary to close in a house. Indeed, a pair of the smallest Fox & Jacobs model—the "Today" line designed as an "affordable house" during the panic of 1975—can be loaded onto the one truck.

The real factory, of course, is out on the site. Each Fox & Jacobs tract, serviced by a crew of workers that may run to three hundred men, is an assembly line on a piece of land, different from the usual assembly line only in that it's the people and the materials rather than the product that moves. The delivery of parts to the work-station is as precisely organized, the schedule is as exact—and the economies derived are as substantial—as anything in conventional industry.

"In 1955 we decided we could market homes at a rate of two hundred a year," says Dave Fox, restless in his windowless small office at the center of the one-story building that stands before the factory shed. He is a legitimate westerner, from Wyoming originally, the hair from a barrel chest rising through the yoke of an open print shirt; he looks about him through steel-rimmed bifocals. "Two hundred is a nice number—that's one house a day. At that rate, we decided, we could style ourselves as housing *manufacturers* rather than contractors. The theory is continuous production, so you have to have the marketing in pretty good shape. You adjust production to sales, underbuild a little so as to get continuity. Our production people have to know twelve months in advance what they're going to need, so we can buy materials."

By the time the lumber and assembled panels go out from the shed, considerable work has been done at the site: streets have been marked off and piping run under them for plumbing, electrical, and telephone services. Lots have been platted on the almost flat, almost treeless Texas earth. (Individual lots will be graded flat where necessary; Fox & Jacobs does not build on hillsides.) Each lot is about 80 by 110, yielding roughly four houses net to the acre, each house oriented so the garage will protect the living quarters from the late afternoon summer sun and the largest window area will open up to catch the morning sun in winter.

At the center of each lot the bulldozer has leveled a pad on which the house will stand. Conduits from the street bring the utilities to a group of low stanchions poking up near the center of the pad. A trench is dug around the pad, a thick polyethelene film is put over it like a heavy tarpaper to act as a vapor seal, and wood forms to hold the coming concrete are anchored in the trench. Woven steel cables threaded at the end are then run the length and width of the pad, through the wood forms. While the concrete sets, nuts will be tightened on the ends of these cables to put them in mild tension and give the concrete a last shake to assure uniformity. Until 1973, Fox & Jacobs used conventional networks of bars for concrete reinforcing, but the cables under tension do the same job with less weight, and thus cost, of steel.

Unlike most builders, Fox handles the concrete work in-house. He has his own trucks, drums revolving, and gangs of lightly trained masonry workers (mostly black) on the site. Concrete is a strange material. The central ingredient is cement, mostly calcium silicates, derived from limestone ground up very fine (with clay added for "Portland cement"), then roasted in a very hot oven to dry it. This is a dirty process and ecologically a bad neighbor. Mixed with water, the calcium silicate grains extend interlocking tentacles in ways by no means entirely understood by the chemists, to create a lattice that is very dense, hard to the touch, and all but impossible to compress because it is already so compressed. The slab that will support the Fox & Jacobs house would weigh in at more than fifty tons.

In addition to the calcium silicates and the water, concrete contains fine sand and "aggregate," which is simply a fancy word for an inert component that may range in size from gravel to rocks an inch or more in diameter. For lighter-weight concrete, the aggregate may be "clinker," the sweepings of a furnace. There is an all but endless variety of combinations, varying in strength, texture, and utility. Plaster and stucco are chemically the same stuff as concrete, without the aggregate. They all share a calcium compound that turns into a solid mass through "hydration," a chemical reaction that occurs fairly slowly (and can be speeded up or retarded by the addition of other chemicals to the mix). Concrete in the process of "curing" must be kept damp (if it dries out prematurely, it will be flaky and weak), and at the same time must not be too wet (excess water will diminish its capacity to bear weight). The process releases heat: let the hydrating concrete get too hot, and it may crack when cooling; let the mix start off too cold in winter, and the hydration may be incomplete. The failure of concrete that had been poured in cold weather was the reason that new apartment house collapsed in Boston a few years ago.

Dave Fox has a set of demands on his concrete. As the anchor of the house, it must be able to take considerable torque when the wind blows and tries to push the superstructure off the foundation: thus, the reinforcing cables. It must be sufficiently monolithic to distribute all the weight of the house (and the car in the garage and the fireplace in the living room) over the dimensions of a pad on Texas soil that expands and contracts. (One of the most common problems in American housing is uneven settling of the slab or footings, causing cracks in the structure.) It must be smooth enough so carpeting can be laid on it directly without ripples (then some of it must be ruffled with sandpaper to help with the laying of tiles in the kitchen and bathroom). And it must be "soft" enough to take and hold nails, for the house will be attached to this foundation entirely by the use of nails.

Once the concrete has "cured" for five days, the wood forms that held it are "wrecked," to use the term of art in the business (actually, Fox & Jacobs disassembles the forms and reuses them on a nearby site, saving lumber and work on it). Having inspected the electrical and plumbing connections before the slab was poured, the FHA examiners now come around again to look at the concrete: everything Dave Fox builds must be eligible for FHA insurance, whether the purchaser uses it or not. The ends of the steel cables are sawn off, the land around the slab is graded to assure that rainwater will run away from rather than toward the house (this takes some planning, as the Texas soil is slightly plastic and may compress as much as six inches under the weight of the house), and the concrete mixer truck returns with a coarser grade of cement and aggregate to pour for a driveway.

Now the product of the factory shed is trucked to the site, and the preassembled frames and panels unloaded beside the slab. A framing team of four men, a supervisor and three assistants, is responsible for framing three adjacent houses in a day. The supervisor has a blueprint in his pocket, and needs it, for there are lots of little variations in layout; he marks on the slab in pencil the placement of the frames for the interior partitions and the exterior panels. One man can pick up and move the heaviest of these panels; some teams work in pairs, some work singly. One supervisor, stripped to the waist in the hundred-degree Texas sun, was starting a house himself with the interior frames (there's no reason *not* to work from the inside out), while his three workmen were putting up outside panels on the adjacent slab. It is by no means unskilled work: all these things have to be exact, and there are on-site chores—like chiseling out channels for one by three diagonal bracing strips—that take both a good eye and a sure hand. This is virtually the only part of the Fox & Jacobs job in which men use traditional hammers, in graceful long-arc swings that pound the nail deep into the slab with a single stroke.

When the framers are finished, the slabs do not, however, look like that forest of sticks familiar from most American tract building sites—partly because much of the outside from the beginning shows the factory-installed brown masonite sheathing, partly because there isn't in fact that much wood. My first visit to a Fox & Jacobs site was in company with three men who were in Dallas for the National Association of Home Builders convention, and as we walked through the mud to a framed slab, a Canadian builder said casually, "Not too many studs." Since the nineteenth century, American home builders have placed their vertical framing members "16 inches on center," which means the studs are sixteen inches apart

measured from the center of the stick. When the studs must support
live loads on a second floor, this close spacing may be important to
the solidity of the house (though the standard text on this subject,
Dwelling House Construction by Albert G. H. Dietz of MIT, says
flatly that "there is no structural reason for the 16″ spacing; it is
merely another instance of traditional usage"). From their beginnings as large-scale homebuilders in 1955, Fox & Jacobs have built
"24 inches on center," contending that in their one-story homes the
savings from a one-third reduction in the use of framing lumber far
outweigh any minimal structural loss incurred. Larry Martin, the
firm's public relations director (more recently moved on to be publisher and editor of *Texas Business* magazine) argued that the first
Fox & Jacobs houses have now been around more than twenty
years, and have come through brushes with tornadoes unscathed, so
construction with studs 24″ o.c. has proved out for Texas. Lumber
is also saved in the Fox & Jacobs house by simply nailing the frames
together at the corners rather than "building up" special corner
posts.

In the old days, and still today in rural housing and in California
(where nonstandard roof lines are the local architects' pride),
builders added a roof to this framing by means of angled "rafters."
Carpenters on site cut an open wedge, a "bird's mouth" to rest each
rafter on the house frame, and nailed the other end to a "ridge
board" at the top. This is hard, skilled, time-consuming work, and
produces a good deal of what the trade calls "wood butchery"—
i.e., spoiled lumber. The substitute for it is the factory-made "roof
truss," a complete triangle with base board and "chords" rising to
hold the roof shape rigid, the whole put together not with nails but
with saber-toothed rectangular steel clamps driven through the
joints all at once in a sort of multiple punch-press machine. Fox
buys roof trusses from a supplier for the simple reason that he found
someone who sold them cheaper than he could make them; on the
day this condition no longer applies, he will make them himself.

Separately stacked at the site, the trusses arrive cradled in the carriage of a snorting taxicab-yellow forklift truck, which drives right
to the slab and deposits them on top of the frame. Two men with
good balance then take the ends of each truss. Carefully walking on

the less-than-four inch ''plates'' that top the frame, they carry each truss to its destined place and nail it to the plate and stud with a power gun which takes nails from an unrolling paper strip the way a World War I machine gun took bullets.

Rafters have not been banished entirely from Fox & Jacobs houses: on all but the minimum house, little nests of them are nailed to the trusses to make decorative fake dormers, almost the only element of a Fox & Jacobs house that isn't planned for use. People want such decor, perhaps with reason: the location of the dormer more than any other factor makes a row of quite similar houses look like a row of individual homes. It boosts profits, too: Eli Broad of Kaufman & Broad told a meeting of New York security analysts in 1976 that he was looking for improved earnings because he had added a few ''new elevations'' to his line, at a cost of about $200 each, and his market research said they would up his selling price by $1,200 to $2,000.

It will be noted, incidentally, that the floor layout of the house must be such that vent pipes can rise straight up between the trusses. Even in an electrically heated Fox & Jacobs house there are many vents—for the hot-water boiler, for the exhaust fan in the hood over the stove, for the plumbing waste pipes, and often for a fireplace: almost three-fifths of the new detached homes in America, and almost half of Dave Fox's minimum-priced line, are bought with fireplaces. ''The equation 'fire + hearth = home' needs no proving,'' wrote architect Robert Woods Kennedy, adding that ''the house . . . contains this one masculine symbol, the chimney.''

Quarter-inch plywood in four by eight squares, lifted to the roof in packets by the invaluable forklift, is then attached to the trusses with nails from the air hammer, and the roofing—vapor-sealing builders' felt and rolls of asphalt covering etched to look like shingles—are successively banged in place with a stapling gun. Down below, the aluminum windows in their attached wooden frames are being slotted into the openings left for that purpose in the panels; and the bricklayers are building the outside wall, carefully moving the string that keeps each ''course'' level, attaching their courses to the house by means of L-shaped corrugated aluminum ''tiers'' nailed through the masonite to the studs, poking into the mortar. On the ends of the house under the triangle of the roof line, on the little dormers and over the garage doors, prefinished wood siding, the vinyl paint baked on in the homeowner's choice of colors, is nailed to the end truss on the studs. The doors pre-hung on their frames are nailed in place; the garage door is attached to the only heavy beam in the house, a two by eight spanning that necessarily wide opening.

With the house locked up, the stuff that might be swiped can be delivered—plumbing and electrical fixtures, heat pump and associated ductwork, appliances, cabinetry, carpeting, tiles. Also the stuff that has to be kept dry at all costs, most notably the gypsum wallboard—"drywall," "sheetrock"—that forms the interior walls of most American new housing, multifamily or single family. Fox buys his "mechanicals"—electrical, duct, and plumbing work— from Frymire, a Dallas subcontractor whose little panel trucks can be seen scooting about the site at all hours. In the smallest of the Fox & Jacobs homes, this work is all in one place. The kitchen and both bathrooms back against a common wall to minimize piping (Fox's fresh-water plumbing is copper; waste-water plumbing is vinyl). The wiring layout was designed by computer for the shortest run of wire—including 220-volt circuits for clothes dryers and shop or ceramic equipment, even in the cheapest houses, and special circuit breakers for things like the heat lamps in the bathroom ceilings. Some of the wiring, including a socket for the floor in the family dining area, is already in the slab.

Temperature control in the Fox & Jacobs house is via a "heat pump," a refrigeration unit that exhausts heat into the house in winter and out from the house in summer. Fox mounts it in a housing below the roof trusses and thus under the insulation that will be blown to cover the area above the ceiling. The ductwork is a rectangular fiberglass tube, also suspended below the insulation, which makes the ceiling low in the halls, where all of it runs. Heat or air conditioning arrive in the rooms through vents over the doorways; the "return" to permit the circulation of air is a small grillwork vent in the hall. The savings are double: the concentration of ventilating equipment and ducts means virtually no wastage of heat to a long run of piping, and having the ductwork in the insulated rather than the exposed section of the house unquestionably reduces energy loss still further. Fox & Jacobs began installing their heat pumps this way even before there was an energy crisis. It's a wash transaction in the building process—the use of a compresser unit about a tenth less powerful than would be needed if it stood in an uninsulated place saves about as much as the dropped ceiling costs—but it's legitimately valuable as a sales tool. The purchaser of a Fox & Jacobs house probably spends 10 percent less per year for heating and cooling than the purchaser of a house of the same dimensions that puts the compressor on a slab outside and runs the ductwork over the insulation in the attic.

Insulating "bats" of fiberglass are hung to fill the entire space between the two by fours, and then the gypsum wallboard is nailed to all the frames to make the surfaces of the rooms and closets. Gypsum is another cement, calcium sulfate. Like the cement used in concrete, it is ground, roasted at high temperatures, and then reconstituted through hydration; but in this case all that work is done away from the site. What comes to the house is a reconstituted heavy board in a paper jacket, usually four feet by eight feet. Carrying each of these to the studs that are to hold it is a two-man job. For the installation itself, one of the carpenters puts on a pair of stilt shoes which lift him to a level at which he can sink the top nails straight into the studs (again, using a gun rather than a hammer).

Gypsum wallboards are not notched together in any way, but simply "butt" against each other, and the seam is covered with tape for painting. The painting is done, once more by a man on stilt shoes, with a spray gun rather than a brush or a roller; one coat covers. Some rooms are wallpapered, again by a man on stilt shoes. Gypsum does not stand up well to moisture. Fortunately, our low-paid Asian allies have learned the art of manufacturing ceramic tile in big sheets easily glued to gypsum lath; and that once-luxury product can now be part of everybody's home. For the kitchen, Fox & Jacobs purchases a metal tile with a baked-on pattern; the housewife gets a choice of designs.

Wood floorings are not available in a Fox & Jacobs house; kitchen and bathroom floors are vinyl tile, the rest is carpeted wall-to-wall, the vapor-sealed underlay resting directly on the concrete. Choices of colors and patterns are available for just about everything—siding, roof, front door, countertops, paint, wallpaper, tile, carpeting. The buyer at every price level gets air conditioniong, dishwasher, disposal, outdoor patios (only a "porch" on the cheapest house), and two trees; he supplies his own refrigerator (this is considered a matter of individual taste in Texas; there is a plumbing connection ready for one of those gadgets that produces ice cubes), his own washer and dryer, and his own lawn. Fox & Jacobs used to supply grass, but found that people took much better care of their lawns if they planted the grass themselves. An underground sprinkler system is available as an option with the house.

The schedule for all this—and it is absolutely inviolable: "the men find it a millstone around their necks sometimes," Fox comments—shows the house and its contents essentially complete seventeen days after the first piece of framing is nailed to the slab; thirty-five days after the stakes are driven to tell the back-hoe operator where the trench will be. Four more days are given to touching up and cleaning up, final inspections, and the installation of the water and electric meters. Then the owners, who have probably come around periodically during the month to watch their house being built—Fox & Jacobs welcomes the visits, encourages people to bring their friends—take the keys and perform whatever ceremonial rites they find fitting.

In early 1976, on sites that ranged from eight to twenty-five miles from the center of Dallas, Fox & Jacobs offered the average American home of 1,540 square feet of living space (plus two-car garage), at a price just under $28,000, everything included. The smallest home in the line, at 1,230 square feet with a one-car garage, still two baths, retaining, touchingly, a little entrance foyer to save appearances, cost $20,950—and could be purchased on an 8 percent Veterans Administration mortgage, no downpayment, for $197 a month including taxes and insurance. At that time, the average price for an American detached house—presumably an average-sized house—was $43,000; Farmers Home Administration houses with a single bathroom, no air conditioning, and 1,000 square feet of living space were coming in at $22,000—and even Turnkey III houses with less than nine hundred square feet of space and no garage at all, no selling costs whatever to the builder, cost the Winston-Salem housing authority more than Fox & Jacobs charged their individual customers for the bottom of the line.

It is even more instructive to look at houses that cost less. In Phoenix, for example, John F. Long— "who came out of the Navy," to quote a local banker, "with a hammer and a nail and a wife with a strong back"—has built up more than twenty square miles of the city, virtually its entire northwest quadrant, almost thirty thousand homes. When he first reached heavy volume in the mid-1950s, he organized a company like Fox's, to do nearly all the work with his own employees ("We needed a twenty-acre lot," he recalls, "just to park our rolling stock"), but during the first market collapse in Phoenix in 1962 "I converted everything to subcontracting, got the overhead off our back, sold our equipment to our foremen. The experience was useful—the fact that I built it all myself means that when I'm talking to a sub I know whether his prices are competitive or not." A brisk little man with a rather tentative smile under a large nose, Long works out of offices even less fancy than Fox's, a trio of converted mobile homes locked together.

Though he donated land for amenities of the kind Fox ignores (including a golf course, school sites, and innumerable small parks), Long had always built for the very bottom of the market, boxes virtually without trim originally on straight streets (they now wind). In the glory days of the early 1970s, he began to upgrade, meeting competition from the astonishing gaggle of California and midwestern corporate builders who suddenly appeared in Phoenix and then suddenly disappeared (some of them, from the face of the earth) when the market collapsed in 1974. By 1975, Long had cut back to "basics" (or the bone), and was building literally the same house he had built in the early 1950s—940 square feet for three bedrooms (one of them nine feet by eight feet, six inches) and one bath. His house, on a lot about one-eighth smaller than Fox's, cost $2,000 less. What was *not* in Long's house that was in Fox's included the second bathroom, the garage, the air conditioning (no air conditioning in Phoenix!), the dishwasher, the disposal, the carpeting, the bathroom tile, the range hood and exhaust, the brick exterior, the civilized little entrance foyer—and three hundred square feet of living space. What Long's buyer got that Fox's did not was a location reasonably close to the center of town and access to a few more community facilities.

There is no disposition here to criticize John Long, whose profits correctly calculated were undoubtedly lower than Fox's. (He says he can keep his prices low because he is self-financing and need not pay interest to the bank—as though part of his real costs were not the interest foregone on the money he uses for construction.) Long's heavier use of subcontractors involves some sacrifice of economy, but *House & Home* ("McGraw-Hill's marketing and management publication for housing and light construction") has cited him for unusual efficiency in building. Moreover, his pride is involved in his efficiency: told that Kenneth Hofmann had said he could tell whether a piece of work was in proper rhythm the moment he walked onto a job, Long said scornfully, "I can tell just by *driving past* a job." (There is a technical term for this—"ratio delay study," defined by two accountants from Coopers & Lybrand as "observing large numbers of workers to determine the incidence of idleness.") But Long is clearly no Dave Fox.

Indeed, with the single exception of Pittsburgh's Ryan Homes, Fox does not have a competitor for the champion's belt among home builders for lower-middle-income families. For the last few years, the National Association of Home Builders has held its monster annual convention in Dallas, and one of its features has been the daily morning tour of the Fox facilities, always well attended by marveling colleagues. Any reference to Fox's operation is sternly brushed aside by other builders as irrelevant to a serious discussion of housing costs or prices. Fox builds on flat land, which means little cost for grading; he builds on a slab, eliminating the costs of foundations and load-bearing wooden floors with joists; he builds on low-priced grazing land. Rumor in Dallas says the 2,500-acre tract for The Colony was acquired for $2,000 an acre; Fox says his land cost is "more typically" $5,000 an acre.

Nobody else has Fox's continuity, which is terribly important in cost control. (A help to continuity is Fox's status since 1972 as a subsidiary of Centex Corporation, which can keep funds flowing when needed.) Nobody else dominates a market the way Fox does, giving true economies of scale. And nobody else, it might be added, relies so completely on an executive staff drawn from people with no experience in homebuilding—a field production boss from Frito-Lay, a controller-treasurer from Texas Instruments, a staff production organizer from Procter & Gamble.

Fox operates nonunion, and stresses that "we design our house and our process so you don't need real high skills. We don't need a man who's learned how to cut roofs and joists, how to build a window and order wood and do all the things a journeyman carpenter needs to know. Same thing for paint—we don't need a painter who knows how to mix paint and stain. They don't need—because of the way the work is laid out—to ask a whole lot of questions." In fact, he gets the best of both worlds, because many of his crew are union-trained people, taking available jobs. Enough goes wrong even on a Fox & Jacobs job to demand the resourcefulness of the professional construction worker, who expects problems and takes care of them.

And in some trades Fox is protected by one of the more demanding state licensing systems for tradesmen. An Educational Testing Service study was particularly impressed a few years ago with the Texas testing program for plumbers, "administered in a very modern testing facility especially designed for the purpose. Several large rooms contain miniature pipes and fittings, and a candidate can be required to do the plumbing for an entire house. . . . Candidates are told to examine the entire house plans; to assess the need for materials; and to make a complete list of the pipes, fittings, joints, and fixtures that would be needed in order to plumb the model. . . . When the materials arrive, the candidate proceeds to perform the necessary tasks. . . . The grading procedures used . . . have been standardized considerably."

Dallas is a low-wage area, and even at rates close to double the hourly wage for industrial workers in that city Fox's labor costs are perhaps two-fifths of what builders in New York, San Francisco, or Toledo have to pay. He also avoids restrictive work rules, permitting him to set demanding production schedules. Nobody can do today, though, what William Levitt did in the late 1940s: "We worked on a piece system," he recalls. "If a fellow was smarter than another fellow, or if he worked harder, he made more money."

Actually, featherbedding in unionized home construction is by no

means so common or so serious as employer propaganda pretends
(except in the Northeast), and Charles Rutenberg of U.S. Homes
reports that his Minneapolis subsidiary, fully unionized, has been
able to establish daily production quotas not much different from
Fox's. Outside New York, management incompetence is a bigger
cost item than union-protected malingering even where the unions
are strongest. One of the few comparative studies in this business
looked at housing labor costs in unionized Ann Arbor and nonunion
Bay City, Michigan. Wages were 23 percent higher in Ann Arbor,
but labor costs on the same "model house" were identical: the
union work force, better trained, had been more productive, and
management, forced to pay more, had used labor more efficiently.
"The only reason we go to union bricklayers is efficiency," says
English immigrant John Fitzwater in Clarksburg, West Va. "A
union bricklayer can lay four times as many bricks, and he does it
properly: he learned his trade through an apprenticeship program."
This doesn't work, of course, in places like New York, where the
union contract limits the number of bricks a man can lay in an hour.

"Ask your sub why his bid is so high," Jesse Harris of Dallas
suggested. "He may tell you you have a supervisor out there who's
so disorganized the sub has to build in an inefficiency factor. The
way to get the sub to do your job on schedule is if he knows that
when he shows up on time he'll be able to do his work—you're
ready. You can be sure your competitor isn't." A reasonable estimate of what Fox & Jacobs purchasers save through the builder's
use of nonunion labor might be from $1,000 to $1,500; and the
workers probably get $500 to $750 less.

Fox's real secret, however, lies deep in the heart of Texas. At the
colony, his biggest site, he was the Master as developers simply
cannot be elsewhere in America today. The only government there
was a sanitary district, which he controlled, having built the sanitation system. He gave no land for schools: the county board of education has to buy its land. So do the churches. He provided no golf
courses or swimming pools or tennis courts or community centers or
even parks: the residents can arrange to build such things after they
move in. He built the roads he thought the site required. He disposed of his spoil where he saw fit. He needed no permissions from

planning boards, zoning commissions, code bureaus, environmental protection agencies: he paid few sales taxes, and the homebuyer paid none.

In Phoenix, by contrast, John Long's purchaser has to pay state sales taxes of about $530 and city sales taxes of about $150; about $80 in fees to public agencies for approval of zoning, site plan, paving, sewer, water and other utility plans, $60-odd in building code fees, and about $600 per house in the costs of government-specified curb and gutter and sidewalk installation. Activist governments at all levels keep finding new features that must be in the house, like hardened burglar-resistant safety glass in windows near a doorway ($75 a house) and a high fence between the homes and any arterial road to keep out dust. "People drive around, look at existing subdivisions, and don't think they're pretty enough," says Gary Driggs of Western Savings. "It makes for constant pressure to upgrade zoning." In Phoenix in 1975, the area office of the FHA and some ambitious lads at the Environmental Protection Administration got together to demand a water softener in every house. The ecology department at neighboring Arizona State blew a gasket: dumping the salts from water softeners down into the aquifer would poison the city's water supply within a few decades. But bureaucratic *amour propre* was now involved, and the only solution that could be found on appeal to Washington was one that compelled the homebuilders to install *attachments* for a water softener ($70 a house) but not the thing itself. Thus the dignity of the regulator was preserved, at the expense of the homebuyer.

As long ago as 1967, the Urban Land Institute felt it necessary to warn that "the public engineer . . . will be doing the home owner a great disfavor if he takes the naive attitude that any extra improvement costs he causes are somehow paid for by the developer. Such costs, plus overhead and profit, are passed on to the home buyer." But nobody was listening. Dick Crowell of the Goodyear-sponsored Litchfield Park, a private New Town near Phoenix, observes that "One of these days the consumer is going to wake up and say, 'I just can't afford all this protection I'm getting.' "

A good deal of this protection represents the setting of luxury standards for moderate-income housing; after all, it costs the government nothing. A Canadian builders' group comments, "The majority of the most active participants in the political process in every jurisdiction—and practically all the top decision-makers in every government—are homeowners. They are not likely to become unduly alarmed by rapid escalation of property values. Instead, they might view the rising values of their homes as a confirmation of their foresight and a just reward for their hard work."

"Street standards," Oregon engineer David Evans told a builders' convention, "are set by traffic people, fire department people, maintenance people. And they don't care about costs." In Baltimore County, Maryland, where curbs and sidewalks must be built even when housing densities are one to the acre, the Regional Planning Council reports that the cost of street development in 1973 was $1,600 per acre—while the cost of street development in Howard County, which had no such requirements, was only $600 per acre. (Baltimore County also charges 30 percent of these additional costs as an "overhead fee" to the county for imposing the costs.) In a situation like John Long's, the added cost of curb and sidewalk is an almost exact trade-off against the added cost of air conditioning. Many of Long's customers can't afford both. Is it really sound public policy to require low-income newcomers to Phoenix—many of them old people who suffer in hot weather—to take sidewalks rather than air conditioning?

In California, of course, such costs go even higher. ("Every time a city council meets," Los Angeles consultant Sanford Goodkin says disgustedly, "the cost of housing goes up.") There builders may be required to install storm sewers large enough to prevent even minor flooding during a storm so severe that one can be expected only once every twenty or even fifty years. The cost of trucking out rather than burning vegetable matter bulldozed on a building site can run into four figures per house; to prohibit such burning in areas without smog worries is to load a big burden onto new homeowners for tiny benefits to their neighbors.

Almost everywhere in California, and increasingly in other states, tract developers must deed at least a tenth of their land (apart from the streets) to the municipality for public use (thus increasing the land-cost component for the houses by 11 percent). Streets inside developments, which for safety reasons should be narrow (nothing stops speeding like a twenty-foot roadway) are mandated at thirty-foot and even forty-foot widths, and the required roadbed in frost-free locations where six inches is ample may be set at a minimum of eight inches.

Some juridictions require builders to put up a school, and put the costs on that house they're selling. It is not generally understood how badly this works out. Assume a $2 million school for one thousand homes. Paid for in the householder's thirty-year mortgage at 9 percent, annual interest and amortization for the school is $193 per home; paid for by a forty-year public bond issue at 5 percent, annual real estate tax for the school is $116 per home. There are lots of potential homeowners for whom that lost $1.50 a week is a real penalty.

In San Diego, the City Council approved a prize-winning design for development in the north end of the sprawling city only after the developers agreed to build *all* the public facilities—schools, water plants, sewage plants, police stations, firehouses, libraries, parks— on their own (or, rather, their homebuyers') money. Even so, Mayor Pete Wilson was against it, partly because he thought the requirement was illegal, partly because "even if the property owners give all the facilities, we will have to extend the services, and their cost will aggravate our revenue gap by two and a half million dollars a year."

Proposing *A Growth Management Program for San Diego* in December, 1976, that city's planning department began from the premise that "the prime responsibility for the provision of community facilities is the developers'." Among the criteria for approving "additions to existing urban areas" the "program" listed "an analysis of the cost/revenue of the proposed development or redevelopment based on objective studies and to cover total expenditures, including but not limited to school districts." Also "the extent to which the proposed development or redevelopment accomplished physical environment, social housing, and economic goals of the city as expressed in adopted Council Policy, ordinances, and resolutions." In other words, private property in San Diego should be developed to achieve public purposes. This is an attractive idea in

some ways, especially if "Council Policy" really is the right thing
to do just now, but the Fifth Amendment does proclaim among its
prohibitions the words "nor shall private property be taken for
public use, without just compensation." Perhaps the Constitution
doesn't apply in San Diego.

One must sympathize at least a little with the Californians. In
1950, San Diego was a city of 250,000; by 1975, it had crossed the
750,000 mark, and a lot of the surrounding terrain was a mess. "If
we don't control the growth and the smog," says planning commissioner Louis Wolfsheimer, "this place will become like Los Angeles." Some outlying sections like Rancho Bernardo and Mira Mesa
have in fact developed attractively without much planning, the
shops and industry and public facilities following the housing; but
especially near the sea and on the canyon walls there has been an un-
conscionable amount of building on sites that should have been left
alone. The problem is, as Councilman Lee Hubbard insists, that the
one really effective way to control growth is to reduce job opportunities. Wolfsheimer in effect confirms this with a complacent statement that "people now come here from the Midwest, stay five or six
months, and drift up to Orange County."

"Controlled growth" turns out to mean loading all the costs of
growth on the newcomers, and giving all the benefits to the existing
residents—especially those who had the best cabins on the ship
before the political leaders lifted the gangplank. "What's happening
now," Hubbard says, "is that it's great for me, but my son can't get
a job." San Diego's unemployment rate is the highest on the West
Coast. Which makes problems of its own: "If you're going to be
unemployed," said Alan Rothenberg, a young Bank of America
officer who went on loan to Governor Jerry Brown to help work out
the state's welfare and housing programs, "you'd rather be unemployed in California. And you can start getting unemployment insurance in this state the moment you cross the border."

In any event, part of the controlled growth policy has been an
increase in all the fees the homebuilder and his customers have to
pay. In 1972, the City of San Diego charged $5 for a "tentative
map" of a lot; by 1974, the fee was $1,011. That's one of the
reasons Dave Fox can build houses so much more cheaply in Dallas

than Ray Watt, the closest to Fox's opposite number in this corner of California, can build them in San Diego: even within the Dallas city limits, Fox pays no such fees. So working-class housing in America increasingly moves outside the Standard Metropolitan Statistical Areas, and when Census Bureau studies show this movement everybody speculates wildly about the return to the soil. Sure.

Homebuilding is a world full of characters. At the University of Connecticut, they talk about Olson, the best builder in the area, whom you can't reach on the telephone and who won't build to anybody's order because he doesn't want the hassle of dealing with people. In Winston-Salem, Billy Satterfield, dressed in golf jacket and turtleneck shirt, mane of gray hair over bright blue eyes, gives his philosophy:

"I've been a leader in this town. I built the first bowling alleys, the first apartment houses, the first motel, the first shopping center. The Lord has given me *foresight*, because he's my partner. He gets ten percent of all I make—and when I die he's going to get another ten percent of what's left. The Lord gave me a gift for making money. That's why I could never retire; it would be cheating Him."

It was Billy Satterfield who launched the new Bermuda Run Country Club and housing estate outside Winston-Salem by selling the first lot to Arnold Palmer. "That was Arnie's own money, too—I took everyone to the bank so they could see that Arnold Palmer spent his own money to buy a lot on my Bermuda Run Country Club. Wasn't that nice of Arnie?"

Indeed it was.

"And all he had to do," a rival bidder suggests, "was guarantee to buy it back from him for double what he paid."

Grandiosity is not uncommon. "This business," says Gary Driggs of Western Savings & Loan in Phoenix, "attracts the entrepreneurial, self-confident, hard-driving, their-own-man kind of person." They live high on the hog when the market is strong and disappear in large numbers when it turns. Driggs's competitor George Leonard at First Federal Savings in Phoenix reported in 1975 that "there are three or four of these construction businesses that basically we're running today, because these guys got bigheaded and

bought themselves boats and airplanes.'' On the higher reaches of
the business, men like Sam Lefrak and William Levitt stand ready
to solve the problems of the world. Lefrak dabbles in foreign exchange and has plans to take over the rock-and-roll business. Levitt
even has a guaranteed solution for the problems of New York City:
''Give all industry ten years to get out—manufacturing does not
belong in the city; prohibit all commercial traffic between 8 A.M.
and 8 P.M.; and prohibit parking.'' This proposal is not in truth very
convincing, but Levitt believes it.

For many years the biggest homebuilder in America, Levitt sold
out to ITT, which ran Levitt & Sons right into the ground. Now he's
in business again for himself, on the second floor of a two-story
bank building on Long Island, his personal office furnished like pictures of Buckingham Palace, occupying about as much space as the
banking floor downstairs; seventy years old but looking younger, a
small man with expansive gestures. ''I see no future in America for
the volume builder,'' he says, ''but there's an unlimited future in
the Third World for the volume builder.'' In practice this means for
Levitt, as for others, playing footsie with the liegemen of the Shah
of Iran, who can be persuaded to contract for houses in 10,000-
home bunches (it's always 10,000 homes) and then somehow don't
have the money when the time comes to do the work.

In Melbourne, Florida, Robert L. Cochran, who is both a contractor for others and a developer for himself, works in an office of
his own design with dark cork walls, his desk standing in majestic
isolation on a carpeted podium while his visitors sit in an area
below, consoled by the comfort of reclining swivel chairs. ''Our
business,'' he says, ''is a business and a hobby, too—that's my
wife who let you in downstairs. My son, who's twenty-one, is in
charge of the heavy equipment. I'm from Charleston, West
Virginia; my father was a contractor and my grandfather was a contractor. I started with a strip-mining operation, so I could learn
heavy equipment, then I was in service a couple of years, then I
worked in Baltimore as a mechanical contractor. When I was
twenty-seven, I started in business for myself. My brother and I
were in floor and wall coverings, subcontractors, and I ran a service
station for Texaco in the winter months. I had three operations, I
worked from six in the morning till twelve at midnight, I decided I

was approaching thirty and the weather was bad, it was time to go South where the weather was warmer. I told people I was going to Florida; they said, 'Where?' I said, 'I don't know.' We just happened to stop in Melbourne.

"I cultivated the doctors. For a while here we built ninety percent of the doctors' homes in Melbourne. We'd meet with them at night, design the house, and then build it during the day. Commercial is easier and quicker. This year [1977], I'm doing a $13 million shopping center, a $10 million industrial park, thirty-three homes at $100,000 and up, thirty homes $80–90,000, a 150-unit condo project, some $47,000 townhouses on Satellite Beach, a PUD on Merritt Island, $26,500 for a thousand-square-foot home, that's selling fast. We've got land out west of town, we'll give people a house where you can add to it, an acre of land, well and septic tank because it's a full acre, and we can do it for $30,000. That'll be interesting. There's a need for some rental apartments around here, too, so we'll build them. I don't have to worry about money—I don't even have to go to the S&Ls, they're here at the office every day, asking if I have any projects. . . ."

Away from the cities, relations can be quite informal. "Nobody around here deals in contracts," says John Fitzwater of Clarksburg, West Virginia. "It's all verbal. Subcontractors won't sign the AIA form. They say, 'If our word isn't good enought for you, to hell with it.' "

It's a different world.

The construction worker's world is different, too. "The thing you have to keep in mind about the construction worker is that he doesn't like you," says Gil Wolf of the National Plastering Institute, who like many others has been both a journeyman and a contractor. "The tradesman considers himself in an artistic field. He takes a bunch of nothing and he builds something from it. And *he* does it, too. A building is not built the way it is on paper—the architects' lines always measure out, but it's not going to be that way when you start to build. The concrete and steel have a tolerance level, but plastering doesn't have a tolerance level. The plasterer judges by eye—it's all touch, touch and eye. Plasterers are very innovative, they're always inventing things.

"A plumber is just an assembler," Wolf continues, warming to

the conversation; he is a big man, large head, long hair, who went to
Tulane from Cincinnati on a football scholarship, then decided he
wasn't interested in college. "Plasterers and masons and bricklayers, the people in the trowel trades, always feel that the people in
the mechanical trades just make us a lot of trouble. We work with
wet matrix, which means we have to feel responsibility to the material; if you don't put it on that wall it's going to set. The contractors
used to exploit that—they'd come around and load up the boards at
a quarter to four, knowing the plasterer would keep working until he
finished what was on the board.

"Plaster is a luxury item today, but it shouldn't be. People are
disturbed about drywall in their homes; it's not a durable product.
It's our fault. Plasterers will go out and work in the shacks when
there isn't any commercial work, but when the commercial wakes
up—who wants it? You've got to walk through mud, they don't
have water—you've got to go to a cistern—and in the shacks you
never get away from the *angles*. It's hard to work out the closets,
the bathrooms; always *angles*. The commercial guy does those
long, straight walls. And out in the shacks it's cold, the trucks don't
come, the wind howls . . .''

To a large extent, homebuilding has become nonunion by default.
"Prior to World War II," says Don Danielson of the Carpenters,
"we were maybe seventy-five percent organized. World War II
came along, we switched to defense construction, there was no
building of houses, but everybody was working. As we came out of
the war, there was a substantial demand for housing, but also demand for commercial and industrial construction, which paid better.
And a whole new way of building houses came into being. Tract
building: the developer. In most instances, he doesn't employ a carpenter. He's the promoter, he puts the package together, and he
subs out the actual building. Carpenters became independent contractors—a carpenter would say, 'I'll frame 'em up for so much a
piece, I'll take eight or ten houses.' Then anything goes: one guy
and four cousins, anybody can throw 'em up and sell it.''

A lawyer in Winston-Salem says, "You can get into your car, go
out into the country, and find somebody there who is or has been
employed in the construction industry. We're full of carpen-

ters—they like to live in the country and come in. All they want is that during the hunting season you don't ask them to show up.''

The carpenters' union has 800,000 members, more than three-quarters of them journeymen who have been through at least a four-year apprenticeship program which these days includes quite a lot of formal instruction. There's no doubt that the skills of a journeyman carpenter, who can do millwork and cut doors and plan as well as execute joists and trusses, are wasted when the task is framing; given a chance at other kinds of work, few union carpenters want to do framing. In fact, a lot of them are overqualified for carpentry itself. ''Most of them,'' Danielson says with mild hyperbole, ''become supervisors or contractors within five years.'' Once a union carpenter, however, always a union carpenter—it's like admission to the bar. ''I've been in negotiations with a tough contractor,'' Danielson says. ''I'm crawling up one side of him and down the other, and suddenly he'll pull out his book and throw it on the table: who the hell do I think I am, talking to one of my own members like that . . .''

What's hard to grasp (because the minds of writers and scholars and book publishers and indeed book readers don't run that way) is that this has been mostly a *pleasurable* occupation. Homebuilders get addicted to it. ''I make my living,'' says Kenneth Hofmann of Walnut Creek, ''not by telling people how to do it but by building the house. I'm second generation born in California, I was a very poor guy when I started, and now I'm a very wealthy man. But I'm here every day, I don't play golf, and I don't take European vacations. And there are a thousand of us.''

But the number diminishes. Part of the malaise of American housing today is that all the parts of it have become less personally satisfying. ''This was an interesting business in 1955,'' says Dr. Oliver Jones, who came out of the Federal Reserve System to be executive vice-president of the Mortgage Bankers Association (and took an early retirement in fall 1977). ''The government spent a great deal of time and effort trying to get these markets to perform better. What we've seen in recent years is the great expansion of front-end costs, all the delays that result from Congress extending itself into the writing of administrative procedures. And you know the money

market: every day you lose is a cost. It takes all the fun out of the
business; today this business *isn't* fun any more; it's a pain in the
ass.''

Mismanagement of national policies accounts for some of the
problem: the little guys were driven to the wall by the unavailability
of money in the big inflation of 1973, the big guys went down the
tube in the market collapse of 1975. "The banks can't talk about
anything but their troubles," John Crossland, Jr., of Charlotte said
in fall 1976. "You break down the doors to get in and then all
they'll talk about is their troubles." His competitor John Broadway
pulls out a "Parade of Homes" brochure issued by all the thirty-two
Charlotte builders in the year 1956; only two of these thirty-two organizations survived twenty years, despite the immense expansion
of the city. This sort of thing makes people uncomfortable and
greedy: might as well get it while it's there.

Gil Wolf traces the malaise farther back, to the 1950s in Cincinnati and a hospital job that was the first thing he ever worked on, as
an apprentice plasterer. "The engineers laid out that building two
and half inches out of square," he recalls, "and everybody went
broke on the fourth floor. Up to that time in American building, all
contractors took responsibility for the job as a whole: if you had to
do something over, you did it. On that building for the first time,
guys began back-charging each other. And they all do it today. Suddenly you began to get down-time and up-time, and the unions expecting hours when nobody was working."

Then the main course of a very good lunch arrived, and Wolf
brightened considerably. "Still," he said, picking up knife and
fork, "it's the last frontier. You can just go out and buy a truck and
get into the construction business. . . .''

Chapter 11

Technologies

The basic component in Pessac, which was determined by the standarized design, was a cell measuring 5x5 meters, which . . . enabled us to use standardized windows. . . . The floors and roofs are all based on the same design and consist of concrete T-beams 5 metres long throughout the whole of the estate. The same applies to the staircases, chimneys, etc. . . . From a structural point of view, we know that we will be able to achieve complete industrialization and complete tailorization of the building site, thus ensuring both quality and economy. . . . We approached the problem in the way in which many similar problems have been approached by industry: airplanes, motor cars, etc. . . . are machines to fly in, machines to drive in. We have tried to produce a *machine to live in*.

—LeCorbusier (1926)

Imagine an automobile assembly line where each step along the line is undertaken by a different company with its own financial interest and separate labor union! . . . Present practice is impossible. The client asks an architect to design something specifically for him. In making drawings the architect will specify various components out of catalogues. He is nearly always restricted to elements that are already manufactured. Then the contractor, who has usually had nothing to do with the design process, examines the drawings and makes his bid. Industry supplies raw materials and components and has little contact with the contractor. The various building material manufacturers make their components totally independent of each other. . . . It is an absurd industry.

—Moshe Safdie (1970)

Nathaniel Rogg, while executive vice-president (i.e., boss) of the National Association of Home Builders, liked to tell the story of a visiting delegation of Czechs, one of whom got on the plane to go home with a long rectangular parcel wrapped in brown paper. He was taking back with him, he said, the greatest technological invention in the history of homebuilding: the two by four stud.

A date can be given to this one: 1833. That was the year when Augustine D. Taylor, a carpenter newly arrived from Hartford, Conn., got the commission to build the first Roman Catholic Church in the city of Chicago. If he had built it the old way—with thick timbers ten inches square, cut in a mortise (a keystone-shaped wedge at the end of the beam) and tenon (a keystone-shaped slot in the post), the job would have taken at least six months and the time of a number of skilled carpenters, then in short supply in Chicago, and would have cost at least $800. By using a frame of long sticks nailed to a ''sill'' at the bottom and a ''plate'' at the top, Taylor and three helpers brought the job in for $400, in three months. The system was sarcastically labeled the ''balloon frame.'' Siegfried Giedion in *Space, Time and Architecture* acclaimed it in 1941 as ''the point at which industrialization began to penetrate housing. Just as the trades of the watchmaker, the butcher, the baker, the tailor were transformed into industries, so too the balloon frame led to the replacement of the skilled carpenter by the unskilled laborer.''

The significance of the change was great. ''Early visitors to Chicago were astonished,'' Daniel Boorstin writes, ''at the speed with which baloon-frame houses were built. In one week in April, 1834, seven new buildings appeared; by mid-June, there were seventy-five more. In early October, 1834, a writer noted that, although a year before there had been only fifty frame houses in the city, 'now I counted them last Sunday and there was 600 and 28 and there is from 1 to 4 or 5 a day and about two hundred and 12 of them stores and groceries.' '' Another contemporary observer argued that ''If it had not been for the knowledge of the balloon frame, Chicago and San Francisco could never have arisen, as they did, from little villages to great cities in a single year.''

The precondition of the balloon frame was the development of a nail-manufacturing industry, which brought the price of a pound of nails down from twenty-five cents at the beginning of the nineteenth century to eight cents in 1828, five cents in 1833, three cents in 1842. During the same period, sawmills began to turn out "dimensioned" lumber, all pieces cut to the same size, vastly reducing the work to be done at the site. From New England days, the American house already had its clapboard siding and shingle roof; now it had an easily transportable, easily assembled frame. "In the art of construction," the early architectural commentator George Woodward wrote, the balloon frame "is one of the most sensible improvements that has ever been made."

Technically, the modern two-story wooden house is not a balloon frame house except in a few places like Detroit where traditions survive. In balloon frame construction, the studs run all the way from the sill that rests on the foundation to the plate that supports the rafters, and the second floor is nailed onto the vertical supports. Thus all the weight rests on the full length of the outside frame. Most modern American building uses instead a "platform frame," in which the studs that support the roof rest on the floor of the second story rather than running all the way down. But the principle of the rigid frame of sticks survives: there is a sense in which it is true that we still build homes in America as we did 140 years ago.

Among the triumphs of this sort of construction was the adaptability of the American house to advances in creature comfort. The wood frame wall left a hollow for the introduction of inside plumbing, electric wiring, insulation, and vapor seals. Because few of the interior walls were load-bearing, rooms could be enlarged and house layouts changed to meet changing tastes, at relatively little expense.

The persistence of this technology has unfortunately created an unusually durable myth, that American homebuilding is inefficient, suffering from retarded productivity. Like most myths, this one has some warrant in history, specifically in the 1930s, when the combination of a unionized work force and fear of unemployment created restrictions on the introduction of new methods and materials (and many of the restrictions were written into building codes by political pressure). But the truth seems to be, as Christopher Sims of Harvard reported to the President's Committee on Urban Housing in 1968, that "since 1947 the price of construction industry output has risen no faster than other prices in the economy. Since 1947 labor productivity in construction has clearly begun to rise."

From 1948 to 1966, on-site labor's percentage of the total construction cost dropped by about one-third. By 1970, according to the National Association of Home Builders, only 17.3 percent of the *total* cost of a typical new single-family house was labor cost; and by late 1974 the proportion had dropped to 15.6 percent. (But the reason, alas, was that the "hard cost" of the structure had dropped to only 48 per cent of the selling price; labor was about 36 percent of actual construction cost.)

In yet another report to the Urban Housing Committee, Ralph J. Johnson of the NAHB Research Foundation made a list of about 150 "technological innovations in home building in the past two decades." They included:

Split rig trusses
Component wall panels
Annular ring and spiral shank nails
Single-layer siding-sheathing
Improved paints
Prefinished siding
High-pressure melamine laminate counter surfacing materials
Wall-hung water closets
Raised bottom bathtub for above-floor trap
Washerless faucets
Electric heat pumps for cooling and heating
Stress-rated lumber
Ready mix concrete site delivered
Self-sealing shingles
Epoxy coatings for plywood
Prehung doors
Prefabricated stairs
Polyethlene vapor barriers under slabs and crawl spaces
Substantially increased use of power hand tools

It was, in fact, an impressive and convincing catalogue, forcing assent to Johnson's comment on the search for a technological fix to increased housing costs: "I know of no reason to believe that radically different materials and systems would replace existing systems and materials, except blind faith in the ability of technological research to produce startling results. Such faith is more justified when the product does not have to be 'sold,' such as in space and weaponry research."

The rationalization of housing production proceeded along two rather different tracks, of which the better known is the one associated with Levitt & Sons, a fairly small firm of custom builders begun in 1929, emerging into the big time in 1941. Their first large venture was a tract of 750 single-family detached houses in Norfolk, Va., for rent to naval personnel, followed by a project of 1,600 row houses in Portsmouth, for construction workers in the navy yard. In 1946, the Levitts began picking up land in the vicinity of Roslyn, Long Island, at $1,000 an acre; and early in 1947 they began the construction of the first Levittown, just under forty miles from Times Square, convenient to the burgeoning light industry and aircraft manufacturing plants of Nassau County.

"We devised a little homely system of twenty-six steps in building a house," William Levitt recalled not long ago. "We tried to copy the automobile industry. A fellow would come to one house and go bang, bang, bang, and then go over to the next one." There was no factory, except perhaps the sawmill in California, a Levitt property, which precut all the lumber precisely to size. "Prefabrication," Levitt says, "is a dirty word to us." But there was an enormous amount of ingenuity—the plumbing tree with associated radiant heating coils assembled and put in place before the slab was poured, the hot-water-heater-and-oil-burner in a counter-high package to be installed in the kitchen, backing against the fireplace in the living room, the Plextone paint of separate colors coagulated and mixed in the spray gun, the asbestos siding and the sandwich panelling that provided insulation as well as interior walls—all of it either original with Levitt or used on a large scale for the first time in a Levittown.

Life was a lot simpler in the immediate postwar years. Levitt built "nine swimming pools, nine village greens, and I can't remember how many playgrounds. I put in the roads, the drainage, the water lines. But no sewage then, just cesspools." (Actually, there were concrete mortuary vaults converted to septic tanks by the ingenuity of Levitt's engineers; and this was nowhere near adequate sanitary practice for a forest of houses on sixty-by-one-hundred-foot lots. Today the people who live in Suffolk County, in Levittown and the imitations spawned further out on the island, are painfully raising about $1.5 billion to put in a sewer system.) "Later, in Levittown, Pennsylvania, we put in sewer systems." All said and done, Levittown was big value for the money: less than $8,000 in 1949 for the Cape Cod house with its expandable attic, someone's home now for a generation, and for sale at five times what it cost when new. Eric Larrabee observed admiringly at the time that Levitt "has done the impossible and made it pay."

But what Levitt did (and Dave Fox does) was really custom building on a tract. The economies were in large part simply the result of the builder's size. The parts of one Levitt house were, if all went well, interchangeable with the parts of another Levitt house, but their standardization was within the confines of one company. For the industry as a whole to become more efficient, there had to be standardization beyond the dimensioned studs and the nails. Parts from different manufacturers had to fit together, so a builder could order what he needed from a catalogue, and it would fit where it was supposed to go—windows, doors, trusses, header assemblies, prefabricated floors for platform framing, millwork, much else. The performance of the readymix concrete had to be standardized; a man had to know what he could expect from different kinds of wood for floor beams; from the thermal insulation values of what he hung in the wall cavities and placed under the roof; from the sound insulation values of the doors and wallboards he installed.

Some of this standardization has been very tricky. Different trees yield lumber of different characteristics, and different parts of the country have different trees. Like all living things, a tree is moist (more than 25 percent water by volume); by the time a stick has been inside a warm house for a year or so, it will have dried to a water content of about 12 percent. As the wood dries, it shrinks. The builder's two by four has been only a "nominal" two by four for more than fifty years, because these dimensions are those of the "green" stick as it comes out of the sawmill. In 1925, when Herbert Hoover was Secretary of Commerce, the government set up standards for raw lumber in interstate commerce, giving official status to the distinction between air-dried lumber (mostly from the Pacific Coast) with moisture content as high as 19 percent, and kiln-dried lumber (mostly from the South) with moisture content no greater than 15 percent. By the time the sticks had been dried and their surfaces smoothed, the two by four was down to 1⅝ by 3⅝, probably to something a little less in the case of air-dried lumber. Lumber from each species is "graded" according to quality (knots, etc.) by inspectors from the association of growers of that species, supervised by the American Lumber Standards Committee of the Department of Commerce; very few localities will permit the use in housing of lumber that does not have a Commerce Department stamp.

The most complicated grading system involves the "span" of a joist—i.e., the unsupported length of a beam supporting a floor—that can be counted on to carry safely a live weight of forty pounds per square inch. Part of the purpose of this system is to give the builder a sense of equivalents, so he can shop the lumber market efficiently and buy the performance his job requires at the lowest price. If his house is twenty-eight feet deep, for example, calling for two fourteen-foot spans between the sills and a center "bearing plate," a contractor looking up the standards can take his choice of No. 2 grade nominal two by ten beams of Balsam Fir, California Redwood, Western White Pine, or Western Hemlock; or he can use No. 1 grade nominal two by eight beams of Douglas Fir-Larch or Southern Pine. Or these requirements can be eased a little if the "underflooring" nailed to the beams is a strong epoxy-glued plywood.

The 1925 standards survived for forty-five years, less because everybody was happy with them than because there was no way to get agreement on anything new from the various interests involved—the builder, the lumber company, the lumber yard, the academic, the government, perhaps even the consumer (though basically the consumer gotta believe). "Each softwood region developed its own

grading rules," says Gerald F. Prange of the National Forest Products Association, "and each tried to make the best of the characteristics of its own species." A group mixing the NFPA, the American
Lumber Standards Committee, and the American Society for Testing Materials (plus some FHA and some university people) wrestled
with fluctuating intensity and persistence for the better part of a
quarter of a century ("I lost most of my hair in the lumber standards
battle," says Prange), and finally in 1970 produced new standards
that, among other things, reduced to 1½ by 3½ the dry size of the
"nominal" two by four and required Western producers who season
their lumber in the open air to start off with a somewhat larger green
stick. "Now," says Prange with some satisfaction, "you give the
mill an incentive to kiln-dry his lumber that he didn't have before."

And for better or worse—maybe the latter because lumber prices
have gone up so drastically—the permanent emplacement of standards and equivalencies has made possible the development of a fu-
tures market in construction-grade lumber, permitting builders who
feel they have enough cost instability in the rest of their business to
assure themselves the wood they will need to meet their plans at
prices established long before they need it. The system is essentially
the same as that by which Kellogg's assures itself of corn for Flakes
or Hershey a supply of chocolate for Bars well in advance of processing, at commodities market brokerage commissions that are considerably lower than the interest rates on loans to carry a high inventory. There is some suspicion that the system is being rigged—a
group of members of the National Association of Home Builders
has brought an antitrust suit against the big producers—but it's
probably better than no system at all.

Most of the standardization in American homebuilding, however,
has been done without any government intervention. The "Uni-
corn," the standard American building module of the four-inch
cube, is traditional rather than official, though every full-time
builder uses it. For the great range of special products, from steel
bolts for the foundations to skylights for the roof, American homebuilding relies almost entirely on a private publishing enterprise:
Sweet's Catalogue Files, part of the McGraw-Hill empire.

"The existence of the same reference work," says George E.

Drake, manager of the architectural/engineering group at Sweet's, "is a prerequisite to successful communication." Actually, Sweet's Catalogue Files is a series of reference works, one for each section of the industry—general building, light construction, mechanical and sanitary engineering, etc. The largest of the Files, for general building, runs thirteen very fat hard binders full of manufacturers' catalogues, all sorted according to the Uniform Construction Index worked out by Sweet's with the American Institute of Architects, the Construction Specifications Institute, the Associated General Contractors of America, and the Consulting Engineers Council.

There are sixteen "divisions," described as "constant in sequence, few in number, and short in name," expressing "the relationship of units of work"—Concrete, Metals, Wood & Plastic, Thermal & Moisture Protection, Finishes, etc. Within the divisions are about two hundred "broadscope sections," which do express "units of work"—Expansion & Contraction Joints, Hardware & Specialties, Ceiling Suspension Systems, Incinerators, etc. There are a Specifications Format, a Data Filing Format, a Cost Analysis Format, and a Key Word Index directing the browser to the broadscope section that includes that product. New sets of Files, some thousands of manufacturers' catalogues in each, are issued every year, and one or another is sent off free of charge to about 50,000 builders, architects, engineers, subcontractors, universities, libraries, and government offices around the country. The choice of recipients is made by Sweet's analysis of the reports from its brother subsidiary of McGraw-Hill, F. W. Dodge & Co., the major source of information about American construction. You can't buy Sweet's Catalogue Files; either Dodge's computer says you do enough business to be worth the manufactuerers' attention, or you don't get it.

Sweet's has offices in sixteen cities in the United States and two in Canada, but the New York operation is the center. Its sales manager, Irving Gross, gives the company's slogan: "Maintain availability and accessibility throughout the entire market at the moment of need." Gross says there are about three thousand manufacturers of building components in the United States who do enough business to justify including their catalogues in Sweet's, which costs them about $9,000 to $15,000 per catalogue, depending on its length and the set of Files in which it is to be included. That's not much; but of course it's for a distribution service only: the catalogues must be delivered to Sweet's in the right quantities at the right time, ready to be bound. McGraw-Hill will do the printing if you wish, and most manufacturers do wish, but it's not required.

Whether or not McGraw-Hill does the printing, the Sweet's team of fifteen or so architects and engineers is available for consultation and advice on how the catalogue should be organized and what it should say. "The catalogues before Sweet's sees them," says Arthur Douglass, one of the consultants, "tend to confuse product data and sell. But you shouldn't talk about sublime and beautiful in Sweet's Catalogue Files; you should talk about extra weight gauge for extra strength. And you should tell people the problems. We have to say, 'What are the limitations?' Tell people about salt air, temperatures, needs for coating. Installation information. Try to show your product in place, in different parts of the country so the local architects can go and look if they wish."

The planning of a building is a collaborative effort, and a lot of it gets done piece by piece, with telephone calls. "Both the engineer and the architect have to have the same catalogue," says Irving Gross, "and the owner must be involved. 'What doors have you selected?' 'Williamsburg.' 'Well, I'd like to check it.' 'Sure, look in section eight point one of Sweet's, under Williamsburg . . .' And it's there for that unpredictable moment of need, because the need arises as the job proceeds." Good salesman.

More important than the value of Sweet's to the individual project, however, is its value to the industry, as the consultant group at Sweet's patiently, drawing on the best collection of information in the country, tells the manufacturers what specifications they must supply, what dimensions they would be best advised to make, how their product could fit most conveniently with other products. The standardization that government committees could never achieve just happens under the crushing weight of market forces, sufficiently informed. It is through Sweet's, if at all, that American builders will achieve the transition to metric measurements—the traumatic move from the four-inch to the ten-centimenter (3.94-inch) module mandated by Congress and pushed by the increasing importance of foreign markets. A Sweet's engineer chairs the national committee on metric measurements in the building industry. At the architecture schools they grumble that their graduates, instead of designing unique objects as they were trained to do, "just assemble those buildings out of Sweet's Catalogues."

This is absolutely beyond the ken of Europeans, who proceed in such matters entirely through the development of official *agrements* (both the French term and the accords themselves are now matters of law throughout the Common Market). But then you can't *get* agréments, because somebody's ox is always being gored, and the result is that all "building systems" in Europe are proprietary: if you're buying a wall structure from one manufacturer, you have to take his windows, too, because other people's windows won't fit. The French housing ministry, trying desperately to promote competition among suppliers of building components, has been fighting to force common standards on the manufacturers, to date with virtually no success at all. Part of the reason, unquestionably, is that existing links between suppliers and customers are strengthened by familiarity with each others' needs and products; new suppliers find it hard to make contact with the market, and even established makers find it hard to make contact with potential new customers.

A question about Sweet's nags: what if the information is inaccurate? "Well," says George Drake, "publishing a notice in Sweet's is like publishing it in the paper. We can't police who gets in. The AIA [American Institute of Architects] made an attempt to have qualifications of materials, but there are antitrust aspects. . . . Most of the things that have to be done have to be done by the manufacturers themselves. If one manufacturer objects to a catalogue from another one, we can ask for evidence that the product meets the standards. Not proof to us, but proof to the complaining manufacturer. Sometimes when two people get in a fight, we throw them both out."

It should be pointed out in passing, however, that decisions on components purchasing and even design elements are not necessarily made objectively. "Best time I had as a builder," says a man who has gone on to other things, 'was being entertained by the utility companies. The electric people were peddling Gold Medal

Homes. They took you out to the best restaurants, and if you wanted a blonde or a redhead or a brunette, that was available, too.''

Into this industry with its increasingly efficient builders and increasingly interchangeable components came the doctrinaire policymakers of Lyndon Johnson's Committee on Urban Housing and the Nixon Administration. The technical studies performed for the Committee had all warned that there were no major cost savings to be expected from efforts to ''industrialize'' homebuilding or even apartment-building in the United States, but the report itself called lyrically for the creation of ''a high-technology housing industry.'' Among academics and bureaucrats, it was almost an article of faith that homebuilding in America was backward by comparison with other industries, and that the introduction of large manufacturers with experience in the more advanced industries would make all the difference. HUD Secretary George Romney quickly launched what he called Operation Breakthrough and proclaimed the arrival of an age when, as builder Marvin Gilman of Wilmington put it, ''we were going to produce three million housing units a year, each with headlights.''

A fully developed industrialized housing system had been tried out in the United States as early as 1936, in Jersey Homesteads (now Roosevelt), New Jersey. ''The houses,'' writes Edwin Rooskam, ''were going to be the first prefabricated concrete dwellings, with whole walls precast in a slab plant. The scheme was ambitious: the entire side of a house was going to move through a series of buildings, with all the equipment in place, the window frames, the pipes and the electric lines, and with the concrete curing as it moved. There was only one trouble with the idea: it didn't work. The walls wouldn't stand up, they couldn't take the strain.'' Later years saw a profusion of miracle cures for the housing headache: metal houses, plastic houses, skins stretched on Buckminster Fuller's geodesic domes. Paul Litchfield of Goodyear, trying to keep his Phoenix factory going after the war, proposed a house to be built by inflating a rubber balloon, spraying gunnite on it, then collapsing the balloon and cutting holes for doors and windows in the gunnite. Senator Joe McCarthy first put his foot on the path to his perdition by accepting a $10,000 "fee" from a failing homes manufacturer called Lustron. All gone, mostly fast.

What was wrong with Operation Breakthrough was revealed in the arrogance of the name. Even in the euphoria of the aerospace age ("If we can put a man on the moon, we can————''), it was unlikely that newcomers could find inexpensive shortcuts to the creation of a product as complicated as a house in an industry as big as housing. The profits from successful volume building were so much greater than any "set-asides" or "seed money" the federal government could offer that promising approaches to major costs reductions would surely be tried out somewhere even without government urging. No doubt the Research and Development budgets of the homebuilders were small; the reason for that was likely to be that the returns did not look promising. The Committee on Urban Housing argued most unconvincingly that housing lagged because "the innovator bears all the costs of developing the innovation and gaining consumer acceptance and public approval for it, only to find that his competition can almost immediately share in the benefits." Guarantee profits to large companies willing to think bold thoughts, and there would be a brave new world: "The only certain characteristic of a high-technology housing industry is that it would be better able to produce rapidly and efficiently whatever kinds of dwellings the American people want to have." Few statements could have meshed more neatly with the mind-set of the incoming Nixon Administration.

A musical comedy could be written about the results, if we still had a Cole Porter to do it. On the heels of the urban riots, many of the nation's big corporations wanted to devote some of their money and skills to solving social problems—especially if money could be made at it. The list is long: American Cyanamid, General Electric, Inland Steel, CNA Insurance, Philip Morris, Boise Cascade, Warner Communications. . . . From every specialty, the wise kings came with their gifts to help the infant housing industry in its muddy manger. A few new ventures arose independently, like Stirling Homex, where stackable plastic rooms came off an assembly line and were hidden away in the woods while stockholders and governments were told units were being shipped off to sites to be plugged

together and piled high to house the people. It wasn't until 1977 that
the officers of this operation actually went to jail.

The odd thing was how few of these innovative systems supported with HUD money even promised to reduce the cost of housing. Michael Goss, a young Captain Kangaroo of an engineer now a
sales executive for Kingsberry Homes in Atlanta, remembers his
years as a project manager on a Boise Cascade Operation
Breakthrough program in Macon, Georgia. This was for steel-frame
single-family housing—an odd choice for a lumber company. "I
did some rough calculations before we started," he says, "and I figured that using the steel frames would cost about a dollar a square
foot more in building the house." Inquiry was made about why anybody should use steel frames under those circumstances, and Goss
said, "I don't know." But he defended his employers: "General
Electric," he said to show how much farther off the mark others had
gone, "had this cast plaster system that was just inevitably *much*
more expensive."

Very little is left of the Operation Breakthrough programs. The
companies that were prefabricating large sections of single-family
housing for distribution to local builders suffered severely in the
1974–75 bust. In the Southern areas where they compete, Jim
Walter Homes can sell its small houses conventionally framed by
local contractors cheaper than Kingsberry can sell its small house
partially panelized at the factory. (The Kingsberry house is probably
a better product, though.) There are no economies from standardization at Kingsberry or at Toledo's Scholz, which supplies panels
(shipped complete with windows and doors) for more expensive
homes. Neither stockpiles components: work on a house is begun at
Scholz, says sales manager James M. Sattler, only after receipt of a
"DPO," for "Dealer Purchase Order." Scholz sells not economy
but design and safety. "We can beat the locally designed house,
which is usually by the builder himself," Sattler observes, "because he won't hire an architect. We say to him, 'We've got an
engineered house for you, with a controlled cost.' And we help him
with marketing. We're trying to make him a better merchandiser;
we know we can't make him a better builder." In the small towns
where many of these panelized homes are sold, builders find the

mere completeness of the package its greatest recommendation, because house components as standard as door hardware may be hard to get from the local lumber yard without expensive delays; and the little small-town builder doesn't qualify for Sweet's.

The major beneficiaries of the early days of Operation Breakthrough were the mobile home makers. They were already *there:* in 1968, they had shipped 318,000 units. "The most aggressive, market-minded drive to date to produce low-income housing is that mounted by mobile home producers," Elsie Eaves reported to Senator Paul Douglas in 1969. She foresaw a future use of mobile homes stacked in apartments and linked in rows as well as spotted on pads in parks. By 1972, the mobile home makers were shipping at a rate of 575,000 a year.

In those days, this was an easy business to enter, and a fair amount of defective merchandise, some of it highly susceptible to fire, came on the market. The shakeout of the mid-1970s, which saw mobile home production drop to just over 200,000 units in 1975, eliminated most of the marginal operators; and the 1976 HUD "Construction and Safety Standards," occupying forty three-column pages in the Federal Register (and well written, by the way), have considerably increased the capital costs of entering the business. The standards also knock on the head once and for all the "mobility" of the mobile home, for they require different insulation qualities, roof strengths (to hold a weight of snow) and wind resistances (for safety in hurricane zones) according to the section of the country where the thing is to be used. No modern mobile home is a "trailer," says Walter Benning, president of the Manufactured Housing Institute. "There's no private vehicle made that can haul a twelve-[foot]-wide, let alone a fourteen-wide."

Technologically, the mobile home is not interesting. It is a stick-built construction job with two by three studs, a lightweight truss roof with a rounded top, usually finished with aluminum sheets; the floor joists are two by six plywood beams; the interior walls are normally a thin gypsum board; the exterior walls are normally aluminum or vinyl, something light. The steel frame with wheels on which the mobile home will ride to its place of implantation may or may not be integral to the structure.

With few exceptions, mobile homes come twelve or fourteen feet wide, the latter being considerably more expensive. (The old eight-foot trailer now exists only as a self-powered ''recreation vehicle.'') Until 1976, several of the more important states for mobile home purchasing forbade vehicles more than twelve feet wide on their roads, but as of 1977 California stood virtually alone in forbidding the fourteen-foot width. A ''double-wide'' fourteen-foot mobile home (a pair of structures made to be linked together at the site) will give about as much living space as the average new single-family house.

Whether or not mobile homes are inexpensive housing is a matter of some dispute. There are undoubtedly economies from the fact that the entire thing is built indoors—no loss of days to the weather—with lighter and normally cheaper materials. MHI figures claim an average retail price of $13.09 per square foot of living space, as against $23.50 per square foot for conventional home construction in 1976. And it can be argued that the circumstances of construction indoors compensate for the use of lighter materials by permitting a tighter fit of components than is likely to be achieved on an outdoor site. Mobile homes are easy to buy, too: the things are on a lot, like automobiles, and they come furnished, fully equipped with appliances, ready to plug in on a pad. ''Instant housing,'' says Walter Benning.

On examination, however, the price question becomes more complicated. The need to anchor the mobile home to a site, build steps to the door and add skirting to the structure, brings the real cost of the mobile home in place to about $15 a square foot even on MHI's calculations. In most of the sunbelt areas where mobile homes are concentrated, the cost of conventional home construction is considerably less than MHI claims; Fox & Jacobs builds for the same money per square foot that it costs to build a mobile home. And the institutional structure of the mobile home market adds very substantially to the buyer's costs.

Mobile homes are usually financed on chattel loans, like automobile loans, not on real property loans; and the average interest rate on a chattel loan runs two or four percentage points higher than the interest on a home mortgage. The average loan on a mobile home is written for something less than eleven years, as against almost thirty years on a conventional home. And more than half the mobile homes sold are placed on rented land, with an average rent in 1977 of something more than $75 a month. This rent includes, of course, the lot-owner's real estate taxes, water, and sewer service.

Let us now compare the average 970-square-foot mobile home at $14,500 in place against a Farmers Home style house of slightly larger size at $25,000, assuming a 10 percent downpayment for both, interest rates of 11 percent for the mobile home and 9 percent for the conventional house, loan durations of eleven years for the mobile home and thirty years for the conventional house, and a real estate tax of $400 a year on the conventional house.

The mobile home owner will pay $246 a month ($171 on the loan, $75 in rent); the conventional homeowner will pay $214 a month ($181 on the loan, $33 in taxes).

At the end of eleven years, the mobile home owner will have title to his dwelling free and clear; the conventional home owner will still owe almost $19,750. Assume now a well cared for house and an inflation of 5 percent a year in the price of existing homes (actually below the experience of the last two decades). The $25,000 house of 1977 will sell after eleven years for $42,750, or $40,200 after a 6 percent real estate broker's commission, leaving the conventional homeowner with an equity of about $20,500. Assuming a well cared for mobile home, what will the owner be able to sell for?

To which the answer has to be: nobody knows—but under the present institutional structure, it will be a great deal less than $20,000, almost certainly less than $10,000. At least half the increase in price we have calculated for our conventional house represents an increase in land values, which our mobile home owner won't enjoy, because he sits on rented land. Even more important is the lender's valuation policy for the price of any home, standard, co-op, or mobile.

On conventional homes and co-ops, the bank or S&L will send out an appraiser, decide what the house ought to sell for (mostly by comparison to other houses in the same area), and then write a loan for some percentage of that. On mobile homes, as on automobiles, the lender will look up "current resale values" in a blue book exactly like the book a dealer uses in appraising a used car; and like the used car figures, these book numbers for mobile homes go steadily down as the chattel ages. Even if the owner could find a buyer willing to pay him $12,000 for his mobile home, which arguably would leave him proportionately as well off as his conventional homeowner friend (remember, he "paid" only 60 percent as much to begin with), the bank or S&L or finance company will probably be willing to lend only $5,000 on such security, which means his total sale price is not likely to be more than about $6,000. At the end of eleven years, in other words, the mobile home buyer will have put in about $3,300 more than the conventional homeowner, and he will take out about $14,000 less. As the system works today, then, mobile homes are not very attractive as an answer to the housing problems of the poor.

This is not to say that there are no advantages. At their worst—and this is what you tend to see from highways, for the better mobile home parks like the better neighborhoods of conventional homes are not placed on main highways—the locations set aside for mobile homes turn into dusty collections of boxes. At their best, though, mobile home parks offer many of the attractions of the Planned Unit Development—community swimming pool and recreation hall, a staffed office to handle problems, plantings, and peer pressures organized to maintain appearance and neighborly manners. Like high-rise housing for the poor, mobile housing works best for older people. "They're still independent," says Julia Herron, long-time public relations director of MHI, who is of an age with the people she's describing, "but they're watching over one another. You know somebody is watching your house."

This observer sat through a meeting of about a hundred delegates from mobile home parks around Vallejo, California, all seated in the recreation room of one of the parks and eating each others' salads and cupcakes as they discussed the comparative values of rent control and looser zoning in keeping down the rents of sites. A New Yorker couldn't help comparing this much despised "life style" with the real conditions of old people of roughly the same incomes in the jungles of the West Bronx and the Lower East Side.

The failure of mobile home production to recover from the 1975 recession has created soul-searching, not to say personnel turnover, at the Manufactured Housing Institute. Walter Benning is a new president, fresh as bread from the sophisticated legislative liaison office of ITT, and among his early actions was to move the headquarters from a remote suburb of Washington (near a rather unattractive mobile home park) to an office convenient to both Capitol Hill and HUD. ''Our people used to sit back while bills were coming through the Hill,'' Benning says, ''and then they'd say, 'How does this affect us? We're not even mentioned.' Congress took mobile housing out of the office of the Assistant Secretary for Housing at HUD and moved it to the regulatory office. . . .''

Actually, the 1974 housing amendments extended considerably the authority of the FHA to insure and the VA to guarantee mortgage instruments on mobile homes, and the problem the industry faced in 1977 was that lenders had suffered serious losses on their mobile home loans made during the glory years of 1971–74.

Delinquencies in 1976 ran about 5 percent of the outstanding loans, as against less than one quarter of 1 percent on conventional homes. ''The lenders got burned because of sloppy underwriting and minimal downpayment,'' says John H. Maguire, MHI's young financial v.p. ''The owners of the homes were in a negative equity situation, and it was easier for them to mail the keys to the lender than to sell the house. In well-kept parks today mobile homes are not depreciating, they're maintaining value—but too often the owners are not allowed to sell their homes on site. Some park operators have a policy that if a person moves out of a park he takes that home with him. Then it ends up being a distressed sale on the back of a lot, like a used automobile. It makes a terrible image problem—a self-fulfilling prophecy.''

The hope is to establish mobile park condominiums, where the homeowners own their land, and in rental situations to bully the park operators, many of whom are also home dealers, to provide the safeguards lenders need if they are to write mortgages rather than chattel loans. An MHI publication aimed at developers claims gallantly that such places offer ''permanent, gracious living and the owner takes pride in his home and assumes his share of the responsibility to ensure that the surroundings measure up to his personal tastes and requirements." In Arizona, MHI claims, there has been a rapid and trouble-free expansion of traditional FHA mortgages on such properties. On that basis, subsidizing mobile homes would indeed make more sense than subsidizing elderly apartments at $600 a month in decaying areas of big cities that offer old people a daily ration of terror and not much else.

Perhaps the most potent of the myths that supported Operation Breakthrough was the one that acclaimed European industrialized construction as a model for the American future. High-rise apartment building in the United States was moving from its steel-framed tradition to reinforced concrete, an art as well as a construction form in which Europeans were preeminent. And by the mid-1960s one could see on every skyline evidence of the value of European technology, in the form of the marvelous counterweighted Danish crane that lifted itself up by its own heels and stood in the center of the construction rather than on the side. Since the middle 1950s, the European governments had been claiming motion toward universal adoption of a standard module—thirty centimeters, almost exactly a foot in English measure—that would be used by all manufacturers to assure the compatibility of parts. The Jespersen and Larsen-Nielsen systems in Denmark, both involving the use of prefabricated load-bearing interior walls, were visibly successful in stimulating interesting architecture, and seemed also to be moving Danish housing into an era when space and amenities would meet American standards.

Industrialized housing had been established in Denmark in 1961, when a new Buildings Act mandated it upon all apartment building in the country—both publicly sponsored and privately financed apartment houses. "By the introduction of provisions of modular coordination in the building legislation," two Danish architects wrote four years later, "it has been possible to break the vicious circle which is set up when everybody waits for somebody else to begin: architects and engineers would like to apply modular building components and only await their appearance in the market; the manufacturers would like to produce such components but wait for a sufficiently large demand."

An engineer named P. E. Malmstrøm was the designer of the Danish system taken up first in Scandinavia, then increasingly in the rest of Europe. (The Finnish B-E-S system, which the Russians have been trying rather unsuccessfully to imitate and the French have considered adopting, is in fact a rip-off of the Malstrøm plan; the Danes can be voluble on the subject.) In 1960, in anticipation of the new law, Malmstrøm formed a partnership with the firm of Jespersen & Son; and their product has been known since as "the Jespersen System." Its components are produced in an interesting factory out in the Danish countryside, a big shed perpetually wet underfoot, very quiet except for some clanking as the concrete dispenser rides by on overhead rails and the forms are pushed into the furnace room to start the curing cycle in a steam bath.

In the Jespersen system, the vertical load is carried down to the ground through a series of precast concrete walls that do not in fact rest directly one on the other or on the floor slabs between them. The wall slabs are corrugated at their ends, and the floor slabs are both corrugated and cut on a diagonal at the ends, so that the weight is carried on a bolt system cast into the walls, and the floors can be joined by reinforcing bars that run through the joints. Once the metal is in place and everything is level, a concrete mason fills the vacant area with mortar. Diagonal supporting bars hooked to wall and floor help keep the walls plumb while the mortar sets. The load-bearing walls run across the width of the building, with special sandwich walls containing insulation for the two ends; and the façade the length of the building is then free for anything the architect wants to hang on it—Jespersen can offer a choice of glazed surfaces, as desired, or can hang somebody else's tiles. Door openings are cast into the concrete (bathroom plumbing is cast into the floor slabs, electrical wiring is in the walls alone, which means no ceiling fixtures), windows are pre-installed in the façade pieces. A floating wood floor system—the Danes absolutely demand parquet—is installed two inches over the slab on each floor, the air space permitting wires and heating ducts to be drawn where needed and also improving vertical sound insulation.

Interior finish is made easier because construction tolerances are close; kitchen cabinets and built-in bureau drawers are easily installed, the walls may be painted or wallpapered. Jespersen estimates 105 days from the completion of work on the foundation to initial occupancy. One of the useful things about the Jespersen system is that apartments can be finished, and tenants can be moved in, one floor at a time, while the rest of the project is still in process of construction.

How relevant all this might be to the American scene was and is obscure. The Committee on Urban Housing in 1968 took note of some "four hundred industrialized building systems available for licensing throughout Europe," but also noted that "It is not wholly evident that the highly industrialized multifamily systems are more efficient than sophisticated on-site [American] methods." Elevator apartment construction, moreover, is only a small piece of American homebuilding—never, not even in the high-production years of 1971–72, as much as 10 percent of the nation's output of housing. Still, it's by definition very visible; the city planners have been requesting greater density to legitimate land values and real estate tax assessments, and the zoning commissioners are demanding that builders cover less and less of their sites with structure, forcing the buildings higher to achieve similar densities. And in the unionized cities—especially after 1966, when the construction workers won contracts that would previously have seemed unimaginable (in New York, the cost of public housing rose by *40 percent* per square foot between 1966 and 1967)—one aspect of prefabrication was vastly appealing: "The Jespersen System," its promotional book proclaimed, "does away with the traditional site operations which rely heavily on skilled labor, and replaces them with a highly mechanized factory production process coupled with site erection procedures largely carried out by unskilled labor."

A major source of economy in the Jespersen scheme is the stockpiling of components. Walls are all the same height (280 centimeters, just over nine feet), slabs are all the same width (270 centimeters, just under nine feet). Wall lengths are available from one foot to eleven feet, with different electrical and plumbing conduits cut in them, or with doors placed in the middle or near the end. Slab spans range from six feet to eighteen feet (for American use, Jespersen got the slabs up to twenty-five feet). Some thirty-nine sizes of cast concrete pieces constitute the entire line, and some buildings (especially office buildings and hotels) may be built entirely of assemblies of only five different pieces. At the factory, then, when business is fast enough to create confidence, Jespersen can pile up concrete pieces the way a lumber yard piles up lumber, with all the savings such continuous process implies. The pieces have to stand in the yard two weeks anyway, to complete the curing process. It's tricky to load the trucks, though, because pieces must go on the truck in exact inverse order to their use at the site: the crane operator at the factory and the crane operator on the site must be working from the same blueprint.

In any event, HUD took an interest. "Half a dozen of your people came over and met with our people," says Andrew Tait of the National House-Building Council in Britain. "At the end, one of our people groaned, 'My God, you're going to make all the mistakes we made'—and by God, you did." (Actually, American builders did not make the greatest British mistake, which was reliance on an *external* load-bearing wall system, an example of which fell down in London in 1968 after a gas explosion in one of the kitchens; "the use of industrialized methods for building high-rise flats," housing official Fred Berry reported in 1974, "received a setback from which it has not recovered.") George Romney himself visited Denmark. Jespersen's manager Jens Holm recalls that he was especially intrigued by the sound insulating qualities of the system, both horizontally from apartment to apartment and vertically from one floor to another. "He had a young man with him," Holm remembers, speaking absolutely idiomatic English (he came to the management of Jespersen from two decades of work in the cement business in America), "and sent him upstairs to jump up and down. Turned out he went to an apartment which was not the apartment directly above; you couldn't hear him at all. . . ."

To develop American versions of industrialized multifamily construction, Congress and HUD established a system of "set-asides," bonuses that could be claimed by a builder who was attempting some process new to these shores. Eight years later, some of these "set-asides" were still unspent. Some projects, like the concrete boxes Moshe Safdie of Montreal's Expo 67 wished to use in Lyndon

Johnson's aborted Fort Lincoln project, turned out to be so expensive that not even the set-asides could make them viable. Others fell afoul of union work rules. Mostly, the builders looked at the specifications, noted that when stuff was prefabricated in a remote factory there was no way local inspectors could inspect it, and pointed out to sponsors interested in the set-asides that the local building code required inspection by local inspectors.

There was, however, one agency in America that could simply brush aside local codes—the New York State Urban Development Corporation headed by Edward F. Logue, Yale and Yale Law School, a man with a Robert Moses-size need to build, who had run the New Haven and Boston urban renewal programs before Governor Nelson Rockefeller brought him to New York in 1968. New York had been running a Housing Finance Authority for some years, passing on to cooperatives and limited-dividend sponsors the reduced interest rates on the state's tax-exempt paper; but the HFA was a reactive agency that could only review applications from sponsors and builders. In the aftermath of the urban riots, Governor Rockefeller and his brother David (whose Chase Bank then considered itself expert on real estate investments) decided that financing support would not be enough: the state would have to initiate housing projects and where necessary set up autonomous corporations to own them. Called in for consultation, Logue told the Governor he would take the job only if the new UDC had power to override local building and zoning codes, avoiding the delay and expense that would otherwise bedevil a state agency as much as a private builder.

This was a very hard sell to the state legislature, which was given the bill to pass on the day of Martin Luther King's funeral but turned it down anyway; and Rockefeller, returning from the funeral, had to keep his surrogates twisting arms in the Assembly all night to get the vote reversed. Later in 1968, New York Senator Jacob Javits persuaded the Congress to add state agencies to the list of eligible recipients of Section 236 interest write-downs and Operation Breakthrough set-asides; and UDC was off to the races. By mid-1970, according to the State Moreland Act Commission that later investigated UDC, Logue and his staff (which would eventually number 550 people) were deep in planning more than 45,000 apartment units to be built all over New York State.

Planning at UDC was different from planning elsewhere. "Since UDC is the supervisory agency for its own projects," staff economist Frank Kristof wrote in a panegyric to the agency about a year before its default and collapse, "many of the essential approvals required for project processing are internal. Once a project has been accepted and has passed preliminary market and financial feasibility tests, four main lines of activity take place simultaneously . . . the architectural activities . . . agreement on the construction budget, setting up of a financial structure, and arranging for required subsidies . . . arrangement for a contractor . . . a contract must be negotiated, and the architectural and construction work has to be supervised. Legal activities include arranging for site acquisition, the construction contract, mortgage closing, and subsidy contracts. . . . Enormous time savings have been realized. . . . In many instances, UDC did not own the land, lacked a construction contract, and had no working drawings but simply foundation drawings; nonetheless, the construction start date was met."

Though Logue now remembers that he was not especially interested in industrialized housing, prefabrication fitted neatly into these "fast-track" arrangements, and a number of contracts went to manufacturers—the Shelley system, the Campsi system, Buildings Systems, Inc., of Cleveland (which went broke on Roosevelt Island, a UDC project in the East River in New York City), and others. A man who worked closely with Logue in New Haven, and did business with him as a contractor to UDC, says "his aim was to make Ed Logue, not George Romney, the great producer of industrialized housing."

A Rochester builder was then in negotiations with Jespersen, which had gone partners with a family of investors in Toronto to form a Canadian subsidiary. The Canadians had financing from the Chase-Manhattan Bank to set up a Rochester operation, but Logue intervened: HUD set-aside moneys under Operation Breakthrough were available only for American companies. Logue made a marriage between Jespersen and the Kinney-National division of Warner Communications (a conglomerate within a conglomerate), promising six thousand units on Roosevelt Island and in Rochester. The Penn Central Railroad had just gone broke, leaving its cavernous box-car factory and repair shop in Rochester available for purchase. UDC arranged for the new Jespersen-Kay Systems company to take it over and start pouring concrete.

Everybody—Jespersen, National-Kinney, and Logue—remembers the three Jespersen apartment projects in Rochester as disastrous. HUD, meddling, wanted twenty-five-foot spans, walls seven inches thick, and especially cast nonbearing concrete partitions inside the apartments, all of which required adaptations of the machinery. The tolerances required by the system were closer than those American manufacturers were accustomed to providing for housing projects, and the workmen were not used to handling the materials. It is not impossible that there was some sabotage. Meanwhile, the designs for Roosevelt Island were inhospitable to construction with load-bearing walls, and Logue went off to find a builder who could work with a poured-in-place concrete frame and precast lightweight exterior walls. There was little point even trying a system of factory-installed electric and plumbing conduits in New York City: the construction unions had made it clear that they would insist on taking all such stuff out of precast walls and reinserting it by hand.

Jespersen did not receive the scheduled third payment in its royalty agreement with Kinney-National, and a man was sent to investigate. He found the Jespersen machinery piled on the side of the yard at the factory, which had been converted to other uses. He made an appointment to see Paul Millstein, a fat yet shadowy figure now the *eminence grise* of United Brands but then the executive vice-president of Kinney-National who had negotiated the Jespersen deal, only to learn that "the previous Friday Mr. Millstein had simply put on his hat and walked out of Kinney-National, and nobody knew where he was." Kinney had closed down the Toronto operation; and thus the most highly respected of European industrialized housing systems vanished from the North American continent.

The memories it leaves are bitter for everyone, for reasons that do not have much to do with Jespersen. The technologically advanced UDC buildings in Rochester have been plagued with electrical fires. The state assemblymen who fought against a law that would give UDC the power to override local building codes turn out to have been more right than they could have imagined. The monument to

Nelson Rockefeller and to Ed Logue, now dismissed from UDC—and to George Romney, who as secretary of HUD oversaw the award to UDC of more than half of all the Section 236 contracts given to public agencies—stands today as thousands of units of handsome but uncomfortable and shoddy and perhaps unsafe housing.

The vice was inherent in the system Frank Kristof extolled. As the Moreland Act Commission reported, the feasibility study for an apartment project at UDC ''involved filling in the line items on the . . . forms used by developers applying for HUD Section 236 subsidies. . . . Schedule B listed the various revenues of a project (including the 236 subsidies) and the maintenance and operating costs. Once the total estimated rental income was determined on these 236 forms, the total mortgage amount that the project could carry could be calculated, since rental income would have to be sufficient to cover debt service on the mortgage and maintenance and construction expenses of the project. The maximum allowable construction cost or so-called 'construction residual' could then also be estimated. . . .'' Run by someone who is not by training a builder—whose attitude toward merely technical people is indeed contempt—UDC was *primarily* engaged in making the ''residual'' figures look right so construction could begin.

''You can't look at these things on a balance sheet,'' Logue said the other day in a rare defensive moment. ''You've got to want to get it built. If you don't want to get it built, it won't get built.'' Logue was committed, too, to distinguished or at least ambitious architecture, which meant that he wound up hustling ahead with men who were amateurs in the housing field, because there hasn't been any business for distinguished architects in multifamily housing. Roosevelt Island was designed originally (what was built is considerably different) by Philip Johnson, who says with his marvelous pixie honesty that he knows nothing about housing: ''It's the duty of the developer to train the architect. A *good* developer has people in house who ride herd on us in most abominable ways. UDC thought they knew as much as Fred Rose about building apartments. [Rose is the engineering brother in Rose Associates, which did *not* get the contract to build Roosevelt Island.] Well, they didn't.''

Architecture to impress the passer-by will always add costs; even after he made optimistic projections of rent receipts and operating expenses, and wangled set-aside money for construction innovations, Logue was on the edge of what HUD might approve. And he was in a hurry. Money had to be saved on the quality of construction: both design and execution were slapdash or worse, and corners were cut. At Roosevelt Isalnd, for example, the heating-and-insulation package was deliberately designed to keep the apartments habitable only down to twenty degrees, though the city's building code called for zero degrees. (In the hard winter of 1977, pipes froze inside the buildings on Roosevelt Island.) Electric heat was used because it's much cheaper to build that way (then much more expensive to operate—but that's for the tenants or the government subsidizers to worry about). Leaky aluminum windows were installed (with some difficulty in one of the buildings, where the drawings had omitted the specifications for sleeves in the light-weight surface walls). Some of those windows slide to open, and have an eight-foot length, which means they cannot be washed from inside. And they can't be washed from outside, either, because the specifications did not include the window-washer's anchors required for excellent reason in the building code.

None of this would make the housing actually unsafe, but there was another area of design where UDC tailored its plans to cost considerations. At the time of the UDC construction phase, aluminum wire was much cheaper than copper wire. The problem with aluminum wire is that it "creeps" with use and age—that is, it tends to work loose from its fastenings. When it does, the result is not the loss of service, which happens when copper wire disconnects, but the formation of an aluminum oxide that does conduct electricity through a high resistance that builds up enormous heat. The resulting fires are hard to fight and quickly destructive. The hazards can be mitigated, and perhaps eliminated, by the use of compression clamps rather than conventional lugs at the terminals. UDC, over the objections of its engineering consultants, specified conventional lugs rather than the more expensive compression clamps on its aluminum mains. In at least one case in Rochester, aluminum wire was used not only for the main carrier to the apartment but also for the connection of electric stoves. Where else in the UDC buildings aluminum was used in a hazardous way is not known: no city inspectors looked at the UDC buildings, because UDC was above the code and was empowered to issue its own certificates of occupancy. And the New York Board of Fire Underwriters did not inspect the UDC buildings because it was not asked to do so—UDC could get fire insurance without an underwriters certificate.

Five of the eleven apartment houses in Rochester have had serious fires. The fire department there believes that at least three of them were related to the misuse of aluminum wiring; the state inspectors disagree. In the aftermath of these fires, the State Division of Housing began to require the managers of UDC projects to submit them to inspection by the Board of Fire Underwriters. Meanwhile, at Roosevelt Island and in perhaps ten thousand other UDC apartments, there are horrifying hazards behind the walls. In summer 1977, Roosevelt Island had its first fire, reportedly in a telephone cable conduit. The fire spread because the builder had failed to install the usual fire stops in the conduit; and nobody had ever inspected. Called to the scene, the fire department learned with horror that the skip-stop elevator design had left whole floors without public corridors by which firemen could gain access to apartments—and duplex units with exits on only one floor, a trap for the inhabitants if fire broke out on that floor.

On the wall of Logue's office hangs a parchment with the text of an award from the American Institute of Architects to the Urban Development Corporation: "Its concern for a livable environment is reflected in its support of imaginative site planning, attractive design and responsible management. Its innovative approach to economically integrated housing has had nationwide impact."

In reality, the default on a series of UDC notes was what triggered the financial collapse of New York City, very nearly destroying the credit of the state and all its municipalities. The total subsidy required for the average UDC project—counting federal aid, local tax exemption, and the income tax break on the bonds—adds up to more than $1500 per room per year—$500 a month of subsidy for a two-bedroom apartment, mostly for middle-income occupants after they pay 25 percent of their income for rent. And the

housing is unsafe. The most remarkable accomplishment of the
UDC operation, says Roger Starr, former city Housing and Development Administrator who sits on the board of Roosevelt Island, is
that "they made building codes respectable again."

The news has not yet reached Washington. In spring 1977, distressed by the difficulty of finding entrepreneurs to go partners with
the federal government on Section 8 housing for New York, HUD
in effect put UDC back in business as a builder via a deal which encouraged banks to lend to the busted state agency on the security of
forty-year Section 8 contracts. The reason given for the choice of
UDC as a preferred Section 8 instrument was its ability to override
building codes. The New York City building code now requires
compression clamps for aluminum wiring; the state code, which is
the only thing UDC accepts, does not. So there will be more innovative and unsafe housing for New York, courtesy of the federal government.

Perhaps the old ways were best.*

---

*Some of the material in these pages appeared in a series of three stories by the author and
Martin Gottlieb in the New York Daily News in November, 1977. In response, UDC announced that its future construction in New York City would be governmed by the city code.
There will still be no city inspections, however, and the meaning of the announcement is
obscure.

Chapter 12

The Consumer Perspective

In the bath-room must be the opening to the garret, and a step-ladder to reach it. A reservoir in the garret, supplied by a forcing-pump in the cellar or at the sink, must be well supported by timbers, and the plumbing must be well done, or much annoyance will ensue.

—Catherine E. Beecher and Harriet
Beecher Stowe (1869)

A work [of architecture] cannot express emotion or touch our inner sensitivity unless its form has been dictated by a genuine intention. And Mr. X (whom neither of us knows but who will one day become the owner of one of these houses) will only respond to that intention if we have invested it in the building. This intention is the care that we will have taken to entice him on to this little plot of ground by giving him all the good light that he needs, by excluding harmful draughts from his house, by planting his flowers and his fruit trees in the sun, siting his kitchen with expertise, fitting his front door in plumb and facing the garden path, placing his windows to provide a good view, his bedroom where he will not be overlooked by the neighbours, etc., etc. . . . . If we did not spend such loving care on each house, we would be turning out miners' cottages, in which case our system of serial production and standardization would have failed because the dwellings would not be good to live in. Standard components are letters; with those letters, and in a particular way, you have to spell out the names of your future house owners.

—LeCorbusier (1926)

The proportion of purchasers who complain is higher in the South-East of England, where houses are more expensive, and probably on average better finished, than in the North-East of England. The

283

reason for this is no doubt partly that those who pay more expect more for their money. It also seems to be that, on average, purchasers in the South-East are more skilled in the use of words, and perhaps less skilled in the use of tools, than those in the North-East. . . . A house is the biggest purchase that most people make, and it is the place where they spend much of their lives. Many people, therefore, have a special feeling for their house, which can vary from great affection to considerable distaste. While these feelings are governed by many factors outside the builder's control—including, as has been remarked, the state of the marriage—the good builder is one who recognizes the importance of his contribution, not purely in building a good house, but in introducing the purchaser to it in a friendly way and explaining clearly to him what he should do if anything should go wrong.

—National House-Building Council
(Britain, 1973)

On October 9, 1975, the Secretary of State for Housing of the French government held a press conference to announce the formal adoption by the government of the Qualitel system to rate apartment houses for the benefit of renters and condominium purchasers. It was, as might be expected, a flowery pronouncement:

"Qualitel: the name alone is a program. And it is a fact that Qualitel is a tool in our policy to promote real quality in housing.

"The necessity for this policy now seems to me universally perceived. The Economic and Social Council heard me yesterday on this subject, and next week I shall be in Clermont-Ferrand to award to a private builder an acoustical label of three stars—another important element in this policy, which we follow with determination.

"The hour has struck for Qualitel!

"Permit me meanwhile to offer a brief history. March 26, 1974, my predecessor M. Christian Bonnet signed in the name of the State, with the Association for the application of the Qualitel method . . . an agreement providing for the delivery of ratings of the quality of housing.

"This signature marked a choice on the part of the State—a choice not to intervene by the heavy hand of regulation, but on the contrary to establish a balance between supply and demand. . . ."

Qualitel "is a decisive element in our action to sensitize and in-

form consumers about the real, basic qualities of their housing. . . .''

What is going on here is at once the most ambitious and the least
effective effort by any government to control the design of housing
through market forces. The 236-page guide to inspectors is the
property of the state, with reproduction (''even in part'') prohibited.
It provides diagrams and decision trees to help the inspectors give
numerical rankings from everything from the parking layout (*Mau-
vaise* if the space is less than 12m² for head-on parking or less than
10.5m² for diagonal parking, *Bonne* if the dimensions are larger
than that, point ratings for M and B according to the number of
spaces and placement of the lot) to the plumbing capacities (the hot
water system is *Très mauvais* if it won't supply water warmer than
95°F, *mauvais* if the hottest is under 125°F, *bon* if the hot water runs
over 125°F (or 134°F: the guide is a little uncertain here). Interior
noise insulation may be *Médiocre, Satisfaisant,* or *Bien.* Environmental measurements include convenience to transportation, the
number of trees (''an old tree is worth 6 points, a newly planted tree
is worth 1 point, five bushes are worth 1 point''). For a family of
four, the state has already established a minimum apartment size of
604 square feet, rising by 106 square feet for each additional occupant; small even by European standards. Points are now awarded for
space over and above those figures, to reward separate toilet and
bath rooms (the French don't like the idea of elimination and cleaning up in the same room), for closets (the French don't build closets
much), and for lights inside the closets. . . . It goes on and on.

The purchaser or renter gets a purple and aqua folder called Profil
Qualitel. It includes a little bulls-eye map of the ''environment'' of
the project, with Michelin-like symbols for industrial zones, sports
grounds, places with a good view, parks, nursery and other schools,
horseback riding trails, churches, post offices, hospitals, subway
stops, etc. Next comes a ''level of quality'' chart which rates the
structures on a scale of 1 (bad) to 5 (good) on one count of Access, three counts of exterior construction and design (including
wall and floor coverings in public areas, service facilities, elevators,
etc.), and nine counts for the apartments themselves (possibility of
installing kitchen and laundry equipment, quantity and usefulness of

space, wall and floor coverings, plumbing and electrical installation and acoustical insulation from—separately—interior and exterior noises, temperature control in winter and summer). Then there is a further rating, still 1 to 5, on four categories of cost of occupation: maintenance of the halls and open spaces outside, elevators and heating. And the attention of the consumer is drawn to certain other possible cost factors, not rated, that are present in this building (i.e., air conditioning, swimming pool, house phone, need to burn light bulbs during the day in stairwells and halls—much French construction provides windows to stairs and halls, and most French prefer things that way).

It's easy to make fun of this—those Latins do go *ad astra per aspera,* pretty fast—but in fact the Qualitel profile or something like it would be useful to American apartment shoppers, if you could get people to agree on the standards. The problem is that you can't, and though the Association Qualitel is formally sponsored by no fewer than twenty-three organizations—from The Women's Civic and Social Union and the National Institute of Consumption through the architects, builders, engineers, technical schools, and even something ominously called the Truth Bureau—they haven't been able to get the French to agree, either. Pierre Schaefer of the real estate developers' federation comments testily, ''We say the Frenchman is capable of deciding on his car, his clothes, his wife—but not on his house. Suddenly he becomes stupid when he looks for a place to live.'' An assistant argues specifically about details, among them, not unreasonably, Qualitel's insistence on rating the size of each room: ''That's subjective. They should measure just total space. I want a big *salle de séjour,* you want a big bedroom—each of us should have his choice. The difficulty is that Qualitel is information not only for the purchaser but also for the builder, and it will control what he builds.'' But that, of course, is the purpose.

The theory behind Qualitel is that the shortage of housing in France is coming to an end, market forces can now be called upon to promote improvements, and since markets work best when information is best the government should help spread the word. ''Help'' is the operative verb—the system is voluntary, and no builder, even the most heavily subsidized, can be required to secure a Qualitel rating for his product. The Association has been distributing literature urging apartment shoppers to demand a Qualitel profile—"a true technical identity card of your future home." But it's the builders who must make the decision; and a year after the optimistic press conference only about 10 percent of new French apartments were being put up for rating. "The professionals have been against it," says Qualitel director Claude Trenin, an earnest young engineer whose bureau is housed in a mousy converted right bank apartment house and employs fewer than a dozen people. Still, the professional organizations have found it politically wise to pass resolutions endorsing Qualitel, and the system may yet get off the ground, if it can be made less complicated and time-consuming—as of 1976, the construction of a Qualitel profile took a trained man ten working days, which the architects and engineers who make up the Association's part-time regionally based inspection service cannot afford to give for this purpose.

There is, moreover, a possibly fatal flaw in the Qualitel process, which looks only at the designs for the apartment houses, not at the end product. "The *Conseil Juridique* wants Qualitel to look at the work in progress," Trenin admits; "Qualitel says, 'We *can't.*'" Policing of construction practices in France is done first of all (as in every country) by the office of the architect who signed the plans; then by the government, centralized in France, which in fact sends examiners only to a 10-percent sampling of each year's projects; most importantly (and this procedure is uniquely French) by inspectors from the insurance companies which will write the policies on the buildings. What interests the insurance companies, however, is only part of the complex of construction standards that will affect the lives of future residents. Schaefer says the builders are responsible, they can be sued, but the customer's chances at that sort of thing are no better in France than they are in America.

In one part of the world there are, in effect, no standards for the design of single-family homes, yet the buyer *can* sue (more exactly, force arbitration) to remedy defects, with an excellent chance of success. That's Britain, where an opinion poll of new home purchasers a few years ago indicated that a guarantee from the National

House-Building Council was the most important single aspect of a new house: 59 percent said it was "absolutely vital," and 26 percent said it was "very important." The second most important aspect, incidentally, was "a well designed kitchen"; sixth came "large living room"; twelfth, "central heating"; thirteenth, "large bedrooms"; sixteenth, "two toilets."

The British government has, in fact, promulgated design and space standards for housing, adopting in 1969 the recommendations of the parliamentary Parker Morris committee of 1961—but these standards apply only to construction by tax-supported local housing authorities; private builders may, and often do, offer less. ("The public sector," a private builder says a little grimly, "is not cost-conscious the way the private sector is.") The force operating to bring the private builder to more ambitious design standards in the crazy social organization of modern Britain is the tendency of the local housing authorities to sell off their properties to their tenants. The force operating to keep British private builders turning out mean little boxes is, as the National Home-Builders Council self-descriptive pamphlet explains, "the willingness of home buyers to pay extremely high prices for small houses of very ordinary workmanship. Clearly, this relates back to a shortage of good houses."

Home building was the only sector of the British economy to get out of the Depression in the 1930s, partly because the banks, at government urging, made money available for houses and partly because it was, as in America, a business virtually anyone with a hammer and a saw could enter. Many corners were cut, and the product was widely considered unsatisfactory. Andrew Tait, executive director of today's National House-Building Council, reports that "the builders went to the government and said, 'What shall we do to be saved?' " The government recommended a voluntary self-policing organization with participation on the board by architects, consumers, and "building societies" (in American terms, S&Ls), and a National House-Builders Registration Council was formed in 1936 to award seals of approval to builders who met what were rather rigorous standards of workmanship. As late as 1960, "registered" builders were responsible for only about 15 percent of the new houses for sale in Britain. Members had to offer a two-year warranty against defects in the homes they built.

Between 1945 and 1955, 78 percent of all new housing in Britain was built by the public sector, but then private building began to revive—with, if anything, even more public displeasure about low quality. The industry decided, Tait says in his fine Scots burr, "to impose self-discipline and make further government regulation unnecessary." In 1965, the Council began to issue ten-year guarantees on the structural parts of its members' houses, and also purchased insurance to pay off any victimized purchasers whose builders had gone bankrupt. Membership was opened to all builders who would permit the Council to inspect their work in progress, and at the same time closed to builders whose track record seemed likely to damage the Council's good name or its insurance record. "One of the first things we found," Tait says, "is that builders don't know what they're selling half the time, just as the buyers don't know what they're buying." The building societies were approached and asked to agree that they would write mortgage loans for new houses only on buildings by registered builders; and slowly but steadily, they came around. Representatives of the building trades unions came on board. By the early 1970s, 99 percent of new private homes in Britain were being built by NHBC members, and the secretary of state for housing in the Department of the Environment was charged by law with appointing the chairman of NHBC.

Today NHBC employs a staff of 250 to 300 full-time inspectors, who take a look at virtually every house built for sale in Britain (the only exceptions are constructions supervised by an architect or surveyor employed by the purchaser). The claim is that the good builder is pleased to be inspected: he "knows that in the last analysis he is dependent on his worst foreman on his worst day. He welcomes anyone who can help him in this difficult job of quality control." The list of defects the inspectors find is an interesting one: "misplaced damp-proof courses, dirty wall-cavities, undersized roof and floor timbers, and poor finishes (in paint, plaster, woodwork, or brickwork). All such defects, of course, are required to be rectified as soon as possible after discovery, and before the issue of a certificate. The most common serious fault is failure to compact infilling." Given the limited time the inspector can be at a site—NHBC estimates one to four hours per house, scattered among several visits—not much attention is paid to finishing defects, "which purchasers can after all see and complain about for themselves."

Complaints must be in writing, to NHBC, which passes the letter on to the builder in question and opens a file. In 1974, the last year for which figures are available, about one house in fifteen was the subject of some complaint, but three-quarters of the complaints were settled by the builder and his customer without further recourse to NHBC. If a second letter arrives, NHBC asks its sender to fill out a formal complaint notice and make a deposit of ten pounds which will be refunded if the inspector finds there is in truth a defect. Such inspections were running at a rate of between four and five thousand a year in 1976.

"Most opinions given," NHBC reports, "are mainly in favor of the purchaser, and usually the work recommended is carried out"—the builder gets thirty to sixty days to take care of the matter, depending on its gravity. If the purchaser disagrees with the inspector's finding, he can demand a hearing before an arbitrator chosen not by NHBC but by the national Institute of Arbitrators. The finding there is essentially binding.

Qualitel concerns itself almost entirely with design; NHBC concerns itself almost entirely with construction. Measuring construction quality turns out to be a harder job than was expected. *Which?,* the British version of *Consumer Reports,* once sent six experts to report on a single tract house, and got back six different opinions of its qualities and defects. The sort of defect that seems worth complaining about varies considerably from purchaser to purchaser. "When an electronics engineer buys a house," Tait says, "you know you can open a file: he'll have endless complaints." (The American version of this troublemaker is the commercial airline pilot, who has a lot of time around the house and works in an equally artificial, zero-defect environment.) *All* houses have defects; as Tait argues, "It's a very difficult thing to put together houses in the open air and get everything exactly right." The most common source of trouble is shrinkage, as the wood, bricks, and concrete dry out, which can cause warpage and cracked plaster, and leaks. "The Council's general rule of thumb," its booklet reports, "is that where the shrinkage is sufficiently minor to be remedied without difficulty in the normal course of redecoration, it will be regarded as normal: but even this rule may require modification in certain circumstances."

Recent experience has been somewhat disconcerting for the British builders. It was expected that once a house got past its second birthday the remainder of the guarantee would be for advertising purposes, but troubles in five- to ten-year-old houses have been cropping up beyond expectation. "We're short of good building ground," Tait says, "and we build brick houses, they're heavy. It's a problem you Americans don't have, because you build frame houses, lighter, more flexible construction. So we've barred certain kinds of foundation that proved an undue risk. Then, we've been getting dry rot in window framing, so we went to treated wood, which you in America had been using for years."

And, of course, inflation played hell with the building business in Britain, multiplying the problem of insurance for the sins of bankrupt builders ("though it's not always the bad builder who goes under; sometimes it's the good one, trying to maintain his standards") and escalating the price of repairs. "There have been cases," Tait says, "where a house was built for three thousand pounds and we've paid eight thousand pounds on a claim. Inflation reached the point where no prudent insurance company could take the risk. So now our home buyers are told that they're insured up to the actual cost of their house—but for anything over that, they'll have to buy a top-up insurance for themselves."

Another change Tait has been promoting is an experience rating for builders. In theory, the Council can expel a builder if his product turns up too many defects, and sometimes the deed is done; but since the building societies have stopped lending to houses without the NHBC certificate, expulsion from registration has been too much like a death sentence. Instead, Tait has arranged a penalty rate for builders with a bad claim record—up to 75 percent more on the premium. Last encountered, he was hoping to balance the penalty for the bad guys with a reward in the form of a rebate for the good guys; but that's the sort of elitism modern Britons find troublesome, and it may not go.

These things are simple in Finland. The lenders withhold for one year the last 10 percent of the construction loan. "Then there is an inspection," says Åke Granholm of the National Housing Board. "Have the doors swelled and won't close? If so, he has to repair. On felt roofing, the subcontractor is responsible for ten years."

In the United States, the lines of defense for the home buyer and renter have been the Minimum Property Standards of the Federal Housing Administration and the assorted building and housing codes of the states and municipalities. The first of these is required of all buildings to be mortgaged under a federal insurance program or financed with federal subsidies. Because there are so many programs—and because builders may wish to have their houses qualify for FHA insurance even if they will in fact be financed by conventional mortgages—it is hard to get a grip on the porportion of new construction actually built with reference to FHA standards. Something like 30 percent would seem a reasonable estimate, but that 30 percent is heavily concentrated on the low end of the market where the temptations to cut corners are most severe.

"Minimum Property Standards," says Duane Keplinger of FHA's architectural and engineering office, "is a compilation of materials and methods that have a history of satisfactory use under conditions of residential use. We don't look on it as a code. We have three goals—to protect the Secretary [of HUD], the consumer, and the future. If we have a forty-year guarantee, we want to be sure the house is going to be serviceable for forty years."

The Standards unfortunately mix design and construction quality with site evaluation (size of side yard, setback from street, etc.), infrastructure (roads, sewers, etc.), and even geology. One of the factors impeding the rehabilitation of properties in St. Louis is FHA insistence on significant earthquake bracing in the structures. Keplinger, a mild man with long head, thinning gray hair, and gold-rimmed glasses, becomes aggressive rather than defensive about the St. Louis rules: "St. Louis is 120 miles from New Madrid, site of one of the most severe quakes that ever hit this country; it changed the course of the Mississippi River.'' But with some hundred of years between earthquakes in the Mississippi basin, city dwellers looking to renovate homes might rationally wish to gamble that the structure that has survived soundly for the last seventy-five years would be good enough for the next twenty.

In the 1976 edition of the FHA code, Minimum Property Standards have begun to move from prescriptions (how many nails in a stud) to performance requirements (''demonstrated resistance to applied loads''). There are seventy-six HUD/FHA area offices reporting to thirteen regional offices, and they are allowed some flexibility in modifying standards for local needs—but not much. The tendency is still to write specific requirements of a papa-knows-best variety. George Drake of Sweet's, as chairman of the National Advisory Council on Research in Energy Conservation, complains about the developing FHA standards for home insulation: ''They're talking about eight, nine, ten inches of insulation, but most ceiling joists are six inches. And the prescriptive standards about the size of a window in a wall are just crazy, they don't permit the maximum entry of solar heat. Energy design has to start from the ground, where you put the building on the site, not from specifications of the structure.''

Modern government regulation, dominated as it is by middle-class desires that everything shall be nice, tends to keep adding requirements. (''We make a little progress every year,'' says Keplinger contentedly.) Some of these are clearly related to the safety of the residents—smoke alarms in private homes and sprinkler systems in apartment houses, for example; it is hard to sympathize with George Sternlieb of Rutgers when he classifies requirements for smoke detectors among the ''costs of over-regulation.'' Others clearly do involve imposing upon people who would like to economize on their housing some costs that express the preference schedules of people who consider housing a prime good—a washroom in addition to a three-piece bathroom, for example, in all three-bedroom houses or apartments. If the people buying the house would rather have another couple of beers a week—which is about the cost factor on the mortgage—it's hard to see what business it is of their government. The absence of the washroom certainly will not reduce resale prospects sufficiently to jeopardize the insurance on the mortgage. And it makes little sense for the government to deny insurance to mortgages on very inexpensive Baltimore row houses which people are breaking their backs to rehabilitate, just because FHA thinks such houses are too narrow.

A reduced rent apartment house for moderate income families was proposed not long ago for a site beside the East River Drive in New York, near where I live. FHA refused approval unless the building was redesigned with special windows and much greater acoustical insulation—the noise of the cars, which no doubt informs the rhythms of these pages, was something the government could not bear to see inflicted on the poor. The project was then dropped because the numbers wouldn't work with the addition of the extra cost. Cyril Harris, the acoustical physicist who wrote the sound insulation standards for FHA, was appalled but ignored. Any dispute with the Environmental Protection Administration runs into that agency's willingness to falsify data because its goals are so noble. Perhaps the builders and the EPA deserve each other.

Raising the standards to protect the consumer will not accomplish much, of course, unless the standards are enforced, and there is reason to believe that the policing of construction quality leaves much to be desired in many areas of the country. A builder who wishes FHA approval of a single-family dwelling must notify the local FHA office in time for inspectors to make three visits, one when construction begins, one at the framing stage and one when the house is completed. The inspectors are federal employees, but they are mostly not very well trained or very well paid—and the quality of their supervision may be suspect too. "In my office quite a few years back, in Kansas," Keplinger reports, "everyone on a supervisory level was a registered architect or engineer, but we don't have that luxury now." Many area offices, in fact, have *no* architect or engineer in the architecture and engineering section. Normally, the job doesn't require great technical expertise: the Commerce Department lumber stamp, a shield from the American Society for Testing Materials on a vapor seal, an Underwriters Laboratory certificate for a panel box, substitutes acceptably for personal inspection. But things are not always normal. If a carpenter forgets to put builder's paper between the window frame and the wall, the inspector like the homeowner will learn about it when the window leaks. A builder who wants to hide defects—especially soil and foundation defects, which are the worst problem in both frequency and severity—probably can do so even if the inspector is entirely honest, which is probable but not guaranteed.

FHA does not inspect construction at all on multifamily housing with more than sixteen units. The rules for such buildings require the presence of a supervising architect or engineer, and he certifies to the FHA that everything has been done by the numbers. The proposition that this does not always work right has been fascinatingly presented by John D. MacDonald in his novel *Condominium,* in which a retired engineer who has bought into a new apartment house "couldn't inspect the underground pilings, but he could inspect the visible cast-in-place concrete and make a judgement of the piling work from that. . . . He found construction joints badly located, impairing the strength of the structure. He found one where the bond at the joint was faulty. . . . He found a hairline crack in a joint, and when he found two places along the crack too deep for the blade of his penknife, he had returned with a two-foot length of stiff leader wire and satisfied himself that the joint had been carelessly prepared in addition to being badly located. . . .

"In a bearing surface area where he knew that the specifications had called for class-A concrete, he found a wall in the garage portion where the pour had been skimpy, where he estimated concrete at four sacks per yard instead of six. . . . In genuinely sloppy concrete work, as this seemed to him to be, there comes a point where the accumulated goofs eat up all the safety factor, and then if there is enought stress on any portion, enough to crumble it or crack it, the deflection is transmitted to other portions of the structure. They in turn crack or twist or crumble, and the whole thing comes down. . . .

"He made a mental list of the things which could go wrong with all the reinforcing steel. Too long a wait—over an hour—before the tension reinforcing of the pilings. Steel with grease on it, or too much rust, or with mill scale on it. Bad welds, too few dowels from footings to walls. Undersized bars. Brittle tie wire. Unstaggered splices in adjoining bars. Bending and field cutting of bars around openings and sleeves. Fast sloppy pours that left voids under and around the reinforcing, or knocked the bars off the chairs, unnoticed. . . .''

This was not even fictionally an FHA-insured project, but it could have been. The architect-engineer who signed the certificate that this work had been done properly could have filed it with the federal as well as the local government, and nobody would have known better. In multifamily housing as in the hospitals, the country is at the mercy of the sense of responsibility of its professionals, which can be dangerous in a permissive era. Our means for policing professional performance are as feeble in the construction industry as they are in the education industry; the cures commonly advocated—increased regulation and the employment of lots of lawyers to bring lots of lawsuits—are in the end worse than the disease, which is a strong saying but true. The great threat to the quality of life in America is not so much the things that get all the publicity as the erosion of the doctrine of personal responsibility for individual actions. Among writers, too.

Local building codes are more detailed, usually more demanding, and cover a much larger fraction of both new and rehabilitated homes. They tend to be intertwined with zoning codes which require minimum lot sizes, side lots, setbacks, not infrequently minimum house sizes—like 1,400 square feet of living space for a three-bedroom house. On the structural side, they are notoriously a source of extra cost and inefficiency in American building—except that every time a researcher goes off to document the damage they do he finds that the indefensible sections are not enforced. The authors of the report on *Efficiency in the Housing Industry* done for the President's Committee in 1968 decided that the average unnecessary cost of building code restrictions was less than 2 percent of the cost of the average house, and even that was probably temporary; the report cited a survey by *House and Home* concluding, ''any innovation that really saves money without impairing quality will be accepted by codes and labor—eventually.''

Some of the code provisions academicians regard as pure fuddy-duddiness look to practical builders like reasonable precautions—Dave Fox, for example, who can write his own codes, uses copper rather than the cheaper aluminum wiring even in his least expensive house, and uses plastic piping only for the waste-water, not the fresh-water, plumbing. The use of gypsum wallboard rather than something more sturdy in apartment house construction, which is still not permitted everywhere, has imposed costs in noise transmission and (where the wall between a common hall and an apartment is gypsum board) loss of safety that a reasonable man could consider greater than the benefits of lower prices. There are, of course, code restrictions that protect union featherbedding practices without benefiting anyone—prohibitions against plumbing trees, prewired panel boards, large bricks, and roof trusses twenty-four inches are the most common—but even here the costs added by preposterous fringe benefits and overtime rules are much greater. (In New York, for example, an electrician is guaranteed fifteen hours of doubletime pay a week, even if he works only fifteen hours—but of course that's a formality, because the contract also requires that electricians stay on the job, getting paid, as long as there is a light bulb burning on the site.)

In most communities, in any event, the codes aren't applied on the site. In many cities, certificates of compliance are literally purchased—not from the government, but from the individual inspector. A man who has been a building inspector in northern New Jersey reports that "most building inspectors have a company in the city that does business with the contractors; they're in it for the payoff on a no-show job. Lots of housing code inspectors are home-improvement salesmen, out for business." In New York, notoriously, change orders in construction or approvals of building alterations require a payoff. And even where the situation is completely clean, and the inspector is really working fulltime for the city, the fact remains that he is in the building business. He sees his job as facilitating, not impeding, legitimate construction projects. "We've got a very good group of building inspection people," says Ed Shelton of The Fortis Company in King, North Carolina, praising the employees of Forsyth County. "They're the type who if you make a mistake, they try to help.'' What else are they supposed to do?

The upshot, then, is that the code protects the consumer by assuring the quality of components specified on the plans submitted by the builder to get his permit—but it doesn't do much to guarantee the quality of the house itself. Thus American homebuilders in the 1970s found themselves in a position not unlike that of the British homebuilders in the 1930s: if they didn't shape up and find some way to take care of complaints, the chorus of outrage from their customers would find its way into the halls of Congress, and they would get caught in some spiderweb of regulations. The National Association of Home Builders decided to copy the British model, brought over Andrew Tait as a consultant, and in early 1974 launched a Home Owners Warranty (HOW) program to insure American home buyers against defects in their new homes and also against any negative consequences from the bankruptcy of the builder who put it up.

HOW came in at a tough time, when the industry was sliding into its worst recession since World War II and everyone in it was cutting costs in the search for the ''affordable house.'' The cost of HOW (presumably to the builder; obviously, the customer pays) seemed low enough, only two dollars per thousand dollars in the price of the house, as a one-time charge, but that was $60 on a $30,000 house which had been got down to that price only by a lot of $60 cheese parings. Moreover, the builder buying into the program was of course buying himself a contingent liability stricter than what the law demanded: by joining HOW, he guaranteed to his colleagues that he would repair free of charge any defect they found in his house when they checked up on a purchaser's complaint.

In its first two years, despite aggressive promotion by NAHB executive vice president Nat Rogg, HOW enrolled only 50,000 homes. Tait in England, who continues his relationship with the American program, was concerned about whether HOW could achieve a critical mass. Rogg persuaded some of the institutions in the secondary mortgage market to ask a question about HOW coverage in the forms they used to determine whether or not to buy somebody's paper, but the American S&Ls refused to follow the lead of the British building societies and compel builders to write a warranty to get a loan. In a few areas, most notably St. Louis and Colorado, the builders went enthusiastically for the program, and two-thirds or more of the new homes were covered; in other areas, HOW never even got started: New York, San Francisco, and Houston were out of the picture entirely. Still, in 1976 HOW contracts began to arrive in increasing numbers: the second 50,000 homes took only eight months. Probably because only the better builders opted for the warranty, experience was cheerful: only about two hundred complaints that were not satisfied by the builder himself in the first 100,000 homes, and only twenty complaints that went to arbitration. Then the Federal Trade Commission decided that what was good enough for England was not good enough for the United States, and everybody had to go back to square one.

It is a peculiar story. Congress had been concerned for some years about the all but meaningless ''warranties'' a number of manufacturers gave out with their appliances, and in 1975 passed the Magnuson-Moss bill, setting up stiff requirements for manufacturers of consumer goods who wished to say they gave a warranty. Housing is not usually considered a consumer good, and the Federal Trade Commission is not usually concerned with housing—there is a Cabinet-level department for that—but its mission to find employment for lawyers does of course cut across many aspects of American government. Lawyers being much less important and numerous in British society than here, the program HOW had copied made almost no provision for their participation. The FTC found this failure intolerable, and administratively rewrote the Magnuson-Moss bill—over the loud protests of Congressman John Moss as well as HUD—to rule the HOW warranty illegal.

Like the British program, HOW had required a deposit (twenty dollars) from complainants who wished to have NAHB investigate the failure of their builder to repair their defect (in fact, every one of these deposits had been refunded in the first thirty months of the program, because every complaint had been found at least arguably real); Magnuson-Moss had forbidden manufacturers to charge fees when a customer invoked his warranty. HOW called for an initial conciliation procedure with other builders acting as conciliators;

Magnuson-Moss had forbidden such participation by colleagues of the manufacturer. Worst of all, HOW like its British model took both parties to binding arbitration if conciliation failed, and Magnuson-Moss forbade binding arbitration. The FTC negotiated out with NAHB a new warranty system that would enhance the participation of lawyers, though the builders did win an exception that permitted the lay conciliation service to start the process. It was a large headache, because HOW as an insurance scheme had already gone through all the state insurance regulators; "I hope they know," said Richard Canavan, a veteran architectural engineer and former FHA Assistant Commissioner who was director of HOW during the depths of the negotiation, "that they are making us change all our documents and get new approvals from insurance commissioners in fifty states." In spring 1977, HOW got rolling again; Fox & Jacobs signed up, adding Dallas to the list of cities where a high proportion of new homes were covered; and by midsummer about 10 percent of the new homes being sold in the United States were under HOW protection. But the risk to the builders had unnecessarily increased, and there was some question whether major further expansion was possible. Too bad.

What makes a warranty procedure so desirable in housing is the importance and impossibility of finding out even from trained examination whether there is anything wrong with a house. Even an American house is very heavy: a modest frame structure will weigh some scores of tons (asphalt roofing alone for a 1,200-square-foot house will probably weigh five tons). If it is built on landfill without footings that reach below the landfill, it will settle unevenly, warping the wood, cracking millwork and finishes, making windows stick, and creating leaks. But once the house is up, there is no way at all to find out what it rests on. (Settling problems can be even worse than the landfill horror: the most intractable disputes HOW has had related to houses built above abandoned coal mines in Pennsylvania and Illinois, where the law of "mine subsidence" is a quagmire in itself.) Other qualities of the soil may be difficult for a purchaser to uncover: Kenneth Patton of the New York Real Estate Board returned from a visit to Washington disgusted at the placement of some of that city's new and expensive row houses: "The percolation of that soil is so bad that you put in sewers and point them at your enemy. . . ."

What is in the cavity between the outside and inside walls can greatly affect the comfort, durability, and even safety of the home. All codes call for fire-stops—heavy blocks of fire-retardant material (usually a treated wood)—to be placed between the studs at intervals to prevent the open area from acting as a flue; but there is no way for a purchaser to know whether such fire-stops are there. Concrete walls and brick facings are poor insulators. One of the reasons for the popularity of wood houses in northern United States was the Puritan dislike for the "sweating" that occurs when warmer and moister air inside the house makes contact with cold masonry surfaces. (This is also, of course, why single-thickness windows fog up on cold days: glass is a conductor rather than an insulator of heat.) A well insulated house with a wood exterior will show a severe temperature difference between the inside and outside walls on a cold day, and if vapor can pass between the two, paint blistering and rot are likely results. There ought to be a vapor seal in the cavity, and there probably is—but if there are any holes in the seal there will be trouble. The buyer can't find out.

Nor can he learn that his roof was nailed on with aluminum nails that corrode in his climate or with wire nails insufficiently coated with zinc, or just plain too-short nails. It's not impossible that a mineral-fiber shingle was used with a coal-tar based underlaying, starting a chemical reaction that will discolor and perhaps harm the water-resistance of the roof. An architect recently said that the only hope for getting a waterproof built-up flat roof on an apartment house was to have a man there all the time, watching the workmen. A builder who is also a masonry contractor told a National Association of Home Builders meeting that the only way to be sure the workers on the site don't add water to the readymix to make it easier to handle is constant on-the-spot supervision by the builder's own foreman. Renters and buyers can't possibly know about such things any more than they can know about the quality of the elevator machinery, the capacity of the waste pipes, the adequacy of the heating or cooling equipment. Everything about a new apartment house is a gamble; the best advice Florida engineer Harrison Rhame could offer was never to be the first tenant: wait till the place is half full, and go talk to the residents. The HOW program charged special, higher fees for builders who wished to offer warranties on elevator-type condominiums.

Having neither training nor experience as builder, architect, or engineer—and not much talent even at the usual do-it-yourself chores—I am reluctant to offer guidance to readers many of whom have better information than I do. But from three years of wandering about and asking questions, I have some notion of where builders may try to save money in ways that add to occupants' costs and troubles. Prospective buyers might benefit by asking some of the following questions and performing some of these tests:

In a house built on a slab, feel along the floor where the carpet meets a window wall; if it's cold down there, the builder has skimped on the insulation at the edge of the slab.

In a house on a foundation, poured-in-place concrete is likely to be better than concrete block—less likely to leak or shift. Any crack in a foundation will be big trouble; look and see if there is one.

Avoid a house in which the framing lumber (especially floor joists) is "green"—i.e., more than 19 percent moisture content. Lumber with only 15 percent moisture content is better than lumber with 19 percent. Inquire.

Plywood subflooring that is glued (not just nailed) to the joists will provide a floor less likely to sag. Where wood siding is used, the inside surfaces should be primed and painted (to seal them) before the boards are nailed to the sheathing; if battens are used to conceal and seal the joints of vertical siding, the battens should be primed and painted on the inside surfaces too. In houses with an unfinished basement, the inner surfaces of the siding may be visible. Buying from a tract builder, you can look at the other houses in process.

Where aluminum or vinyl siding is used, stopping all passage of water vapor, holes for ventilation must be left between the bottom of the siding and the sheathing, or the wall structure may rot. A water table is desirable to prevent dripping rain from accumulating near the foundation. There must also be air vents below the eaves of the roof to permit air circulation and avoid rotting up top.

"Split" shingles and shakes will last much longer than "sawn" shingles and shakes.

Though most of the heat loss from a house goes upward, through the roof, the chimney flue, and the exhaust hood over the range, the insulation of walls and windows is extremely important. Rooms feel cold not only because the air temperature inside is low, but also (perhaps more so) because there is a large temperature difference between the walls and the skin of the people in the room. Avoiding this loss of body heat through radiation may permit major savings in fuel bills. This is how radiant baseboard heating became popular, despite the great waste of fuel in electrical heating. (If a house is to be heated electrically, a "heat pump" is much more efficient.) To keep room surfaces reasonably warm to the touch, the heat loss through the walls of a room should not be greater than one-tenth of a British Thermal Unit per square foot of wall space per hour and ideally (the specification would be R-19) only half that. Double-glazing, as in the Thermopane system, will be required to keep windows from being cold to the touch. If the window sash is metal rather than wood, even double-glazing can do only limited good unless the frames are insulated themselves.

Still on the insulation game, weatherstripping is essential around all windows; it is by now widespread but not universal.

Make sure you can open and close all the windows of a new house before you buy it.

The easiest way to test the sound insulating values of gypsum wallboard is to send one person to the adjoining room to talk through the wall. If he is easily heard, you've got problems. This will happen if the sections of board do not come together exactly and have been joined by tape before painting and wallpapering. Heavy and sold in big sheets, gympsum board is not bad as a sound insulator, but sound travels easily through the studs to which it is nailed. If the boards in two rooms are nailed to the same studs, sound insulation will be greatly reduced. The solution—obviously, in more expensive homes—is the use of staggered studs, half extending toward one room, half toward the other. If the boards are not nailed to a common stud, their sound-retarding qualities will be maximized.

Bring a set-square when visiting a house for sale, and place it in the corners and on the floor against a wall. Exactly rectangular corners or meetings between floor and wall are not to be expected; but anything that is off by more than one inch or so at one end of the set square is likely to be en route to causing its purchaser trouble. Squareness of the outside walls is a measure of the quality of the foundation and the workmanship.

If any part of a bathroom or kitchen wall is plaster, the plaster should be based on Portland cement, which after hardening does not react chemically to water. Keene's cement, a very hard-finish substance, is sometimes used in these rooms, but it *does* react chemically to water, and will deteriorate. Gypsum board itself is vulnerable to water vapor, and must be sealed with plaster or covered with a tile or metal or plastic to stand up in a moist room.

Beware the one-piece extruded acrylic or fiberglass tub-shower-wall covering; it looks pretty when new, with its integral coloring, but the gel-coat can scratch in two or three years if cleaned with any abrasive cleaner (which means, in normal cleaning). Some manufacturers have licked this problem: molded bathroom materials carrying the NAHB seal should be safe. Ceramic toilet bowls *are* more durable than plastic toilet bowls, but there is no advantage to the ceramic water tank above the bowl and no reason why a plastic tank cannot be attached to a ceramic bowl. Nor is there any virtue to the traditional cast-iron bathtub—indeed, there is some vice, because it's hard enough to frame the floor of a bathroom to hold the weight of a tub of water without adding in the extra weight of cast iron.

Request the brand names of the significant household equipment—furnace, plumbing, windows. They are at least as important as the refrigerator, stove, and washing machine (for which everyone asks brand names). Make sure some local building supply company carries the brand installed in your house, so repairs or replacements will be simple if needed.

No. 12 wire will carry a greater electrical current than No. 14, and is preferable. Aluminum branch wiring—wiring beyond the fuse or circuit-breaker box—is extremely hazardous and should be avoided. If there is aluminum wiring behind the walls, all sockets and lighting fixtures should have copper connections to their terminals, with the aluminum attached through "pig-tailing" (screwing together) with the copper. In any house where aluminum wiring is directly connected to receptacles or fixtures, even the handiest do-it-yourself householder should leave *all* electrical work to professionals. Connection of aluminum mains to the fuse or circuit-breaker box should be by compression connectors (no visible wire outside or inside the box). The question to ask is whether any aluminum wiring smaller than #4 has been used: the larger the number, the smaller the wire.

Stairs should rise at the rate of about seven inches per step, and the treads should be about ten inches wide. Saving space by building a steeper rise is a way to promote accidents in the family that buys the house.

In row houses, the walls between the homes should be fire retardant with at least a two-hour rating, and ideally they should rise far enough above the ridge line of the common roof to prevent the spread of fires through the roof. In some states, these high separations are required by code.

In both apartment houses and row houses, the walls separating units should be much more than wallboard on studs, and should not be used for distribution of utilities services. A common electric cable which feeds to outlets on both sides of the wall will make a hole in the acoustical insulation large enough to drive a whisper through; a common pipe that attaches to bathroom fixtures on both sides of the wall will guarantee that people hear the sounds from each other's bathrooms all over their homes. There is nothing wrong with gypsum per se; it's heavy enough, and fire retardant enough. But six-inch (at least four-inch) gypsum block, not hung gypsum board, is what's necessary for sound insulating purposes. The surfaces should then be plastered, with a plaster hard enough to resist a pressure of at least a thousands pounds per square inch. Failure to plaster gypsum board walls between apartments and common halls will make occupancy unsafe in slum neighborhoods, because people denied entrance through the doors will simply break through the walls.

For the house buyer, however, there is over the long run no substitute for some kind of warranty program, run by people knowledgeable enough to judge whether defects are sufficiently out of the ordinary to require reparations from the builder. To demand perfection in an object as complicated as a house will drive prices really beyond anyone's reach, for little purpose; to permit shoddiness in construction invites neighborhood deterioration, quite apart from the losses to individual buyers. Walking that line is not something lawyers and judges are likely to do successfully.

The legal system has already demonstrated its total incapacity to establish implied or express warranties of habitability in apartment houses. Here there is no substitute for an adequate supply of accommodations, and an information system—oh, that the local newspapers would only do their job!—sufficient to identify the real estate brokers who lie to customers, the owners who pocket the profits of deferred maintenance on their structures. Renting or buying in a new building, the best assurance is the continuity of the builder/owner. The investor who puts money into a building expecting to take it out with profits in a couple of years, says a disgusted old-timer, "is just like the guy who goes to Las Vegas." And we all know how sloppy he is. This, incidentally, is the systemic fault in condominium construction: the dominant influence is that of the "investor," who plans to get out with the sale of the units, not the "builder," whose name stays on the structure.

Purchasers of existing homes usually have to fend for themselves, though there are a few one-year warranty programs run by insurance companies: ask your broker. In most states, houses are bought essentially in "as is" condition—and you can't even send the termite man around to inspect without the consent of the present owner. HUD recommends asking the local housing code compliance office to inspect the wiring; you can see for yourself if there's rust around the hot water heater or damp marks in the attic or on foundation walls. Often enough, people sell their houses because they're moving to another part of the country, and even if the buyer had a case in law he might have terrible trouble finding the seller. Recently, the courts in some states have begun to hold brokers liable for misrepresentations, which has been giving the willies to officers and proprietors of large brokerage firms.

"The man may have asked about the roof," says Jackson Wells

of Coldwell Banker, who for some years ran the residential end of that operation in Los Angeles. "The salesman said it was fine—the seller told him it was six years old; and, well, a heavy cedar shake roof is good for fifteen years, maybe twenty. But the first heavy winter, the roof leaks. The seller is gone, he's put the money into another house, and the broker is the only person who's left. Now when the salesman is asked about the roof, he should say, 'Put it in your offer that it's subject to inspection by a roofer.' It's hard for a commission-oriented salesman to do that; means delaying the sale. But there are some sales that shouldn't be made."

Amen.

Chapter 13

Profiting

To most people, the greatest objective in buying a home is happiness. A good salesman, while appealing to a home buyer's economic sense, should also appeal to his need for happiness.

—Real Estate Salesman's Handbook

We stuff these newsletters. A woman working on it said at seven o'clock one night, "I never knew it took so many people to sell real estate in San Francisco."

—James C. Fabris, executive director, San Francisco Board of Realtors

I'm a fish out of water at these realtors' meetings. They keep talking about motivating the buyer and the seller, how do you SIGN 'EM UP! How do you sell a housewife a house without a kitchen? There was one session on using hypnosis.

—Kenneth Patton, executive director, Real Estate Board of New York

"We started to sell in March, 1949," said William Levitt reminiscently. "We advertised that beginning the next Monday we would accept deposits. Wednesday night they began camping out. It was bitterly cold; we set up a canteen. One of the women on the line was pregnant; we had to take her to the hospital to have her baby. That Monday night we closed, from seven-thirty to eleven o'clock, fourteen hundred contracts."

Levitt rummaged around in his big Georgian desk and found an old glossy photograph, the original of something that had run in a newspaper. Shot from above, the picture showed three long wooden tables, picnic tables, with dozens of people sitting on wooden folding chairs on both sides of each. The middle table had a typewriter at each place, and women typing. Up top of the picture one could see a large doorway with people crowded in a waiting room on the other side.

"This was the Village Bath Club in Manhasset," Levitt said. "The salesmen were at the first table, showing people the map and explaining the contract. Then they would hand the contract back to the table behind them, where the typists would fill in all the information. At the third table, we had the notaries, and witnesses for the signing." The price of the house, two bedrooms and potentially expandable attic on a sixty by one hundred lot, was $7,990. The money was supplied by the Bowery Savings Bank, with the first 40 percent of the mortgage guaranteed by the Veterans Administration. Levitt took $100 deposits as a binder, but no downpayment was necessary on the house. Those were the days when the VA allowed the banks themselves to "qualify" borrowers for the guarantee; and Bowery, in effect, turned its authority over to Levitt.

These houses were in every way a known quantity. The Levitts had started out in the mass housing business as builders for rent, and had built and rented six thousand homes on adjacent property before this first sales offer to the public (sitting tenants had already been given the chance to buy; Levitt, who needed working capital to build 150 homes a week, sold those not bought by their tenants to The Junto, a Philadelphia educational foundation, which collected rents until the holdouts moved and then sold the residue, at a nice profit). The customers at the Village Bath Club were hungry not just for homes of their own but for places to live; the Veterans Administration no-downpayment mortgages (at 4 percent) meant that houses could be bought as easily as apartments were rented, and in fact for less than the $65 a month rent Levitt had been charging. And in addition to the house itself there were the "built-in" features (a phrase that derives from Levittown)—the modern refrigerator, the washing machine, the television set (actually embedded in the wall, which made it eligible for financing on the mortgage).

Homebuilding is an openly imitative business. ("Everything in this country comes from California, except for a few things from Texas," says Mandell Shimberg, looking back on his days as a builder in Tampa; "so I used to take a trip every year to California and Texas, to find out what I would want to build next.") There had never been a success like Levitt's, ever before. Even at $8,000 a house, 1,400 sales in one evening was more than $11 million of business, the equivalent of $30 million today. Levitt's procedures, especially as refined in the later Levittowns in Pennsylvania and New Jersey, became the prototypes to be copied by tract builders.

Essentially the system involves building and furnishing a small collection of model homes, and ideally grouping them around a community feature—in Levitt's case, a swimming pool—with a shopping center nearby. Customers are drawn by advertising to "inspect" the model homes (Levitt, aiming for veterans, emphasized that the fabled Kilroy would be there), and sales people on the spot sign them up. In Levitt's time, the financing was easily arranged through the scattered-confetti distribution system of the VA guarantee (a system that seemed to invite fraud but was rarely abused: it was discreditable for a businessman to cheat a veteran in those days, as it was discreditable for a gang member to beat up an old lady; *autres temps, autres moeurs*). Now the federal paperwork is horrendous, but by selling long before the house is built developers like Dave Fox can get their closings on time. Most developers do without federal insurance or guarantees and make arrangements with S&Ls whereby their construction loan simply transmutes to the buyer's mortgage, no fuss nor feathers, no bureaucratic intervention, no mortgage taxes.

"Built-in" features are undoubtedly too important. "By and large," says Joseph Ciskowski of Jim Walter Research, "what sells the house is that the refrigerator panels are removable, the cabinets are jazzy, and there's ceramic tile in the bathrooms. Nobody is paying attention to the wiring and the plumbing and the structure." That's almost inescapable when the reliance is on model homes, which are considered crucial because empty rooms look smaller than they are (defects are more visible, too). Mirrored walls are especially popular in smaller houses; a little sign on an endtable informs the shopper that the mirrors don't come with the house, but exactly how much they will cost is explained only much later. Sometimes the furniture is aimed at one narrow market—for Prudential Insurance's disastrous Galaxy apartments on the Palisades across from Manhattan (a multimillion-dollar loser, belligerently overpriced), Gene Dreyfus of Chicago did different small model apartments for the widow, the gay bachelor, the heterosexual bachelor. . . . In the early days, department stores and furniture boutiques were more than happy to contribute to the decor of the model homes: it brought business from visitors as well as purchasers. Such advantages may still be there, especially outside the biggest cities— Scholz Design reports that the local papers gave a large story to some model homes that were the first such ventures by J. C. Penney (which was, not by coincidence, expanding the home furnishings department of its Toledo stores). On the top level, however, designers will no longer work for nothing, and the builder must budget for furnishings in his selling costs. Much of this comes back, of course, if the model can be sold furnished at the end, which can often be done.

The major cost of selling a tract development is the sales force on the site, usually compensated by salary plus a bonus for comparative performance (straight commission makes relatively little sense when the customers are drawn to the site by advertising). In some states the developers' salesmen must be licensed to sell real estate; in others, it's like working behind the counter in a store. Builder-developers—especially those who are continuing to add new sections to a tract—rarely can use the high-pressure tactics routinely employed by the land-development companies to sell lots in the desert or the swamps ("How much can you afford to invest?"—the process is entertainingly and horrifyingly described in Anthony Wolff's *Unreal Estate*). Still, the "we-have-only-four-of-these-left" ploy can be effective in boom areas like Orange County and Denver and the Washington suburbs; and people who fall in love with a house can be as foolish as people who fall in love with other people. And the statement that prices will be higher next year, or even next month, has been true so often in the last twenty years that the salesman can utter it with convincing probity. Speculative builders who are paying 12 percent or more for construction loans may come under pressure to dump inventory by cutting prices (and a bank that has foreclosed on a tract or a condominium sometimes wants to get out as fast as it can), but usually new homes are sold at list price. Plus settlement costs—which until Congress intervened a few years ago the salesman rarely mentioned.

Home sales businesses, unlike other businesses, must be open on the weekends; most are open on holidays. Jack Nixon of Leon Weiner's operations in Wilmington, Delaware, says ruefully that "Leon sold a house once on Christmas Day, so our selling office has to be open every year at Christmas." For many people, it's a part-time supplementary job. The division of sex roles is an interesting one. Women are wanted to explain to the wife the conveniences and comforts of the house (you can see the baby in the backyard from the work stations in the kitchen, etc.), but pretty girls are distinctly not wanted in the bedrooms: the wife does not like the idea of a good-looking stranger in "her" bedroom. Though it's the wife who buys the house, it's the husband who signs the contract, and it's a complicated contract; men often want to have it explained to them by a male rather than a female.

Selling for a tract developer is not very high-paid work, but it's not very hard work either. Counting everything together—advertising, furnishing the models, maintaining the models (which newcomers sometimes forget, to their ultimate cost), and paying the sales force—the selling effort runs developers between 3 percent and 6 percent of the price on routine tracts. California builders reaped windfall gains in 1977 through the virtual disappearance of selling costs—houses were being sold from plans in trailers on the site, with no models to build, furnish, or maintain, to customers who walked in and said nervously, "Are you sold out, too?"

The tract developer's profits if all goes according to plan are 5 percent to 10 percent of the price of the house. If he sells it quickly, he doubles his money, because his own investment is so much smaller than the construction loan; if it sits vacant longer than about ninety days, the interest on that loan will eat him up alive. Hence the importance of selling.

Olympic Towers in Mahattan is a large plain skyscraper on an intersection that also offers Rockefeller Center and St. Patrick's Cathedral: Best's Department Store stood here less than ten years ago. Designed by Skidmore, Owings & Merrill, it has various appurtenances of luxury construction: door frames flush with the cciling, marble bathrooms, nine-foot, two-inch ceilings, windows down to the floor. There are stores (very fancy) and offices on the lower levels. The nearest place to park a car is five hundred feet away.

A co-op, with apartments ranging from $150,000 to $625,000, Olympic Towers was built by Victory Development Co., "a subsidiary of Victory Carriers, president Mr. Aristotle S. Onassis." It was among the more successful ventures Ari left behind him at his death; by 1976 virtually all the apartments had sold, overwhelmingly to foreigners. No more than half a dozen of the apartments were occupied by people who had lived elsewhere in New York before Olympic Towers opened.

"They called me in," says advertising man Alvin Preiss, "after they'd planned a failure." He is a casual, well tailored man of about forty who prowls a little about what is now a temporary New York office (most of his work these days is in Toronto and London). "But there was still time for me to change it around. If they had opened that building as a domestic building you could never have convinced people later that it was an international building.

"What they had done was build a standard New York apartment house in a place where no New Yorker would ever live. They had planned pantry kitchens, and no food service, and they were charging two and three times the going rate. My first question was, 'Who would pay those prices?' The answer was, people in London and Paris—they didn't know you could get three times the space, for less money, ten blocks uptown. They were used to living in a business district; it seemed natural to them.

"So we put in a wine cellar—doesn't cost anything, it was an empty room. Another empty room became a financial wire room, with quotes from markets all over the world. I got publicity by going on a worldwide search for the right concierge. We added a bidet to the bathrooms. I had four designers compete on the model apartments—I gave them each $100,000, but they spent more be-

cause they were competing. We set up a little theater in the Olympic
Airways office, with a screen all around, and we sold by a very
sophisticated slide presentation. The fact that Onassis was the name
behind it meant that it all seemed true.

"When we topped out the building, we had a sit-down lunch for
125 people on the fifty-second floor. There were no walls, but we
had carpeting, we used a silver service—they actually showed our
brochure for the building on national TV. So we opened as a success . . ."

This sort of thing is exportable: Preiss when last encountered was
on his way to Toronto, for the first of three black-tie parties in a club
atop a new condominium on the lake. This place is a little off the
beaten path, "on the site of the Palace Pier," Preiss explains,
"which was a great dance hall in the war. I bought the rights to
Glenn Miller's Moonlight Serenade to be a theme song for the com-
mercials. It's a building where we'll have a staff piano player:
you'll go up to the club, there's a bartender and a maid and the
piano player, you'll go there for a drink before you go home to your
apartment. What we're opening with these parties is the life-style of
the building. . . . There are two buildings, a hundred million dollars
of sales involved, you can afford to have parties."

Different approaches are applied to different communities. For
the subsidized high-rise Starrett City in a remote corner of New
York (rents, $225–$300 a month), Preiss promoted "Brooklyn.
Like It Used to Be." For New Century Town in Libertyville, near
Chicago, townhouse condominiums sponsored by Sears, Marshall
Field, and Aetna Life, the theme was a return to small-town
America. To rent new apartments in the Riverdale section of New
York, Preiss counseled a "rebate" in the form of a $2,000 savings
account passbook given to the tenant after the completion of a two-
year lease; this cost no more than the four months of free rent the
owners had been offering, eliminated the risk of the scoundrel tenant who takes the free rent and runs, and led to news stories that
helped move the merchandise.

For Arlen Realty's enormous Aventura development in North
Miami—a city of high-rises, almost 20,000 condominium units—
Preiss advised an unusually early start on the selling effort, from a
construction shack on the grounds, rough and ready, at advertised

"discount prices" ("we sell apartments wholesale"); and then a change of emphasis as the buildings went up and were occupied. "They'd planned a country club atmosphere," Preiss recalls. "I said, No—Cultural attractions. Make the recreation room a theater. Give Sunday concerts, lectures."

Richard Weiner, who has his own advertising agency but worked with Preiss on Aventura, remembers the lectures: " 'Sex for the Mature Adult.' We brought down Sally Schumacher, an alumna of Masters & Johnson, head of the department at Long Island Jewish Hospital; she knows the mature Jewish couple. And another lecture was called, 'Ask the Doctor.' There was one about money. About half the audience were people who already lived there, but the other half were people who might be customers. And you get publicity in the local press—not in the real estate sections, which are either very skimpy or run by the advertising department, the lowest form of life in U. S. journalism; *outside* the real estate section. For Charter Club in Miami, we brought in Cecil Saxbe, the retired chief of detectives from Scotland Yard, to be head of security. He got on all the talk shows and he was perfect. But in the final analysis, people will not buy an apartment on Biscayne Bay because they saw the retired chief of detectives of Scotland Yard on *To Tell the Truth.*" This was also the building onto which Preiss had tied a huge, ten-ton red ribbon, to illustrate its supposed quality as a "gift" to Miami. It went bankrupt; you can't win 'em all.

"The Apollo program contributed to business in every state of the union," says Jim Guest of Guest Realty in Brevard County, Florida. He knows the program well, having been one of the Grumman Aircraft engineers who worked on the landing module. "Up to the time the satellite was assembled here, one part of the Apollo hardware never saw the other; it was all controlled by drawings. Now, these are high dollar figures. So representatives from each company were making frequent trips to Cape Kennedy to watch over the interests of their part. Such people need a place to eat, and they were on expense accounts. There was entertainment of a male nature, girlie shows; not too-good restaurants, not-too-reasonable cocktail lounges.

"Several thousand construction workers had been needed to build

the base, and when it was finished, that income disappeared. In
1965, when the first satellite went into orbit, the part of the work
that required the visiting was over, and those representatives ceased
to visit Cape Kennedy. The business failures began even before
there were layoffs. It scared the devil out of the oldtimers in this
area.''

Guest had seen it all coming. A handsome man eternally young
with bright blue eyes and styled brown hair, tweed jacket and solid
tie, he seems in some ways a throwback to the 1920s, a cross between a go-getter and a smoothie. A serious-minded salesman.
Grumman had sent him to the Cape in 1965, and he knew he wanted
to remain in this part of the world. By 1968, he had qualified for his
real estate estate license and was ready for the plunge of a year's
required apprenticeship in a broker's office. Early in 1969, he hung
out his own shingle in Cocoa Beach, and he was ready when the
panic came.

''When we landed a man on the moon,'' Guest says, ''it made
sense to cut back on the program. The number of people who put
their homes up for sale even before the first layoff was amazing.
Then the layoffs came. Every street had five for-sale signs and three
more homes that were for sale but you didn't know it. On all the TV
networks there were shows about 'Failure City.' People around here
were very critical of the mass media, and said nobody would want
to come here, but how wrong they were! Because the primary motivation of most people, unfortunately, is to take advantage of each
other.

''What happened here was like wildfire. People came from all
over: 'We've heard there are homes here practically for the asking;
we'll take two or three.' I was new in the business, writing orders
like crazy. There were so many houses for sale that most of them
you didn't even bother to take a listing on—you couldn't possibly
handle all the paperwork. Some people were taken advantage of,
practically gave their homes away. Some homes were saved by renting them for less than the monthly mortgage payments. I'd say to
the owners, 'How much can you afford to lose per year to wait?'
The market turned in a year; people were making profits in a
year. . . .'' It took Guest not quite four years to become the biggest

realtor in Brevard County. At one point in the early 1970s he had ten offices, reaching as far south as Vero Beach, as far inland as Orlando and Winter Park; as things turned down, he pulled back to the six near his home. In 1976, with 108 brokerage firms in the Board of Realtors, Guest did 16 percent of all the real estate brokerage in central Brevard.

Four times a year, Guest runs a training program for people who wish to be real estate salesmen. Fifteen or sixteen people take the course, and eleven or twelve are accepted at the end of it as "representatives" of Guest Realty. ("Representatives"—independent contractors—not employees; if they were employees Guest would have to pay social security taxes.) All told, he has about 125 representatives in the field, permitted—nay, encouraged—to use the name Guest, put it on their cars and their houses, talk it up, as long as they maintain a minimum annual production. Each office has a registered broker, in effect a neighborhood partner, as well as the representatives; nothing is listed for sale unless the broker sees it, and contracts are presented to purchasers only by one of the brokers, not by the salesman, to assure management control. As a normal matter, Guest wants people who are new to the business. "Then," says Dick Schultz, a local math teacher Guest hired to run his training program, "they don't waste our time or theirs telling us how their previous employers did it."

"It's like Arthur Murray teaching people to dance," Guest says. "You break it down step by step."

Schultz is a larger, mustachioed man; he has given up teaching and now heads the Cocoa Beach office as well as running the trainees. "There is no art in selling," he says, "—no art at all. It's a science. A few years ago we had a girl who wore clothes I wouldn't let my seventeen-year-old daughter wear, and she asked dopey questions. With great trepidation we put her out to represent Guest Realty. She sold a million dollars the first year, and the second year she came within a hundred thousand dollars of two million— nobody has *ever* sold two million dollars in residential property in Brevard County. The third year she got divorced, went out to get a new husband, got him, got pregnant—and still sold more than one million dollars. We have lots of people here who know a hell of a lot more than she does about real estate who struggle to sell half a million dollars. She learned what she was supposed to do, and she did it.''

Guest's message is to keep everlastingly at it: ''The more you expose yourself to people,'' he says, ''the more they will expose you to others. In a group, if one man says, 'I'm thinking of selling my house,' somebody else in the group will say, 'I know the guy you've got to talk to. . . .' Real estate is an interesting subject. People don't turn you away when you start talking about it. Everybody knows all the big fortunes were made in real estate. The numbers are always big. Whenever you meet somebody you have to tell them, 'I'm in real estate. I'm with Guest Realty.'

''The important thing is to go after listings. You can do that. The listings are in the geographical area you're in, and the buyers are not. The buyer takes hours of your time, you show him everything in the county, he passes something with another broker's name on it—something you could have sold him—and he calls that broker himself, and you've got nothing. But when you have the listing, you've multiplied yourself twelve hundred times, because there are twelve hundred brokers in Brevard County.''

The best source of listings is of course friends and neighbors. One of the reasons there are so many real estate salesmen is that the weakest member of the fraternity may still have an edge in winning the listing of a home for sale on his own block. An important source of listings is the ''fizbo,'' acronym of ''For Sale by Owner.'' Only about 10 percent of the ads owners put in the paper produce direct sales, and as the owner grows discouraged a broker picks him up. Neighborhoods where nobody in the office lives may be assigned to a salesman for ''farming,'' intensive visiting to find out which houses are likely to go on the market and why. ''You get to know people,'' says Eladia Ganulin, sales manager for Unique Homes of San Francisco. ''It's like selling insurance or anything else.'' Probate courts and lawyers who handle estates are a significant source of listings. Also divorces: like the lawyer, the real estate broker lives off the unlucky and is not loved for it.

Increasingly, as brokerage firms become linked in nationwide cooperatives and franchise schemes, leads for listings arrive from out of town, from a broker who hears about a potential customer coming to his city and calls partly to tip a distant colleague about a property that will be for sale and partly to acquire any information that may be available about the family that's on the move. Some kinds of information may be hard for a broker to handle. He knows that if he brings a black buyer into a neighborhood his chance for future listings there virtually disappears, but the law tells him he mustn't "steer." Usually he takes his chances breaking the law rather than violating neighborhood attitudes. There is some reason to hope, if not quite to believe, that so long as all prospective purchasers are in the same boat—i.e., there is no special government subsidy for blacks—the pressures for racial exclusion are easing.

Brokerage commissions in the United States vary between 5 percent and 7 percent on residential properties (they go up to 10 on commercial properties); and where buyer and seller have different brokers the two split the commission, the buyer's broker getting the larger share, though in some areas fifty-fifty splits are the norm. Once upon a time listings were "open" (a number of brokers might offer the property independently, winner take all) or "exclusive" (to buy this house, the buyer had to see this broker), with "exclusive agency" used for agreements that if the owner sold without the broker's help he owned no commission, "exclusive right" for situations where any sale to anybody generated the commission whether the broker had anything to do with it or not. Today in most cities brokers join a Multiple Listing Service, which gives the broker who submits the listing an "exclusive" but permits every broker to offer a possible buyer pretty much the complete range of properties currently for sale in this area, and to participate in the commission if it's his fish that bites.

The "MLS" is operated in most places by the local board of realtors, usually by contract with a computer service company that generates an updated weekly book of available properties including a great deal of information about the house (typically including a picture) and a coded note about the size and split of the commission. Costs of participation will vary somewhat from place to place. In

Orlando in 1977, a broker paid $14 for each listing placed in the machine and $220 a year for the weekly book; in San Francisco prices were about half that.

A multiple listings book may have two thousand homes, the presentation organized by neighborhood, and it tends to be unwieldy for general questions that ask instead about homes of a certain size or view or with swimming pool, etc. For that purpose in some cities brokers can buy directly from the computer company an on-line terminal that can query a software package. Typing into the terminal, a broker may ask, for example, for all the four-bedroom homes on a hillside with a swimming pool available for less than $55,000 in this community. In Orlando, thirty-nine brokerage offices are hooked into such a service—by leased wire to Realtron, in Detroit. The answer to the sample question above would come back "None," and then the broker can start changing the parameters—eliminating features or adding bucks—until the machine starts printing out some possible properties. The convenience may be more in the ease of convincing the customer that what he wants doesn't exist at his price than in the actual delivery of the information.

Both the Internal Revenue Service and the Justice Department have had severe cases of hives from contact with the Multiple Listing Service idea—IRS because boards of realtors may claim tax exemptions as nonprofit groups while deriving income from MLS operations, and the Justice Department for obvious antitrust reasons. (It is in "open" listing situations that brokers are most likely to cut their commission to offer customers a bargain.) The National Association of Realtors publishes a *Handbook on Multiple Listing Policy* which is an unusually delicate example of walking tiptoe on legal eggs ("any rules requiring members to adhere to a schedule of fees or commissions is [sic] contrary to the Code of Ethics and to the historical policies of the NATIONAL ASSOCIATION OF REALTORS, and is inequitable limitation on its membership" and "If a REALTOR wishes to cooperate with a nonmember . . . the Multiple Listing Service must not discourage or prohibit him in any way. . . .") So far, the obvious nonprice advantages of the service to the public, and the clear need to apply *some* standards for entry to the lists, have kept the government from eradicating it completely.

And the fact that the press is venal in real estate matters probably assures that no Public Interest advocate will be able to raise a major crusade against it.

Getting the listing is of course only one blade of the scissors; buyers and renters have to be found, too—and finding may be the operative word. "It used to be you put up an apartment house on Ocean Avenue and Avenue K," New York ad man Richard Weiner says, "and most of the people who would rent there already lived within six blocks—or they drove by and saw the sign you put up outside. Now you have to get people from all over the country." Most brokers consider advertising ineffectual for this purpose—the ads are regarded more highly as a way to draw listings, from people who see you're on the ball. Word of mouth via banks and lawyers, or a comment from somebody's cousin's husband's best friend, is a more likely source of customers for a broker than the whole real estate section of the newspaper.

Some of this is systematized. In Tampa, Walt Millard is part of a nationwide franchising system called Red Carpet Realtors, and gets a certain number of customers referred to him by others in the same organization, out of state. Then he can talk to them before they come to Florida and he can be ready with a salesperson whose main occupation or personality seems to match well with the newcomers'. He also cultivates contacts with big corporations that have operations in Tampa, to get a line on new executives coming to work in this area. One of his salesmen is assigned permanently to Honeywell.

In Winston-Salem, Mary Ann Parrish of Helms-Parrish Properties says that 60 percent of her business comes by reference from corporations bringing in new people: "In the last few weeks, we've had four promotions to president here, and three of them were people for whom we'd found homes. They'll send any new people to us." At the University of Connecticut, Bryl Boyce, who chairs the real estate department in the business school, says with a touch of hyperbole that he's sure there are three real estate agents in the area who know before he does when a new professor is being hired for his department.

People who live in apartments in an area are often likely customers for houses, especially if they have small children. Miller Nichols of Kansas City sees that his monthly brochure of homes available in Southwest goes to every doctor's office in the area, especially the pediatricians' offices, to be left in the waiting rooms. A broker with a house to sell near a hospital may circularize the staff; if it's also off the seventh fairway of the country club, he may send word of it to the members of the club. Customers do call and say they've seen a sign on a lawn (in 1977 the Supreme Court ruled that under the First Amendment people cannot be forbidden to put a sign on their lawn saying this house is for sale). Every so often a broker runs into somebody socially who has a special requirement and can be sold instantly if that requirement is met. Eladia Ganulin in San Francisco recalls a sculptor who wanted a Victorian house with a studio in which he could work on three- to five-ton pieces of stone.

Apart from highly specific features, worthless to some and priceless to others, houses mostly have to be sold on either value or "life style"—a catch-all category stretching from "PS 6 district" in New York to hints of novel sexual experiences at Marina del Rey in Los Angeles to the medical clinic on the premises in Fort Lauderdale. Swimming pools were big in the 1960s; tennis courts in the 1970s. What is customarily meant by "life style," however, is not much more than what brokers have always called "community"— if you come here, you'll find like-minded people. For what the potential customer seeks, after all, is a place to live.

"We find people not so much a house as an area," says Walt Millard, whose sales force includes about thirty people, some of them teachers from both the school system and the University of South Florida, moonlighting for fun and profit. "Some people come down, all they're interested in is the location of the boy scout troop. There's concern about neighborhood schools, churches, shopping. I take people to schools and have them meet the teachers before I show them a house. If people haven't picked an area, the salesman is wasting his time." Cutting a level deeper, Mary Ann Parrish of Winston-Salem says, "What people want is instant acceptance for their children."

Buying a house is a big decision, a commitment of emotions as well as money. It almost can't be done entirely rationally, and most people are scared of it; they are forever asking for delays, "time to think about it." The salesman and broker, having no source of income but their time, tend to be impatient with this process, and no subject is more common in the literature than ways to act to "close" a deal. The stress on the closing process makes real estate people somewhat more cynical than the run of mankind. Honestly or otherwise (it doesn't matter much), salesmen and sales-oriented service professionals tend to say of their work that "you have to like people"; in real estate brokerage, the line is that "people have to like you." Real estate selling is inevitably a manipulative trade: the NAR *Handbook* explains that *empathy* is good and helps you put yourself in the customer's shoes and study his problem through his eyes; but *sympathy* is bad and may prevent you from making a sale simply because you think these people can't afford this house, or wouldn't like the muddy garden that now looks so green.

Real estate salesmen have to spend a lot of time driving prospects around from house to house, and showing off the interiors. The recommended modus operandi is to show the unfortunate elements first, trusting this evidence of your honesty and the later enthusiasm about the more successful parts of the house to dissolve the risky first impression. In multiple listing situations where the owners are not going to be at home to help with the display of their possessions, brokers usually hang a "lock box" on the house in some position not easily visible from the street. The broker showing the house knows the location of the box and has means of opening it. Inside the box is the key to the house, to be used and put back in place for the next visitor.

During the course of driving about and observing reactions and listening, the salesman is likely to acquire personal opinions about the prospect. Both parties being under some stress, these opinions are not always positive. "I could tell some of the businessmen in this city some important things about some of the people they've hired," says Mary Ann Parrish. Her daughter, who works in the agency, has a sign hanging behind her desk, showing a girl with long hair in fairy-tale robe bending uncertainly toward a frog-like creature on the ground. "Before you meet the handsome prince," it says, "you have to kiss a lot of toads."

This is a business of mind-boggling dimensions. The number of existing homes sold each year is two to three times as great as the number of new homes, and the annual total of money changing hands approached the $150 billion mark in 1977. The National Association of Realtors, based in Chicago, has 800,000 members, 20,000 of whom come to an annual convention; and the Realtors are the aristocracy of this business: in most states, salesmen are not Realtors. More than three-fifths of NAR members specialize in single-family brokerage. Some three million Americans spend at least part of their time selling real estate. In Orlando, there are 2,800 members of the Board of Realtors out of a population of about 100,000. "For every prospect in town on any given day," says Don Saunders of that city, "there are something like forty registered salesmen. But you know the old adage: ten percent of the people do ninety percent of the business."

Residential real estate brokerage has always been a big employer of women. "Ladies who have never worked," says Jackson Wells of Coldwell Banker in San Francisco, "do very well because they have product knowledge—they can interpret the benefits of the house for a woman as a man can't." Others take a more cynical view, pointing out that selling houses requires a willingness to drive around all day in your station wagon with a woman and her four kids (one of whom may throw up in the back seat). There seem to be local variations in the frequency of female employment: Coldwell Banker, for example, reports that nearly all its residential sales personnel are female in Houston and Dallas, but Denver is dominated by men. Nationally, women make up 41 percent of the part-time "Realtor-Associates" (salespeople who are ambitious enough to join the National Association of Realtors). Among full-time Realtors, only 18 percent were women in 1974, and their average income was $15,000 against the men's $25,000. Among full-time salespeople, 37 percent were women, averaging earnings of $9,000 a year, against $15,000 for the men.

To do this job right probably requires a great expenditure of energy and of time. The NAR *Handbook* offers "A Typical Day for a Successful Real Estate Salesman," doesn't get the fellow home until nine at night, and opens the day as follows:

6:00 a.m.   Get up—jog—shower
6:45 a.m.   Breakfast—Read the paper for "For Sale By Owner" ads and competition.
7:30 a.m.   Leave for the office. Listen to motivation tapes in your car on the way to work.
8:00 a.m.   Arrive at the office.
8:05 to 9:30 a.m.   Review daily plan.
                Check your message box.
                Review and prepare the following:
                    Address any direct mailing.
                    Match your buyers with property.
                    Check M.L. sheets and office listings
                    for inspection.
                    Call other brokers regarding their listings.
                    Plug your listings to fellow salesmen . . .
                    Prepare expired listings for evening calls.
                    Make 5 cold calls.
9:30 to 11:00 a.m.   Do all the things that you have just prepared:
                . . . Take out "sold" signs, or "for sale"
                and pick up lock boxes.
                Call 5 "For Sale By Owners."
                Call 3 expired listings.
                Call potential buyers and set up appointments
                to show property.
11:00 to noon.   Inspect new listings. . .

Every so often, some of the leaders of the field decide that a higher level of professionalism might substitute effectively for some of that go-getting. "Ten of us all over the United States decided in 1969–70 that we wanted to bring in a lot of bright young people," says Howard Benedict of Hamden, Conn., who is a real estate commissioner of that state and chaired the mortgage finance committee of the NAR. "But they wouldn't work. They got discouraged easily, and if they were single they would throw up their hands at minor roadblocks. They wanted to do too many uneconomical-type things, and they wouldn't listen.

"I find the business-school graduates are not profit-motivated. They'd much rather have a job in a bank with a fancy title than go out and hustle for a living. All ten of us had them leave owing us money. Now I won't take on any young people who haven't already had two or three jobs, learned what the world is like. I find I do very well with college-educated women who have been homemakers for fifteen or twenty years. A lot of them have the talent to earn eighteen, twenty thousand dollars a year on straight commission." To get a grip on the realities of this business, it must be remembered that such an income probably derives from the sale of only ten or eleven houses a year: these are big tickets.

There can be two honest and informed opinions about whether the services performed by the residential real estate brokers are "worth" the $5 billion or so they cost in 1977." A California sample survey in 1967 found that "fizbo" sales were more likely to be at the seller's asking price and quicker, while sales through brokers were below the asking price and slower (probably because the For-Sale-By-Owner homes for which prices were above the market would end up in the hands of a broker: it can be argued that the seller who sold fast at his asking price was taking less than he could have got.) But in a mobile society brokers' services are all but indispensable. Markets cannot function without information, and as the low return from newspaper advertising demonstrates, the information this market needs must be spread by people. Because the markets are by definition local, it takes a lot of people.

The social function of real estate brokerage is that it sets the price for houses. Brokers use appraisers, who are often brokers themselves (in Florida, only a licensed real estate broker can work as an appraiser), but despite the licensing exams, the "appraisal plants" full of documents, the apparent certainty of the numbers on the standard forms, an appraisal is nothing more than one man's opinion of what something ought to sell for. In some situations rational appraisal would be highly desirable—in Orange County, for example, where the selling price of existing houses has risen far above their replacement cost, arguing an astronomical value for land. But in such places the appraisers cop out entirely, and report as "values" the prices of "comparable" properties recently sold. (And it is no trick to get that information: in most parts of the country, computerized listings of recent sales are available on a regular schedule of publication.) One of the techniques of appraisal is the calculation of value by reference to the rent roll, which is then multiplied by some constant to give an approved price. In New York in 1971, the appraisers multiplied the rent roll by anywhere from five to eight, depending on the neighborhood; by 1975, they were multiplying the rent roll by one-and-one-half even in middle-class and possibly improving neighborhoods like Riverside Drive near Columbia University.

As every businessman knows, the difference between "value" and price is that you sell at the price and talk about the value. The fundamental truth in the stock market is that everybody who sells a share "values" it differently from the person buying it: differences of opinion make financial transactions as well as horse races. The sale of a house is a financial transaction involving a real object for use. There is no intelligent economic theory to explain how people behave in such markets, so it's every man for himself. Institutions arise to protect people from the consciousness of such realities, and an appraisal is indeed a great comforter. Almost every residential transaction involves two appraisals—one for the seller, urged by the real estate broker either as a help in getting the seller's price or (covertly) as a way to make the seller lower an unrealistic price, and one for the bank, urged by the loan committee of its board as a way to protect everyone from later allegations of bad judgment. In some transactions there is also a third appraisal, for the buyer; many have noted that an appraisal made for a buyer tends to indicate a lower value than an appraisal made for a seller.

Appraisals add between $150 and $500 to the transaction costs when a house is sold. Then there's another couple of hundred for title insurance, which has to be supplied twice, separately for the lender and the buyer. This uniquely American industry, which takes in $600 million a year in premiums, is a monument to the unique place of real estate in our legal system and to the inefficiency of local government. The ownership of land is very specifically safeguarded by the state through rules that go back to the protection of widows and pious endowments in the English middle ages; and the transfer of ownership, which means the transfer of these safeguards, can be accomplished only by meeting standards of contract more rigorous than those applied to other purchases and sales, properly registered with the relevant authorities. But the owner of a piece of property may wittingly or otherwise "encumber" it, leasing it, creating an easement (like the use of a path through the property to the beach), selling some part of the ownership rights (like the right to mine coal or drill for oil beneath the surface), or pledging it to a lender (the mortgage) or to a contractor for some repair work (a mechanics' lien). A seller's description of his property has no legal significance: all he can sell is that area described by the metes and bounds or bearings or rectangular survey data on file in the courthouse. (Hence there may also be the costs of a survey, to prove that the seller's description matches his ownership.)

When a man dies, the stocks and bonds and bank accounts and insurance policies he owns go into his estate, for distribution after probate, but his land passes directly to his heirs; and an heir with a 90 percent interest has no right to sell the property. Unpaid taxes of any kind—income, estate, or inheritance as well as property taxes—may give a government a lien on a house, which means that the purchaser may buy an obligation to pay somebody else's taxes if he wishes to keep his home. And there are no effective statutes of limitation here: if seventy years ago a man sold a title to property that he didn't really own, the heirs of the rightful owner can conceivably come around tomorrow and take the house away from its occupant.

Alfred Ring in his real estate textbook tells a story about a New York lawyer who queried an opinion on a title from a New Orleans lawyer who had carried his search back only to 1803. The reply from New Orleans read: "I acknowledge your letter inquiring as to the state of the title of the Canal Street property prior to the year 1803. Please be advised that in the year 1803 the United States of America acquired the territory of Louisiana from the Republic of France by purchase. The Republic of France acquired title from the Spanish Crown by conquest. The Spanish Crown had originally acquired title by virtue of the discoveries of one Chistopher Columbus, sailor, who had been duly authorized to embark upon the voyage of discovery by Isabella, Queen of Spain. Isabella, before granting such authority, had obtained the sanction of His Holiness, the Pope; the Pope is the Vicar on Earth of Jesus Christ; Jesus Christ is the Son and Heir Apparent of God. God made Louisiana.''

Title insurance gives protection against all these possible defects of ownership (though not against all contingencies: the policy covers only the money paid for the house, and if an owner improves his property or inflation increases its market value before the title goes blooey he recovers only his original purchase price in devalued dollars). The procedures vary enormously from state to state—in some states, title insurance companies are prohibited by law, and lawyers give the guarantees themselves—but one rule prevails everywhere: however recently a house was sold, title searched and insured, the whole process must be done again, and paid for again, before a new sale can be consummated. In the twilight years of the twentieth century, with a hundred years of land transactions on record almost everywhere in America and computers available to file and retrieve the data, there must be a better way.

More costs: the lawyers. It was in a case growing out of the purchase of a home that the Supreme Court ruled that bar association fee schedules violate the antitrust laws, but the result through 1977 was to produce confusion rather than competition. Now that lawyers are permitted to advertise, rates for routine real estate closings may go down. For the time being, in most parts of the country, a home buyer must figure that a lawyer to handle the closing will cost somewhere between $250 and $1,000.

Then the mortgagee—the S&L or the bank—comes along and charges a fee for "originating" the mortgage. This fee may cover more than merely opening the file on the home to be mortgaged (the cost of the lender's appraisal and title insurance can be included), but at an average of roughly 1.5 percent of the total loan, it adds up to a lot of money—more than a billion dollars a year for the lending institutions as a group.

There are escrow agents who hold the purchaser's deposit and the funds set aside to cover forthcoming real estate taxes and insurance bills. In some states the escrow holders do much more—in Southern

(but not Northern!) California, they actually take care of the paperwork of closing the deal, saving both seller and buyer the layman's
agony of that last, long meeting. But they cost, too—and it is by no
means clear that they really *need* all the money prospective buyers
are required to deposit with them before the contracts can be signed.
And then, of course, cities and states impose taxes on property
transfers and new mortgages. . . .

Add it all together—brokers' fee, lawyers' fee, appraisers' fee,
title insurance, loan origination fee, taxes—and the transaction
costs incurred in the purchase and sale of a home probably exceed
10 percent of its price. Thousands of dollars. Especially for a first-
time purchaser or seller, the discovery of these costs can be the most
unpleasant single moment in the experience of buying or selling a
house.

Moreover, there has been over the years a lot of backscratching
associated with these costs. The broker, appraiser, lawyer all sit on
the board of the S&L; and people who send each other business
often expect a quid pro quo. There was reason to believe that referral fees and commission-splitting were adding to the rapidly mounting burden of what is in any event a labor-intensive process subject
to rapid inflation. Starting in the late 1960s, bills began to appear in
the hopper at Congress to assert federal control over the service
costs of homebuying.

In 1970, Section 701 of the Emergency Home Finance Act gave
HUD the power to regulate settlement costs of home sales financed
by FHA and VA mortgages, and called on the Department and the
VA to make a study of settlement costs across the country to help
Congress determine if specific legislation was required. Each state
having its own rules and customs (many, as the difference between
Southern and Northern California over escrow holders indicates,
having a collection of disparate rules and customs), the study became an encyclopedia; but its recommendations were focused.
Early in 1972, HUD Secretary Romney told Congress that his people would develop a schedule of ''maximum allowable charges'' on
home sales in different regions of the country. The first schedule,
covering six metropolitan areas, was published as a proposal that
July. Lawyers became concerned about price control on their fees;

bankers began to worry that if customers could not be charged for certain services that were part of traditional lenders' requirements the banks might have to pay the bill themselves; and the title insurers noted with horror that among the recommendations of the HUD-VA study group had been a system of national land registration to eliminate the costs of title insurance.

In two entertaining brief reports to the Alicia Patterson Foundation, Mary Clay Berry has told the story of how the title insurance companies lobbied for a law that would protect their industry and wound up creating a monster. What emerged (after Romney's successor James T. Lynn had withdrawn the proposed schedule and asked Congress for guidance) was the Real Estate Settlement Procedures Act of 1974—RESPA. This long and ill-drafted law sought the traditional American remedy of full disclosure of all costs, and included a traditional American prohibition of "kickbacks." Its purpose, said the Senate Report that accompanied the bill, was to foster "a free market for settlement services."

The new bill applied not only to FHA and VA mortgages, but to all mortgages written by banks and S&Ls with government-insured deposits, which in practice means all of them; and the banks and S&Ls were made responsible for the disclosure statements (under criminal penalties for failure to comply). To make disclosure effective in stimulating competition, HUD was to impose the forms to present the information, publish a booklet explaining it all to home buyers—and require a time period between the delivery of the disclosure and the closing of the sale, so the buyer could noodle over what he had learned and do some shopping if he didn't like it.

This was the real estate industry's bill, supported (though not very enthusiastically by anyone except the title companies) in fear of getting something worse. One of the places where settlement costs had gone well out of line, with big profits from referral fees, was Washington, D.C., and Congressmen often overreact when abuses touch their own pocketbooks or those of their staffs and friends.

In mid-1975, HUD published the required regulations, and suddenly there was hell to pay. The real estate brokers had thought themselves exempted because their commissions were paid by sellers; they learned that an amendment added without fanfare had included them in—and also that the law could be interpreted to make the standard Multiple Listing Service procedure a "kickback." Tract developers learned that their arrangements with the title insurers, whereby they could simply pass the policy given them to their customers, were also "kickbacks" (an interesting example of the way consumer legislation can boomerang to increase costs). Banks and S&Ls found that gathering and presenting the information for the forms, which are several pages long, would require some hours of employee time per mortgage, for which they could not legally be compensated in any way. Lawyers who are paid for real estate work by the piece rather than by the hour found its profitability potentially diminished, as they too were forbidden to charge for any extra expenses required to meet the provisions of the law. (Eventually, of course, the bankers and lawyers would find ways to charge the customer, but at the beginning they had to be careful.) And homeowners who wanted to get their money and leave—or home buyers who wanted to move into their new house quickly— found that they had to wait twelve days for the disclosure information to ferment before they could go about their business. The buyer could waive his right to twelve days' notice, but the waiver was revocable and no lender would accept it. For home buyers, as lawyers Morton Fisher and Gregory Reed wrote, RESPA meant "additional regulatory complications of a settlement process which, to them, was already a Kafkaesque event." Charles Salms of Manchester, New Hampshire, who found himself awkwardly placed as counsel to both the state Realtors and the state mortgage bankers, remembers that their worst fight was about "who would carry the blame in the customer's mind."

Congress was deluged with letters, and responded, liberals as well as conservatives. The panic among legislators who believe in the virtues of regulation may have been caused by a fear that now millions of ordinary voters, home buyers and home sellers, would discover the great truth businessmen have been trumpeting unheard since government grew gargantuan—that time comes cost-free to bureaucrats, who therefore really cannot understand why others value it so highly.

What survived the 1975 amendents and the 1976 regulations was a somewhat simplified version of the HUD form detailing settlement costs (to be delivered with ''good faith estimates'' rather than hard numbers, *one* day before closing), plus the prohibition against kickbacks. All questions about it must be answered from Washington: ''No authority granted to the Secretary under RESPA,'' says the Regulation, ''has been delegated to HUD Regional Offices, HUD Area Offices, or HUD Insuring Offices.'' The global result has been to protect a minority of buyers and sellers against the crooks of the industry at the expense of added costs for the majority. Conceivably, there will be a next step in which settlement procedures are simplified and costs reduced by legislative mandate—but that's theory rather than possible practice, for Congress doesn't want to hear about RESPA again, ever.

Among the immediate effects of the law, Jack Marino, Jr., of the Pioneer National Title Insurance Company told an American Law Institute meeting in fall 1976, was the end of the hallowed practice in New York by which title insurers gave lawyers a 15 percent commission for steering business their way; and reputable lawyers then passed the commission on to their clients because that's what the Canons of Ethnics told them to do. ''The title insurers had been turned down twice by the Insurance Commissioner on a request for a premium increase,'' Marino reported. ''Now RESPA gave it to them.'' Baltimore lawyer Morton Fisher commented, ''It's the new Robin Hood. You steal from the rich, and you keep it.''

Chapter 14

The Sickness of Rental Housing

Where apartment structures . . . are professionally planned, . . . both the quantity and the quality of periodic income, under competent property management, are deemed economically superior to income derived from other forms of monetary investment. . . . In the absence of rent and price control, real property, like a ship upon the ocean waters, floats above its purchasing power-constant dollar line irrespective of depth or rise in the level of prices. It is this ability to hold its purchasing power integrity that has in recent years popularized the demand for shares of real estate trusts and syndicates.

> —Alfred A. Ring, professor of real estate, in his textbook (1972)

It has been said that the economics of tenement house operation were inscribed on the backside of the janitor's wife; it was she who scrubbed the halls (if they were scrubbed), cleaned the stoop and kept the building neat. When she insisted on standing, so the adage goes, the costs of tenement house operation rose with her.

> —Roger Starr, New York City Housing Administrator (1976)

I get people to invest by selling them the depreciation.

> —Richard Ravitch, builder

Historically, the profits in real estate were primarily those of the landlord. An investor paid someone to build an apartment house, borrowed 90 percent of the cost on first and second mortgages, hired a manager or an agent to rent the apartments, collect the rent, and verify the bills, hired a janitor to stoke the furnance and keep things more or less shipshape. The manager worked part-time, and a big piece of his salary was a free apartment. In small buildings a "super" doubled the jobs of resident manager and janitor. Multifamily housing being a pure case of absentee ownership in most cities, quite a lot was riding on the performance of the manager. If he rented apartments to people who didn't pay their bills, or caused other tenants to leave, the economics of the operation would disintegrate: because the mortgage payments and virtually all the expenses are the same no matter how many rent bills are collected, the profits disappear if much more than 10 percent of the apartments don't yield their monthly tribute. If the manager failed to follow up on necessary maintenance work, the owner could face a sudden crisis of leaking walls or windows, a blown boiler, a clogged waste stack. "I can never understand," says broker William Edmunds of Orlando, "how a man can put five million dollars into a building and then leave it to be managed by someone he pays with a free apartment and maybe a hundred and a quarter a week." (The managers themselves make the same point a different way: "Developers and owners think all you have to do is hire some asshole to collect the rent.") In 1976, the median income for apartment house managers in the United States was apparently around $8,000 a year—plus the apartment, which is tax-exempt compensation.

It was assumed that the numbers were simple and steady. Miller Nichols, thinking back to the old days in Southwest Kansas City, reports that an investor-builder figured on ten months' rent for the operating expenses and mortgage payments, one month's rent for taxes, and one month's rent for profit. Assuming a desired profit of 10 percent on investment and a total cost of $6,000 per apartment—both figures reasonable immediately after the war—the rent was $60. That would permit 30 percent of the rent roll for operating expenses and 5 percent interest to the lenders. Placed on the market, this apartment house would sell for 8.3 times its annual rent roll. Note that if the owner could screw a 10 percent rent increase out of his tenants with all his costs unchanged, the "leverage"—the fact that 90 percent of the money spent on the house was borrowed from the bank—would double his profits.

There are still a number of highly profitable apartment houses in America, especially in the Washington suburbs, California, and Texas (most notably Houston, where the absence of zoning laws means that the builder did not have to pay premium prices for land zoned for multifamily use, as he does elsewhere). Older buildings in many parts of the country are a bonanza for owners who have held them a long time. Jack Rosenfield, an earnest young man working his way through law school by helping out in his family's real estate business, speaks approvingly of earnings from "the standard old six-story sixty-unit elevator apartment, which a good super can run by himself"—even in New York, where rent control has been continuously in effect for thirty years and where the average real estate tax takes not one but more than three months' rent.

But as early as the 1950s, experienced investors began to see that the apparent profitability of apartment houses was built on sand. The FHA and VA mortgages, with their low or nil downpayments and with interest rates reduced by government insurance, had made it less expensive for people to buy their own homes in the suburbs, which most of them wanted to do anyway. As noted in Chapter 6, the costs of building apartments began to rise more rapidly than the costs of building single-family homes. And maintenance costs went up faster still: house maintenance, as homeowners well know, is a highly labor-intensive operation. By the middle 1950s, new apartments accounted for less than 10 percent of the housing starts in America. The insurance companies, which had invested heavily in multifamily housing right after the war, pulled out.

What brought multifamily back in the 1960s was a political decision that if the private market would not sustain the profitability of apartment houses, governments would find ways to do so. The housing phase of urban renewal was in essence a way for governments to make apartment house ownership profitable. Capital costs were reduced because the plot was pre-assembled, the land was written down, existing structures were removed—and then operating costs were reduced by abatements of real estate tax on some part of the value of the improvements. The big contribution to profits, however, was given by the federal government in the income tax laws.

Leverage—the multiplication of an investor's purchase by the use of borrowed funds—had proved to yield an insufficient cash flow of profits in residential real estate; under the new tax laws, leverage would be used to create a flow of tax benefits instead.

A building, like all else manmade, deteriorates and decays, a condition expressed financially as depreciation, a gradual loss of value. This depreciation is a hidden cost of operating the building; as a cost, it is deductible from income for tax purposes. The entire building depreciates, the part built with borrowed funds as well as the part built with the investor's own money. but the depreciation is in no way a cost to the bank—only to the investor. So the investor in effect leverages depreciation; he takes as a deduction from his income for tax purposes the depreciation allowance from the entire base cost of the building, not just his own investment. If this depreciation allowance is greater than the income earned by the building, an individual owner or partnership can apply the deduction to income earned elsewhere, reducing the taxes paid on those other earnings.

At this point, an investor can be got interested in building and owning an apartment house even though he earns little or nothing on his investment: the government in effect gives him his "profits" by excusing some of his other income from taxes. "The apartment house business," tax lawyer Adrian Thiel told an upbeat meeting on multifamily housing in 1969, is "an adjunct business and shouldn't be the main business of builders, or the main business of anybody else. It's an ancillary, incidental thing to making money elsewhere and then sheltering it." The Joint Committee on Internal Revenue Taxation reported to Congress in 1976 that in soliciting partners for real estate ventures, "it has become common practice to promise a prospective investor substantial tax losses which can be used to decrease the tax on his income from other sources. There is, in effect, substantial dealing in 'tax losses' produced by accelerated depreciation in real property."

"Accelerated" is the key word for what the government did in the 1950s and 1960s. In theory, after all, depreciation *is* a cost, and taking the bank's share is entirely justified by the fact that part of the bank's monthly mortgage bill is an amortization that (again in

theory) matches the depreciation. Apartment house mortgages, differently drawn, amortize more rapidly than home mortgages. Presumably, when the building is sold, the investor will receive less for
it than he paid, because it has depreciated.

Before World War II, buildings had been assumed to have a
useful life of forty years, and were depreciated steadily at 2.5 percent a year. In 1946, however, an administrative decision in the
Treasury Department permitted investors to depreciate on a "150
percent declining balance" method, and in 1954 Congress permitted the use of "double declining balance." Even in noninflationary
times, this faster depreciation guaranteed that the tax deduction
would be greater than the real loss on resale in the early years.
"When the law was amended to permit accelerated depreciation,"
New York lawyer-developer Eugene Morris says in happy recollec-
tion, "we were able to finish up some tough ones and get out with a
hell of a profit."

Let us look at what these arcana mean in terms of cash money. At
forty-year "straight-line" depreciation, a million-dollar building
will generate $125,000 of tax deduction in five years at a steady rate
of $25,000 a year. At forty-year double-declining-balance (5 per-
cent instead of 2.5 percent, applied to what remains after previous
depreciation instead of to the initial cost), the tax shelter in five
years will be $226,000. Assuming an investor in the 70 percent
bracket and a 90 percent mortgage, double-declining-balance depreciation on the whole building yields an average tax reduction of
$31,600 a year for five years for every $100,000 invested—a yield
of 31.6 percent per year *tax free* to a very rich and heavily taxed investor even if the apartment house makes no money at all.

More: IRS rulings permitted investors to deduct directly from
their income (rather than include in the cost of the building to be
amortized over time) all expenditures for interest and real estate
taxes during the period of construction. Indeed, IRS permitted a
prepayment of interest on construction loans to be deducted from income. Again assuming that the project is financed 90 percent by
loans, an investor in the 70 percent bracket putting up $100,000
might well show a tax deduction of $110,000 in the year before con-
struction began—the government would in effect repay him

$77,000 of his $100,000 investment. Referring back to the previous paragraph, we find that on a *net* investment of $23,000, a very rich man could reduce his taxes by $31,600 a year—a net return of 137 percent, tax free.

What built the multifamily housing of the 1960s, in other words, was not the hope of profit but the certainty of tax avoidance. No wonder so much of it got built—and so much of it later proved unprofitable.

In theory, the government got some of that money back. The greatest advantages from double-declining-balance come in the first ten years, and most of the investors who participated in these "syndications" sold out fairly quickly. When they sold, their profit was calculated not on their original investment, but on that investment less the depreciation they had taken on their tax returns over the years. In our example, the investor who put up $100,000 would have taken $226,000 in depreciation. In the postwar period, that depreciation was a bookkeeping fiction, because the market value of the building was *not* declining—in fact, it was increasing, floating upward (in Alfred Ring's metaphor) on the rising tide of inflation. If our investor simply got back his $100,000 net, he had to show a $336,000 capital gain ($226,000 in depreciation plus $110,000 in prepaid interest, now regained). But capital gains were taxable at a maximum of 25 percent in those days, so the government recovered only $84,000 of the $235,000 the investor had saved on his income tax.

And even those taxes didn't necessarily have to be paid, if market prices were rising, the investor's tax advisers were clever enough, and the bank was cooperative. Assume that the market value of the building increased by 5 percent a year, which was by no means uncommon in the later 1960s, when construction costs were rising 10 percent a year. After five years, the million-dollar building would be "worth" $1,275,000. Assume next the conventional kind of apartment house mortgage for five years, with 20 percent to be amortized over the five years and 80 percent left to be refinanced, a "balloon" at the end of the term. The $900,000 mortgage has now been reduced to $720,000. The bank, being prudent, will not return to a 90 percent loan, but it will go 80 percent. Of $1,275,000

market value—which means $1,020,000. The investor pays off his
$720,000 and pockets $300,000 in cash—paying no tax on it, because this isn't income, it's borrowed money. To recapitulate—on
an investment net after taxes of $23,000, in five years our very rich
investor (only the very rich can play this game) has received tax
reductions of $158,000, plus $300,000 in cash. Oh, to be very rich,
in the revolutionary '60s!

Obviously, the course of true love did not often run this smooth,
but even if the numbers were one-fourth as good as these, it was one
hell of a deal. This is why the Real Estate Investment Trusts were
founded and flourished (briefly), why the banks decided they preferred participation arrangements to straight mortgages, why the
apartment market was overbuilt, and why the Congress finally be-
stirred itself to writing significant tax reform legislation. At the heart
of that legislation were provisions to "recapture" as income to be
taxed at income tax rates rather than capital gains rates the "accelerated depreciation" permitted in previous years. Another provision
eliminated the tax deduction for prepayment of interest and taxes,
requiring that such expenditures be added to the cost of the building
instead and amortized gradually over the years. In 1976, the House
even made so bold as to legislate a Limitation on Artificial Losses, a
prohibition against taking in any one year a tax deduction greater
than the total amount an investor had put in the property—but that
was going too far, and the Senate put in tougher "recapture" provisions instead.

As of 1977, except for people investing in government-subsidized housing for the poor, it was difficult to launder income into
capital gains through the depreciation of property—but it was still
possible to *postpone* the payment of taxes through what the House
so accurately described as "artificial losses." Builders said that
because the new tax law had diminished other tax shelters even
more severely than it hit real estate, there was still considerable
money around for investment in apartment houses. But now the
lenders wouldn't play—they had backed so many losers in the previous boom, they had so many bad loans to multifamily housing on
their books, that they were no longer interested in any venture that
didn't project profits without reference to tax gimmicks. And those
were hard to find.

At the age of 70, Lex Marsh in Charlotte, N.C., is still building rental units, townhouses on a corner of the eighty-five acre property where he lives himself. He figures he needs $450 a month rent to break even, but he rents at $360 a month, on leases as long as two years, with the thought that the rent will come up later. His bank will help: he's a prominent lawyer and mortgage banker as well as a builder, and he already owns two thousand rental units, many of them only lightly mortgaged. His name is on the note, he's been around a long time, and he knows what he's doing.

"Rental housing is still a solid business here," Marsh says. "It's easy to find resident managers—we have a central bookeeping office, they don't have to collect the rent, everything's clean. We pick out a retired or semiretired couple, or perhaps a tenant who's been with us three or four years and wants some part-time work, will take care of one hundred, up to two hundred apartments. A project manager shows and rents apartments, acts as clearing house for service calls, and in general as liaison with tenants.

"Mostly we give six-month leases; after that, they're free to move and we're free to move them. If a tenant has rowdy kids and half a dozen other tenants are complaining about them, we ask them to move—and they do. Most of our apartments are individually metered [for electricity]. Those that are not, we are on the verge of going back and installing individual meters, at a cost of one or two hundred dollars per unit."

But the key fact is that even in this trouble-free market, Marsh is building to show a short-term and perhaps medium-term loss, mostly because all the tax and possible appreciation considerations together make the project look like something his heirs will be glad they have. The Rand Corporation study of Green Bay and its surroundings estimated that rental housing showed a net rate of return on market value of only 2.4 percent, with cash flow positive only because the owners did a lot of unpaid labor themselves. The authors speculated that landlords were happy about their condition in large part because they anticipated an increase in the market value of their properties—but clearly market values will not increase if rates of return remain that low.

As buildings age, moreover, the tax laws which encourage mortgage refinancing for the biggest loan the bank will give must eventually put the owner and all the incautious lenders in the hole. Among the few rules that continue to apply through the looking glass of tax shelter and inflation is the law that aging properties cost more to run. In the Income/Expense Analysis of the Institute of Real Estate Management, the cost of operating rental apartments built between 1968 and 1975 ran half or a little less than half their receipts, while the cost of running apartments built between 1920 and 1930 was two-thirds the receipts or worse. And these figures are without allowance for replacement reserve. Even 50 percent is menacingly high—in our example of a profitable immediate postwar apartment house at the head of this chapter, we allocated only 30 percent to operating cost to give our investor a 10 percent return on his money—and in those days mortgages on apartments were written at 5 percent interest, as against 9 percent or more in the 1970's.

This is a vulnerable business at best. New York's Daniel Rose describes a forty-eight-story apartment tower his family built and owns: "It cost $14.5 million in the late 1960s. We got $12.5 million from the bank, put up $2 million in equity. It generated some return every year until 1974, when we had a negative cash flow of $250,000. We brought the loss down to $100,000 in 1975, and we probably broke even in 1976." Breaking even is not usually considered a satisfactory return on $2 million. "But," Rose added, cheering up, "you couldn't duplicate that building today for $24 million. It means that before anybody builds a new one our return will have to be $1.4 million a year. . . ." And his brother Fred adds, "I just can't imagine that there will never be another apartment house built in New York."

For moderate-income housing with nonprofit sponsorship, the vulnerabilities are even greater. Enoch Williams of the Housing Development Corporation, a subsidiary of the New York Council of Churches, tells horror stories about projects built for and managed by his group. "Our UDC projects are not only master metered [the electricity is paid by management and supposedly included in the rent], they have electric heat. Con Ed [the New York utility] billed us for $97,000 over an eight-month period; we'd had only a few people living there for three months. They bill on 'constant demand'—they charge you for what they have to have available at maximum load.

"In one project we took over from the contractors in the Bronx—you learn this sort of thing by experience—Con Ed hadn't got the meters in on time, or never got them in; anyway, they'd never billed the contractor. You take over, and you get a three-year back bill for $75,000. You go to the Public Service Commission, they say, 'Well, they were three years late—we'll give you three years to pay.' There's an $18,000 monthly rent roll on the building, no way we can pay an extra $25,000 a year back bills. You know what happens: we didn't pay the mortgage."

The fact is that it costs more these days to house people in an apartment than in a single-family home. Savings on heat and hot water are quickly gobbled up by the need to employ managers and maintenance people, to send out rent bills and collect the rent, to buy electricity for the elevator and the lights in the halls and lobby, to maintain the public areas, replace the dishwashers in the apartments, pay the costs of periodic painting and of redecorating vacant apartments for new tenants, absorb the losses from vacancies and delinquencies. Assessors almost everywhere see to it that apartment houses pay a higher proportion of their value in taxes than private homes do.

The worst problem is management. Abraham Goldfeld in 1937 described the desirable qualities in a manager: "Enthusiasm, cheerfulness, unselfishness, calmness, consistency, responsiveness, simplicity, frankness, impressiveness (mental rather than physical), firmness, tact, tolerance and patience, dignity and courtesy. A few qualities, such as intelligence, leadership, and breadth of vision are fundamental." Not surprisingly, it's hard to get such people, especially to manage moderate-income and low-income properties. "Most people looking for management jobs," says Jerome Steinbaum, whose S/K management firm handles 15,000 units across the country, "want to live in Beverly Hills in buildings with swimming pools and saunas. Our buildings are in ghettos, where they have to worry about vandals and about getting ripped off themselves. Most of our managers use the job as supplemental income—you can't afford to pay people for what they have to put up with."

Enoch Williams of New York's Housing Development Corp., an ex-Army officer, finds it hard to recruit because "the maintenance thing has negative connotations—people think of the old resident super with three rooms in the basement, who was drunk half the time, came up and sloshed out the halls once in a while. A lot of community folks figure that if you don't have a desk and a chair— well, that's it. Management is a complex thing to learn—there's a lot of dynamics, from community relations to accounting. The school system isn't meeting the needs. If I train minority people, they're in demand—somebody comes along and hangs that dollar in front of them, and I lose them.''

For middle- and upper-income apartment houses, says Agnes Nolan of the New York firm of Whitbread-Nolan, a lawyer turned real estate agent, ''you're looking for a guy who has the caliber of a good policeman, who can get ahead to a job like the executive level in the police department. You want somebody who knows what the problem is and how it can be handled—that's much better than the guy who picks up a phone to call the plumber at $28 an hour: anybody can do that.''

Robert Cunningham of HUD's property standards office remembers his days in the field, visiting FHA-insured apartment houses. ''If you found a building where the shrubs were trimmed and the lobby was shiny, the halls were clean, all the equipment was polished, you could be sure the first thing the tenants would tell you was that the manager was a son of a bitch. When you went into a building where the shrubs were scraggly and the lobby was run-down and the halls were dirty and there was a mess in the basement, you could be sure all the tenants would tell you what a nice guy the manager was.''

HUD has begun to require ''certification'' of managers in federally subsidized housing, but the training course that leads to the certificate is only one week long because the government couldn't persuade people to come to two-week programs. Sam Lefrak expresses the owner's point of view: ''When you've got your ass scattered out there to the four winds, it isn't simple.''

''Apartment house investment is a bad business,'' says Columbia's James Rouse. ''Even at one percent vacancy on the composite of all our apartments, we don't have a positive cash flow.'' And all the problems are going to get worse. The dollars and cents of low-skilled wages and local taxes have been among the most rapidly rising numbers in the economy, and these costs keep going up through the inevitable years of the real estate cycle when the market is suddenly saturated with new projects and rents cannot be raised on the older ones. With the interest clock running on construction loans at 10 percent a year and up, builders take a fearful bath on new apartment houses if they don't rent fast. "We lost $2 million in operating expense,'" Kansas City's Miller Nichols said of a large apartment house in the Country Club district, "before we filled it up."

In April 1977, *Fortune* magazine suggested that builders were being too conservative in their estimate of how many multifamily units they were going to build that year: "Vacancies may really be the lowest in a generation, auguring rent increases that are likely to make the numbers look better and better for builders in the next few quarters." Before new multifamily makes sense in most cities, however, rents are going to have to go up 35 percent, and that's probably not politically feasible. "People are funny," says Pat Crow of the Florida Apartment Association. "They'll pay increases on the car and the groceries, but not on the rent." And in an inflationary time, of course, rents must rise fast enough not just to maintain but to increase the profitability of the property or its value will decline.

Rental markets have always been troublesome to economists (not to mention tenants), because different places to live cannot be substituted for each other like, say, different brands of soap. The cost of moving is likely to absorb one or two years' savings on rent even if the alternative apartment is a much better buy. Particularly in dealing with tenants whose choices are restricted anyway—people whose work keeps them in places with limited housing facilities, poor people, blacks, in earlier days Jews, Poles, Irish—landlords have been able to indulge in gouging. When housing shortages made inferior accommodations easily rentable, landlords found that money saved on services, or on maintenance, drops right into profit. Herbert Gans reported that in Boston's West End residents of tenements without central heat often felt themselves fortunate: "Kitchen stoves freed the West Enders from dependence on the landlords and their often miserly thermostats." Presently, the tenants organize and petition the local political club for a redress of grievances.

Governments can control rents more effectively than other prices, because the buildings can't move away. Rent control is a subsidy to all tenants regardless of their need voted by governments at the expense of all landlords regardless of the profitability of their investments; and once it has been in place for a while the politicians find it all but impossible to remove. Always and everywhere, rent control eliminates private ownership of for-rent housing—between 1946 and 1976, the proportion of the British population that paid rent to a private landlord dropped from 51 percent to 13 percent. Frank Kristof of the Urban Development Corporation estimates that New York City's rent control has transferred no less than $22 billion from landlords to tenants in the years since World War II.

The landlords take revenge, reducing by all available means the quality of their product. The impact on the city is literally devastating, as taxes go unpaid, mortgages are defaulted, and maintenance is shunned. Little new is built without government subsidy to owners to replace the subsidy landlords must give to tenants under the law; in New York even the conversion of lofts to luxury apartments is so heavily subsidized by real estate tax exemptions that the city pays more of the cost than the developer does. George Schechter of the socialist-oriented United Housing Foundation, which has gone through hell recently in its efforts to sustain moderate-income co-ops, reports that after a generation of rent control New Yorkers think "services come from God, rent goes to the landlord, and deterioration . . . deterioration is something you get used to."

Fred Rose says that "mankind's second most passionate relationship is that between landlord and tenant." Rising rents make hate in one direction; rent control makes hate in the other direction. Louis Winnick of the Ford Foundation reports that when Grand Concourse and Ocean Parkway in New York began to change racial composition and the landlords decided they would have to take as much out as they could before abandoning the properties, they deliberately rented apartments to criminals and disorganized welfare families (welfare paid the rent), to get even with the tenants who had been robbing them, courtesy of the rent control law, for some years. It is a game everyone loses—most tragically, the arriving upwardly mobile blacks, to enjoy a filtering down of better housing than they have ever occupied before, and finding it in a condition of irreversible deterioration. In the final stage, landlords with criminal instincts arrange to have their buildings torched to collect the insurance. To say that these conditions are going to be cured by forcing the banks to make loans into these "neighborhoods" is worse than whistling in the wind.

The modern "life style" is incompatible with property maintenance in conditions of urban crowding: maintenance bills are higher than the investor's planning budget indicated, and resale value is likely to be lower. The glory days of rising market values are gone, because the true value of a housing investment is inescapably a function of durability. Today, delivery men and boys casually vandalize the common areas of an apartment house. Appliances installed by the landlord in a rental apartment are much less well kept than those the homeowner buys for himself. Cockroaches spread through an entire line of kitchens from the messiness of one. Nobody polices other people's children (some don't even police their own). Because heating and air conditioning are on the rent, people leave their windows open; if the electricity is in the rent, they also leave the lights on. And they are surprisingly casual about what they throw down the toilet if it's the landlord who must pay for snaking out the lines.

People in trouble can put the landlord in trouble too. "On one more than routinely desperate occasion," Robert Bendiner of the New York *Times* recalled in his semi-autobiographical account of the Depression, "[my mother] resorted to the extreme device of having one of us enlarge a hole in the bathroom ceiling and then irately demanding repairs before another dollar of rent should be forthcoming." Under rent control, there may even be long-term benefits from such activity, for the courts have held that the presence of such "violations," however caused, deprives a landlord of any rent increases the law might allow him. Departing tenants these days pour cement down the drains, leave the bathtub faucets flowing to flood the lower floors, and set fires on the way out. One way or another, all this—in addition to the rising costs of construction—must go on the rent bill. "Then you've got the rent strikes," says

Preston Martin, former president of the Federal Home Loan Bank Board; "all those lawyers telling people, 'Don't pay the rent—that'll get their attention.''

Conversion to cooperative or condominium ownership is the obvious solution, approved in theory from one end of the political spectrum to the other. (The English radical John Turner has set down as his Third Law of housing that "deficiencies and imperfections in *your* housing are infinitely more tolerable if they are your responsibility than if they are somebody else's.'') The drive in that direction is worldwide, in socialist as well as capitalist countries—in England, France, Finland, Hungary, and Singapore, to take a mixed bag, only the elderly or the means-tested impoverished can occupy new apartments on a rental basis. In New York, the epicenter of rental housing, tenant households decreased by 123,000 from 1970 to 1975, while owner households increased by 51,000. But conversion to tenant ownership has been slowed in the United States by the landlords' reluctance to take reasonable prices for unprofitable properties (they hope for some huge appreciation in urban land values, or some giant federal "community development" program to buy them out handsomely); the tenants' reluctance to invest savings in the purchase of an apartment that will probably cost more in monthly mortgage payments, taxes, and maintenance charges than it now costs in rent. Meanwhile, the newly built condominiums, which are *nobody's* responsibility, turn out to be such shoddy stuff that even an ill-informed public won't buy.

The cant line of the mid-1970s was that inflation has priced the American dream, the little house on a little lot in the suburbs, beyond the reach of the average family. The fact is that the inflation and the general loss of social discipline have priced the modest new apartment even further beyond the grasp of most of the people who might prefer to live in apartments. "New housing which is not federally subsidized,'' New York's Community Development office wrote in 1975, "is beyond the reach of 80 percent of the City's households.'' Rents go up enough to create hardship for tenants in existing buildings, but not enough to stimulate the construction of new ones. The result, despite the fearful experience of New York, is likely to be the spread of rent-control laws. Henry Schechter of the AFL-CIO reports he gets a dozen calls a week from counsel to city governments all over the country, asking him how one goes about writing a rent-control ordinance. Dilapidation follows.

Sam Lefrak blames the politicians: "Cat and dog, black and white. When I was a boy, it was like a family. The contractors would come for the payroll on Friday and my mother would feed them. My father had a rent table and people would come in and pay the rent, say hello. Izzy Waldbaum [now the proprietor of a large grocery chain] had a pushcart; my father built him a taxpayer. Barbra Streisand said when her aunt moved into a Lefrak building in Brooklyn she'd really made it. People took care of the buildings; there was a pride. Now they call you a slumlord, you're a stereotype, a second-class citizen. Well, this is a domestic industry, that stayed in the area. We paid the taxes that maintained the services. Now the legislators have legislated us into the ground. But just try to communicate this. . . ."

No doubt the politicians, the lawyers, the academics, and the editorial writers have been making the situation worse. "Landlords," New York's Community Development program reported acidly to the federal government in 1976, "need assurance . . . that the objective of government housing policy is the resuscitation of buildings for the benefit of their occupants, rather than the punishment of their owners for the benefit of the public conscience." But the situation itself is what's wrong. "We're going to have to socialize rental housing," says consultant Anthony Downs. "Who's going to want to own it?"

Part IV

PUBLIC POLICY

Chapter 15

Money, the Bankers,
and the Government

Real estate development is a function of the availability of money. And the availability of money is a function of the stupidity of lenders.

—A young builder in Atlanta, who upon reconsideration, having decided that he wished to remain in this business, requested the privilege of remaining

ANONYMOUS

The very first fact about housing is that it is built and bought with borrowed money. "The successful builder," says developer Bruce Rozet, "is the man who comes around in his Cadillac when the banks are handing out the money, and when they aren't you never see him at all." If lenders "redline" a neighborhood, the price of existing housing in that neighborhood drops like a stone, and nobody builds there at all. Builders cannot get construction loans unless the banks are assured of the presence of a "take-out" in the form of a mortgage from a long-term lender. Possible customers for a new house where a mortgage *is* available can't buy it unless there is a lender ready to write a mortgage for the family that wants to buy their old house.

Even more impressive is what happens when money rushes into an area—an Adams-Morgan in Washington, a Pacific Heights in San Francisco, a North End in Boston, le Marais in Paris. Home

353

prices can easily double in a year, fed by the sudden confidence of the banks that the property pledged as security for the loan will be worth still more in a time to come. A dramatic illustration of the effect of a change in lenders' requirements came in New York City apartment cooperatives in 1977, when the banks and savings banks reduced their downpayment requirement for the purchase of co-op shares from 25 percent to 15 percent. Suddenly a purchaser with a $10,000 downpayment could pay $66,000 instead of $40,000 for an apartment. The monthly payment at 9 percent interest rose from $304 to $470 (atop whatever maintenance charges were involved in living in the co-op), but here the Law of Inflationary Expectations took care of people's worries: after all, at 6 percent inflation, which the New York *Times* kept saying was "the underlying rate," $470 today would be the equivalent of only $345 in five years, $253 in ten years. Meanwhile, the inflation-fed increase in the market value of the apartment would, it says here, provide a nice nest egg for the family, even in depreciated dollars. "We can do without a few things for a while, John. . . ." So prices rose in a year by 40 percent, pretty much across the board.

The episode that drew national attention in 1976–77 was the land boom in Orange County, California, the southern suburbs of Los Angeles. In December 1975, a survey by the Real Estate Research Corporation of Southern California found 31 percent of the single-family homes in Orange County valued at more than $70,000; by June 1977, eighteen months later, the proportion valued at more than $70,000 had jumped to 87 percent. The *average* price of an Orange County home advertised in the Los Angeles *Times* in summer 1976 was $91,000, and the U. S. Commerce Department reported that the average actual sale of an existing home in the county in 1976 was for $73,000. The right to purchase newly built homes at $90,000 and up—townhouses, mostly, with party walls and maybe 1,500 square feet of space—had been made subject to lotteries at the Irvine Ranch and Mission Viejo, because there was no other fair way to allocate homes among the crowds that gathered as much as a week before new tracts were to be opened. A Home Loan Bank examiner documented the case of an unbuilt 3½-bedroom detached house that sold in November for $104,000, and was offered for resale in January, still unbuilt, at $128,000.

There were, of course, a number of reasons for the boom. The country still tilts southwest. Life in Los Angeles proper had begun to look increasingly unattractive, especially on the school integration front. According to a special survey for the Home Loan Bank, the average family income in the County had risen 50 percent, from $12,000 to $18,000 between 1970 and 1975. The sulphurous smell of land use control that might forbid ordinary people to acquire their own homes was rising in the air over the State Capitol in Sacramento; indeed, a number of usually level-headed Californians were coming to the conclusion that the lunatics had taken over the asylum. (State Architect Sim Van Der Ryn, appointed to the post by Governor Jerry Brown, told an interviewer that he had decided to accept because "Jerry's . . . extremely Buddhist in quite a beautiful way.") A multiplication of environmental controls and local requirements for roads, utilities and schools had already doubled and tripled builders' costs for land development; and large-tract work with its possible economies had in effect been ruled out by a new state law requiring builders to make provision for low-income families in any tract with more than fifty units.

So the boom starts with lower-priced homes, because there can be no new construction at the lower levels to hold down the price. Houses built in the mid-1960s for $15,000 are sold for $35,000. After satisfying the old mortgage, the sellers have about $25,000 that under the tax laws they can keep in its entirety only if they invest in another house. Assuming a "conservative" 25 percent downpayment, they can in theory qualify for a $100,000 house. But most lenders even in Orange County were reluctant to lend any family more than twice its annual income. Assuming the $18,000 annual income that was the Orange County average, the family could borrow $36,000—and could buy a $60,000 house.

The seller of that $60,000 house, built in the late 1960s, had paid $27,500 for it. He has a better than average income—say, $22,000. After satisfying the mortgage, he has $40,000 in cash; so *he* can buy a house at $84,000. Once more: *that* seller had paid $35,000; he now has $60,000 in cash, and an income of $25,000—and he can buy that new townhouse with the great life style at the Irvine Ranch, for $110,000.

This is not theory; it is what happened. "Most purchasers in

Orange County,'' said a Home Loan Bank official, ''are buying their third, fourth, fifth—even sixth or seventh—home.'' The conventional statistical approach to the house a family can afford says that two- to two-and-a-half times annual earnings is the maximum price at today's interest rates. But once lenders start defining what a family can buy in terms of the loan size rather than the house price, the participants in a market like this can easily purchase a house that costs more than four times their annual income. The problem is that this loop is not closed: 18,000 new homes were added in 1977, and Home Loan Bank figures showed less than a quarter of them for sale at $60,000 or less, more than half clustered in the $80–$90,000 range. Not many newcomers can afford these homes, which don't *look* like $80,000 worth to anyone from any other part of the country anyway. (Steven O'Heron and David Parry of the Home Loan Bank describe them as '' 'average' houses at 'luxury' prices.'') So the market becomes dependent on pressure from below, from people pouring into the County to buy its least desirable homes.

And on speculators. The real estate expert of one of California's largest banks said that 42 percent of all the new homes being bought in Orange County early in 1977 were being put out to rent by purchasers who planned to dispose of them in two or three years for double their 1977 price. Economist David Parry at the Home Loan Bank thought the percentage was high—that up to 20 percent of the homes bought through the lotteries at Irvine Ranch and Mission Viejo might indeed have gone on the rental market, perhaps more at Avco's heavily advertised Laguna Beach development. And in a few areas, brokers themselves had been buying up existing properties, using their commission as downpayment, and renting on short leases for resale within a year. But over the County as a whole, the Home Loan Bank believed that no more than 12–15 percent of purchases wound up on the rental market.

The rental market, of course, continues to be dependent on the occupants' earnings, and moved as might be expected in a community where per capita income went up 50 percent—from an average of about $160 in 1970 to about $250 in 1977 for a two-bedroom apartment. Rents over $400 were hard to peddle; rents over $500 were all but impossible to peddle, especially with the glut of fancy homes for rent. Even on new houses that take no maintenance and replacement expense, the costs of carrying property really cannot be less than 10 percent of the sales price per year at today's interest rates. In 1977, then, speculators in Orange County properties were losing $3,000 to $6,000 a year on what they bought. They were still pretty cheerful—several professors of economics encountered in Los Angeles early in 1977 were discussing how they would subsequently invest their gains from Orange County property—but it was the sort of cheerfulness that dissolves like sodium bicarbonate the moment the first upset occurs. The first sales at reduced prices in Orange County, and they will come, probably in 1978, will be like a cry of "Fire!" in a crowded theatre.

In 1976, the Home Loan Bank estimated, there were 97,000 sales of houses in Orange County, 17,000 of them new. "It is interesting to note," O'Heron and Parry wrote, assuming a worst case in which 20 percent of those purchases had been put out to rent, "that, according to this calculation, the housing stock purchased by speculators and investors would be equal to 112 percent of new housing sold in Orange County during 1976, thus representing more than a year's supply of new home demand." This is a little tricky—an overhang of new and old together affects the market for more than just the new—but it's an impressive figure nonetheless.

In summer 1977 the blast-off of prices in Orange County was slowed by the failure of the Governor and the state legislature to agree on a program for real estate tax reduction—real estate taxes being an effective brake on acceleration in this market—but some "tax relief" is certain to be granted in a situation where the state has a budget surplus and middle-income homeowners are really suffering. Meanwhile, speculators can dream about eventual passage of a widely supported 1977 legislative proposal that would be the most restrictive land development proposal yet, convincing even the uncommitted that if they didn't get a house soon they never will.

When the investigators pick over the ruins, one hopes they will direct their most penetrating attention to the role of the lenders and the appraisers who work for the lenders. Aggressive lending by the Los Angeles banks and S&Ls was the initial stimulus and long-term support of this boom: if they had not been willing to accept what were quite clearly transient market prices as the true "value" of these properties, the fiasco could not have occurred. The leading institution, as far as can be learned, was Security Pacific National Bank, the tenth largest in America, which came to mid-1975 with a worrisome portfolio of bad loans, reduced demand from commercial and industrial borrowers, and a felt need to boost income figures by making a volume of loans at high interest rates. Homebuilding in 1974–75 had dropped below the trend lines in Orange County; the demand for homes was clearly there. It was the kind of demand that would push prices if lenders would validate them. Security Pacific wrote a lot of mortgage business.

Meanwhile, money was pouring into the California S&Ls at an unprecedented rate—and at interest rates on savings, especially the longer term certificates, that were greater than the interest the S&L could earn on federal government paper. They had to put the money out. "When lenders are looking for loans," says Stuart Davis, chairman of the board of Great Western, a $7-billion S&L, "there is a tendency for home prices to rise." You bet. "The appraisers must show comparables, but after a while that doesn't work as a check. Normally, the cost of reproducing the house is the upper limit, supposedly less depreciation; but inflation has outrun depreciation."

The fact was that instead of the cost of new homes establishing the ceiling for old homes, which is the norm and the pious appraiser's lodestone, the escalating market for old homes was pushing the price of the new ones. As the houses themselves could not possibly have appreciated so greatly as the appraisers were saying, the "value" had to be in the land: this was a *land* boom. At the Irvine Ranch, where land was carried at a book value of $9,500 an acre, sales to builders in 1977 were actually made at a price of $95,000 an acre. (This is why there was the much publicized fight between Mobil Oil and an investment syndicate including the Irvine heir to purchase the property from the foundation that controlled it.) The builders could pay this price because customers were buying the houses in advance of construction on mortgages the banks had already approved.

"Prices are our number-one concern," Davis of Great Western

said early in 1977. "We talk about it at the board level all the time." But he could find defenses. The Great Western borrowers were occupying the houses they bought: his lending officers, he said, weren't making loans to speculators. (In a quarterly report some months later, Great Western conceded that as much as 25 percent of its mortgage loans in early 1977 had in fact gone to speculators who then put the houses up for rent.) Downpayment requirements were high: except for occasional FHA and VA mortgages written as a public service and instantly sold on secondary markets. Great Western was demanding 15 to 25 percent equity (no great problem, of course, as the Orange County boom actually operated). If families were "qualifying" on the basis of incomes for both husband and wife, Davis said that is acceptable now: "The wife is far less likely to have children if she doesn't want to have children. She *will* continue to work—they look at their employers these days and they say, 'I'm going to be vice-president.' " And the fact was that loan delinquencies were virtually invisible: in Great Western's portfolio as a whole, only about one-tenth of one percent of the loans were more than thirty days behind in their payments.

In January 1977, the net increase in deposits at Great Western was running almost $3 million *a day,* a third to a half of it in high-interest long-term savings certificates. To hold the money in Treasury bills at a less-than-5 percent interest rate would mean that each day's receipts would create annual operating losses of $100,000. Great Western found mortgages to write.

From offices in San Francisco, the regional Home Loan Bank watched carefully, like a child watching a cobra rise from its basket. While they watched, speculation moved north to the Bay Area. "I could make forty percent a year on my house since I bought it four years ago," said the Bank's Les Coplan. "I've been in housing forty years and I can't remember anything like it." New construction for California as a whole jumped to a rate of 290,000 units a year—one-sixth of all the home building in the country. The Bank of America announced that its mortgage and construction lending for the first quarter of 1977 was double what it had been in the first quarter of 1976. Labor shortages in the construction trades were creating chaotic conditions on sites, with subcontractors sending crews wherever that day's price was highest, and carpenters demanding $100 under the table as a bonus for showing up.

In May, the Home Loan Bank finally made a public statement, warning the S&Ls not to bankroll speculation, and announcing an increase in the interest rate the Bank would charge any S&Ls that found themselves in need of help from their federal Mother. The Bank reports that this warning had a salutary effect, and that down-payment requirements (especially where there is a smell of speculation) have gone to 30 and 40 percent at a number of Orange County lenders. Officially, everybody is predicting a "soft landing." Given the stratospheric reaches of the flight and the unsoundness of the aircraft, it seems unlikely. Builders themselves, steadily increasing their Orange County output, seem relatively unconcerned. Informed of this observer's opinion that a drop of 20 percent in Orange County home prices could be expected in the near future, a developer gave a sigh of relief: "That's okay," he said. "I can cut my prices twenty percent and still make money."

For the homeowners and the lenders—and, obviously, the speculators—the bust may be more damaging.

Homebuilding, home buying, home selling are all totally dependent on the availability of credit. We live in an age when the line between credit and money has been erased—the deposit in the bank is identically equal to the coin in your pocket—and (not coincidentally) the great menace to social stability is inflation. Even in socialist countries (where the menace is not inflation but shortages and black markets), governments today can control the creation of money only by manipulating the availability of credit. This can be done either by controlling the assets of lending institutions through direct credit allocation or by controlling the liabilities of leading institutions, mostly but not entirely their deposits. Control by credit allocation turns out to be inefficient because it fails to reward intelligence and requires the government to know much more than a government *can* know. Control by limiting liabilities means allocation of credit by price—interest rates—and in a time of rapid communication and neurotic reactions it places upon lenders who are by no means equipped to handle such problems an obligation to make extraordinarily complicated judgments about the relationship of risks and rewards at various terms of loans for various purposes. What happens usually is that the borrower willing and apparently able to pay high interest rates gets the money, and others do not.

Housing loans are long-term loans, which means their cost to the borrower is grossly affected by changes in interest rates. If the interest on a $5,000 three-year car loan goes from 9 percent to 12 percent, the monthly repayment rises from $159 to $166, an increase of less than 5 percent. If the interest on a $30,000 thirty-year mortgage goes from 6 to 9 percent, the monthly repayment rises from $180 to $241, an increase of more than 33 percent. When demand for loans is high and interest rates are rising, borrowers for the short term can always, almost painlessly, outbid borrowers for housing. Funds flow from the lending institutions that make housing loans to the lending institutions that make short-term loans.

In theory, there are "countercyclical" values in a system whereby money flows away from housing toward the peak of the business cycle when industry is optimistic about investment, then back toward housing when loan demand elsewhere diminishes. The theory is a little chancy in a modern society with its long lead-times and bureaucratic delays, because the impact of money shortages for housing may not actually be felt until everything else is going sour too. (Moreover, as noted in Orange County, the flood of money into housing when other demands diminish can be crazily selective, fueling a speculation that hits people where they live.) In any event, governments make commitments to housing—taking the job of Secretary of HUD, Mrs. Harris said she had a pledge from President Carter *not* to use housing as a countercyclical tool. One of the tasks assumed by modern government has been to develop special institutions that can continue to supply adequate amounts of credit to housing during periods when money is tight.

In the United States we have a truly bewildering variety of such institutions, some of them less than ten years old, but before attempting to sort them out it will be useful to look at the housing finance system of another country, where one very simple institution has in fact—and unfortunately—been able to keep money flowing to housing through the most severe credit crunch. Why this happened,

and why it was unfortunate, offer a background for understanding
our own system, its successes and failures.

The country is Denmark, and the institution is the Mortgage
Credit Fund. There are seven of these institutions, chartered by the
state, each supervised by its own board of directors. At the newest
of them, Byggeriets Realkreditfond (BRF), one of only two that finance housing nationwide, the board is selected from three
groups—the banks and savings banks, the bondholders (including
the insurance companies, pension funds and the national bank), and
the community of borrowers (divided fifty-fifty between the nonprofit government-subsidized Housing Associations, which also get
their money through these institutions, and the entirely private borrowers).

The function of the Mortgage Fund is to take mortgages on property, especially residential property, and issue bonds of corre-
sponding value, guaranteed by the Fund, which can be sold on the
Bourse to raise the actual cash. The money for mortgages, in other
words, comes directly from the financial market, without mediation
by savings banks or insurance companies, though they may participate as purchasers of the bonds. Housing fishes for its money exclusively in the pool that also supplies funds for government bonds,
agriculture and industry. Asked how this unique system began, a
Danish banker said apologetically (Danes are marvelous; they are
the despair of the other Scandinavians), "Well, my country
achieved constitutional democracy only in 1849, and one of the first
things the new Parliament wanted to do was establish a scheme for
home financing. They sent a committee to Silesia, which was supposed to have the best system. Regrettably, the committee misun-
derstood what it saw."

BRF managing director Henning Axel Nielsen, an elegant, helpful, blond young man with excellent English, explains his operation
in the context of housing finance: "You want to build a house. You
go to an architect or probably a builder. Your wife wants to change
this or that, different tiles in the bathroom, and so everything gets
arranged. So you go to your bank or savings bank. Most people
have already bought their piece of land, have held it for three to five
years—that can be mortgaged too. The bank makes a short-term

loan and you go ahead with craftsmen on the project. The bank pays off normally in four installments—when the foundation is finished, when the tile-work or the bricklaying is finished, when the house is closed in, and a final payment. The bank has a surveyor who goes and sees whether the work is done according to the drawings and verifies the quality of the work.

"One day when the house has been finished, you approach us. You have a house, you want to have it mortgaged. We get documentation that you own it, from the locality that it has been legally built, and the architectural drawings. We send an appraiser—we have seventy-five appraisers under contract, they are themselves building masters or engineers. They have technical expertise and know how a house should be built, and what the prices should be for a given construction. And they are local people, they know the local market. And they send customers to us, a quality we put some weight on because we are in business. So the appraiser sends a report to us, and suggests a market value. We use that as a *help* to find our own market value.

"The law tells us we must look at building costs—not the *actual* costs, because we don't in fact have to know what your actual costs were. We did ask about actual costs until July 1976, but when you ask an owner you get very poor information. He expects that if he gives you a very high cost you will grant a bigger loan, so we used to get a lot of nonsense, we were always saying, 'No building should cost that much!' We decide for ourselves on the total of building cost and price of land for your house.

"Say that adds up to 400,000 Kroner ($65,000) in cash money. But we are not going to finance by cash; we finance by bonds with an interest rate of ten percent, which you will sell at a discount on the market. Let's say my type of bond at ten percent has a market value of 70. [i.e., the interest rate on the market for a forty-year bond is 13.5 percent] So I add thirty percent to the 400,000, and that's my value for your property. I say the price should be 525,000 Kroner, and I require a twenty percent downpayment. So on a 400,000 Kroner house, the value is 525,000 Kroner, the loan is 420,000 Kroner, the market gives you 295,000 Kroner for our bond—and you pay 105,000 Kroner cash, which is twenty percent of the value." In other words, an allowance is made for inflation in the valuation of the property to match the market discount on the bonds.

Nielsen continues: "It is a very liberal thing—if you can produce the property you have a right to the loan. In no other country is there such free access to the capital market. I don't know the borrower's age or income or family status. We never ask any questions about the person—but we know *a lot* about the property. Default is almost nothing. Even in these two very poor years [1975–76], we have a total loss of 3.5 million Kroner out of 8.4 billion." That's less than one-twentieth of one percent, and the explanation that the banks have done prescreening does not entirely stand up, because since 1970 the Mortgage Funds, relieving the financial strain on the banks, have been taking over an increasing share of the construction loan function, issuing bonds on the basis of plans rather than completed buildings.

The Funds charge an origination fee of 1.5 percent to issue the bonds (*much* less than banks charge elsewhere in Europe) and service fees ranging from .3 to .5 percent a year. Bonds are redeemed twice a year at random (by the use of a computer-generated number table specially prepared at the local equivalent of M.I.T.). Mortgages are automatically assumed by the buyer when the property is sold; people who wish to prepay their mortgage can do so by purchasing on the market a bond with the same maturity date. Thanks to the trivial loan losses, the Funds are profitable, except that at BRF none of the profits is ever paid out—all earnings go to beefing up the reserves.

On Nielsen's sample figures, the annual repayment obligation of his borrower (semi-annual interest and amortization on a forty-year "annuity" loan) will run 53,000 Kroner, or more than $700 a month. Holy smoke! you say. How many Danes can afford that? Well, Danish income tax rates go to 60 percent on the marginal dollar at about $7,000 a year income, and the interest on a Mortgage Fund bond, like any other interest, is deductible from income for tax purposes. For a family with as little taxable income as $15,000, then, the effective post-tax cost of the loan (while still an enormous, barely supportable burden) is less than $300 a month. And there are no real estate taxes in Denmark. A little more than half the population had incomes that size in 1976. As most Danes like most people elsewhere want to own their homes, the market for single-family housing in Denmark is big—70 percent of all home-building in recent years.

This sytem works like a watch. If interest rates go up, and discounts grow steeper, the numbers are merely evidence of inflationary expectations anyway; and in an era of inflationary expectations people are willing to commit themselves to level future payments way beyond what might seem sensible today. The mortgage and bond sale can be jiggered so that all the discount appears as tax-deductible interest payments, and the government in effect pays 60 percent of the extra cost. The faster the inflation, the greater the demand for loans, the more attractive it becomes for the middle class to invest and borrow for housing purposes.

The nonprofit multifamily Housing Associations suffer, because they have no tax benefits from their interest payments. The government then takes care of this problem two ways—first, with a direct construction subsidy (in 1977, 23 percent of the cost of the building); second, with aid to tenants to help them pay the rent in new Housing Association apartments. Even after the aid, however, workers may have to put up 30 percent of their pre-tax income to live in one of the new Housing Association developments financed at high interest rates—while residents of the older developments, financed some years back lower interest rates, pay only 10 percent of their income for entirely comparable accommodations. And the middle class, once quite willing to rent Housing Association apartments, bleeds out to owner-occupied units to reap the tax benefits.

Industry suffers, because the corporate income tax rate in Denmark is only 38 percent and the banks will not lend to business for less than the yield on Mortgage Fund bonds. Thus the market works to make it possible for homeowners to borrow at a net of 5.2 percent interest on a 13 percent market rate, while corporations must pay a net of 8.1 percent. And the government anxiously tries to find ways to increase industrial investment and output.

Inflation and the sour politics that accompany it are institutionalized in the society as larger and larger numbers of middle-class people are placed in a position where *only* inflation and the chance to pay yesterday's debts in depreciated Kroner can get them out of the hole of extreme housing costs. And even a socialist government reins in the production of workers' apartments because the burden of subsidizing ever increasing numbers of rentals becomes too heavy on the budget.

In 1973, the government froze the Mortgage Fund appraisals to restrict the issuance of bonds for more than current values, and homebuilding collapsed. The government backtracked quickly, and in fact sent the Central Bank and the state pension funds into the market to buy bonds (their purchases were only 17 percent of the market in 1973 and rose to 41 percent in 1974). In December 1975, the government moved to credit allocation by means of an agreement with the Mortgage Funds to lend no more than would be necessary to build 40,000 housing units in 1976. A report from the Mortgage Funds to the European Congress of Building Societies in September of that year concluded wearily with the comment that "the financial system by itself cannot be used as a tool of control. The building activity is depending on a complex of policies in respect of housing, construction, taxation, etc. Therefore a plurality of means should be put to use."

For the first third of this century, home finance in the United States stood on three legs: the owner's personal resources (60 percent of all nonfarm owner-occupied homes were not mortgaged at all as late as 1920), the individual lender (who held 42 percent of the mortgages outstanding in that year) and the savings institution (which held another 40 percent). Real estate lending of any kind has always been regarded as a trap for commercial banks, which might need their money back fast to repay depositors and could not get such money out of mortgages at need—but would always be tempted to make "accommodation" loans on property because the longer term of such loans and their illiquidity meant that they paid higher interest rates and seemed to yield higher profits. Bankers were not, are not, never will be, and probably never *should* be very clever. Until 1926, therefore, nationally chartered commercial banks were not permitted to write residential real estate loans at all.

As the credit union was started because banks would not make small unsecured loans to ordinary people (how long ago it seems!), the "savings and loan" (often "buildings and loan") association was begun in the later nineteenth century to give people a way to channel their savings toward home loans for their friends, their neighbors—and themselves. Meanwhile, in the industrial states where commercial banks were not interested in taking savings accounts on which they would have to pay interest (in farm states the banks had always accepted such deposits, though the interest they paid was often very low), "mutual savings banks" had been formed essentially to provide a place where the money ordinary people saved for emergencies or for anticipated big expenses could be kept safe and employed for whatever purposes guaranteed safe earnings. Such savings banks were not primarily housing lenders—in 1920, the S&Ls had all their deposits and then some invested in residential morgages, while the savings banks used only a third of their deposits for that purpose. But these savings were not like commercial bank deposits; they were long-term money that would not be running out of the bank (which could legally delay repayments without declaring insolvency, if necessary); so savings banks money, like S&Ls money, could be lured into mortgages. By the end of the 1920s, the three "depository" institutions—banks, savings banks, and S&Ls—held about half the total residential mortgage indebtedness in America.

During the 1920s, incidentally, there was also a brief, wild fling at the sort of mortgage credit we met in Denmark, with the bonds issued by private "guaranty companies" rather than by state-chartered funds. Such institutions had been around in New York State since the 1870's; now, suddenly, they boomed. Sometimes on purpose, sometimes because nobody knew what he was doing, the guaranty companies became stock-market swindles; they bought the mortgages the banks didn't want, built no reserves, and simply disappeared, with all but total loss to the bond buyers, when Wall Street laid its egg.

But all the mortgage lenders had a terrible time in the Great Depression; only the very conservatively operated mutual savings banks, with less than half their assets in mortgages, held up. The traditional mortgage instrument of the 1920s had been a fairly short-term loan, three to five years, with a balance left to be refinanced at the end of the term. The lender was safe because the homeowner's downpayment was (or seemed to be) very high: first mortgages were rarely written for more than half the market value of the house. (Much of that other half might have been raised on a second mortgage, but that was supposed to be something the bank didn't have to worry about.) Refinancing was a good thing for lenders, who got a fee each time it happened, but it wasn't common—mostly the mortgages just ran on, month to month, callable at the option of the bank, which could foreclose on the property if the borrower didn't pay when called. This was the standard plot of American melodrama: the lecherous moneyman who demanded the repayment of the mortgage or the use of the pretty daughter. It sounds worse than it was: mortgage loans were overwhelmingly local loans, from a savings association to its own members, and people knew each other.

In the early 1930s, the system collapsed. Farmers couldn't sell their produce; workers lost their jobs or if they kept them took pay cuts, making the fixed burden of the mortgage payment much harder to carry. People had to eat their savings; but the associations had no way to get them their money. As the façades fell away, numerous distasteful spectacles of incompetence and corruption in banks and savings associations were revealed—but the crisis would have come even if the institutions had all been managed by the eleven honest Apostles.

If ever there was a time for government to Do Something, this was it, and the Hoover Administration in 1932 established a Federal Home Loan Bank Board and twelve regional Banks to advance money on the security of mortgages held by savings institutions. The Banks were funded with $125 million from the Treasury, plus payments from the associations that wished to be members, amounting to one percent of their total mortgage portfolio. The author of the law anticipated that the new Home Loan Banks would also (like the Danish mortgage funds) sell bonds backed by mortgages. The most radical section of the law—repealed less than a year later by the first New Deal Congress—required the Federal Home Loan Bank in a district to make mortgage refinancing loans itself to anyone who could prove he had no other way to get the money.

None of this worked very well—nobody would buy the bonds, which were *not* government obligations—and less than a year later the Roosevelt Administration raised the ante enormously. A Home Owners Loan Corporation was formed under the aegis of the Home Loan Bank Board, with $200 million from the Treasury, $200 million from the Reconstruction Finance Corporation, and the power to issue $2 billion in bonds that *were* guaranteed by the federal government (and exempt from federal, state, and local taxes) for the purpose of refinancing mortgages for homeowners. The new HOLC mortgages were to be self-amortizing over a period of fifteen years. At its peak in 1935, after another billion dollars of borrowing authority had been added, HOLC held $2.9 billion in home mortgages. It phased out in 1951 with a bookkeeping profit to the government of $14 million.

The Home Owners Loan Act of 1933 also authorized the Bank Board to charter, supervise, and when necessary lend money to federal savings and loan associations (previously all S&Ls had been state chartered). The deposits in such institutions were not in those days called "deposits"; they were "shares," and what the member "earned" on them was "dividends," not "interest." Those dividends and the S&Ls themselves were declared exempt from federal taxes (the exemption was removed in 1942) and protected from any state and local taxes greater than the taxes imposed on other thrift institutions. They could lend money only to their own members on the security of their shares or to homeowners on the security of mortgages; and their lending authority was essentially restricted to the geographical area within fifty miles of their home office. There is a long and continuing history of legislative changes in the authority of the S&Ls.

With this Act, the United States entered a second third of this century, in which government activity in housing would be competent (i.e., the government knew what it was doing) and mostly, though decreasingly, fruitful. The centerpiece legislation, passed in 1934, was the National Housing Act, which established the Federal Housing Administration and four programs:

(1) Title I, authorizing FHA to insure loans made by virtually any institutional lender for the purpose of repairing or improving residential real estate.

(2) Section 203, authorizing FHA to insure self-amortizing mortgages running twenty years or less, at interest rates of 6 percent or less, in amounts as great as 80 percent of the appraised value of one-to-four family homes. (Later amendments lengthened the mortgage term and provided for 90 percent and finally 95 percent financing.) A fee was to be charged for the insurance, and no mortgage was to be insured unless FHA determined that the project was "economically sound."

(3) Section 207, authorizing insurance of mortgages on multifamily projects for low-income people, to be built by state or local governments or limited-divided corporations, and subsequently supervised and rent-controlled by FHA.

(4) Title III, authorizing the charter of "national mortgage associations" to provide an organized secondary market—that is, a facility to buy and sell mortgages already written, where previous lenders could get their money out and new investors could buy the piece of paper they need.

These programs did not catch on quickly. (In fact, no entrepreneur could be found to start a mortgage association, and in 1938 FHA itself had to form the first Federal National Mortgage Association—FNMA, soon called "Fannie Mae.") The savings and loan associations, having been bailed out by HOLC and started on a separate track with access to the Home Loan Bank when they needed money, resented the "interference" of FHA, its appraisers, its insurance fee, and its paperwork. The commercial banks simply weren't interested. Carl Bimson, later president of the American Bankers Association, had just come down from Colorado to help his brother run the Valley National Bank in Phoenix; he remembers that theirs was the only bank in Arizona willing to write insured mortgages. "The others were already sitting on a lot of mortgage loans. Rocky Mountain S&L had gone broke. . . . When we ran out of money and went to the Phoenix insurance companies, they said the FHA program was worthless—Congress couldn't insure anything beyond its own term, any more than a local city council could bind the next city council. My brother and I went to California and talked to the head of Occidental Insurance, and he made a commitment to buy a million dollars of our mortgages for ninety eight percent. We took a two-dollar loss, but we kept an annual one-half of one percent service charge, and kept the contact with our customers; and we had another million dollars to lend.''

That was the way the FHA system was supposed to work: by appraising properties and approving neighborhoods, FHA would give possible mortgage holders—the insurance companies, the eastern savings banks—a feeling of security about loans on properties in places they knew nothing about and couldn't investigate. Little Valley National became one of the largest producers of FHA-insured paper in the country, and Bimson was called to Washington to help explain to his fellow bankers how they could make the programs work. Progress was slow: it wasn't until 1939 that FHA-insured mortgages accounted for as much as 10 percent of the stock of mortgages on one-to-four-family structures.

Taken together, the federally supported S&Ls and the four basic programs were as brilliant a piece of social invention as American government has ever produced. They were not subsidy programs—indeed, the FHA insurance funds were profitable to the government until the Section 235 and 236 fiascos of the early 1970s put them under water. They sought to influence the allocation of credit to housing not by putting money in bankers' or builders' pockets but by improving the traditional lending instruments to make them national instead of local, less risky and more negotiable—and thus more attractive to new lenders, less expensive for both lender and borrower. People who could scarcely have hoped to own their home before were empowered to do so, at a cost little if at all higher than the average monthly rent. The federal program assumed the possibility of a healthy market in housing finance, and did not seek to achieve public purposes by subverting private judgment. Lenders who made mistakes would still suffer costs (the insurance paid off fairly slowly, and in bonds rather than cash), but they wouldn't suffer ruin.

The nation came out of the war with the banks full of the deposits of people whose consumption had been restrained by wartime shortages, and full of government paper generated by the deficits of the war budgets. One key program was added: the Veterans Administration guarantee, which covered the first half of any losses a bank

might suffer on a mortgage loan to a veteran (up to a certain maximum per house), and thus protected lenders against any but the
most egregious stupidity even on loans for houses purchased with
no down-payment at all. And interest rates were reduced: with government bonds pegged at 2½ percent and quality corporate bonds
selling to yield less than 3 percent, Congress did not wish to see veterans or borrowers in FHA-insured programs compelled to pay
more than 4 percent. The country was eager for housing: residential
mortgage debt outstanding soared from $25 billion in 1945 to $55
billion in 1950 ($22 billion FHA or VA) to $102 billion in 1955
($43 billion FHA or VA). It was not understood how safe these
mortgages really were—"Nearly everyone is agreed," Eric Larrabee wrote of Levitt's $8,000 house in 1948, "that today's housing
values are inflated, and that the collapse will have to come some
day"—but even the pessimists agreed that our veterans had run
risks, entitling them now to others' risks on their behalf.

In the Northeast, the states released their mutual savings banks
from the previous obligation to restrict their mortgage lending to
properties within the state (some of them now wish they hadn't done
so); and across the country the life insurance companies increased
the proportion of their assets that went into residential mortgages.
Some of the insurance companies established extensive field operations to investigate properties—which meant, housing economist
Saul Klaman noted shrewdly in 1960 (some years before he became
president of the National Association of Mutual Savings Banks),
that "regardless of yield . . . these companies are committed to at
least minimum mortgage programs." But the banks and the smaller
insurance companies grew increasingly dependent on a new institution—the mortgage banker. More broker, really, than banker, these
companies would initiate FHA and VA mortgages with funds they
had borrowed from a commercial bank, make up a package large
enough to interest an institutional investor, and then sell the package, retaining the origination fee and a servicing function that
yielded continuing fees. Some commercial banks went into this
business for themselves (the First National Bank of Boston was the
leader), establishing "mortgage warehouse" divisions that supplied
investments conveniently to the mutual savings banks and life insurance companies.

This assemblage of institutions was too successful: the country needed housing, but it needed other things too. The Korean War generated renewed federal deficits. With the Federal Reserve System committed to purchasing low-interest government bonds from the banks and savings institutions at par, it was too easy for the credit-generators to make unlimited quantities of new money. In 1951, a "treaty" between the White House and the Fed freed the central bank from the obligation to maintain the price of government paper, and to prevent interest rates from rising too fast the Fed was given direct control over the terms of credit creation. Downpayments were required on VA mortgages, and considerably higher down payments for all but the cheapest houses were imposed on FHA-insured mortgages.

As the Korean War wound down and inflationary pressures ebbed, this authority was phased out, and the nation returned to credit allocation by price—except that Congress was still protecting the veterans from what it considered excessive interest rates on mortgage loans. FHA ceilings were held at 4½ percent, VA ceilings at 4 percent. The result was that no money was available for VA mortgages; to remedy that, in an action unconsciously pregnant with a new world, Congress authorized FNMA to supplement its "secondary market" function by supporting the market for VA loans, buying them at par to keep the veterans' interest costs low.

This was time-consuming, and ultimately inadequate, because Fannie Mae's resources were limited and because market rates soon rose above the FHA limits too. Experience with home loans in what was now ten postwar years had encouraged savings institutions to proceed with uninsured low downpayment long-term self-amortizing mortgages, and the money began an outflow from government-guaranteed paper to higher-interest "conventional" mortgages. In the middle 1950s, the move away from government guarantees was accelerated by the development of private mortgage insurance companies, charging lower fees and requiring infinitely less paperwork. To keep the federal programs functioning at all, their administrators authorized a system of "points" that would make government-insured mortgages competitive.

Recall the Danish system: to borrow 400,000 Kroner on a 10 percent bond when the market rate on bonds was 13.5 percent, the

Mortgage Fund had to sell a bond with a face value of 525,000 Kroner. What happened in America was the same process, except that it occurred inside the bank. Let us do it with numbers. Assume a $12,000 house to be bought with a $600 down payment and a twenty-five-year mortgage for $11,400. The federal government says the maximum interest rate is 4.5 percent. The market says the interest rate on insured mortgages is 5.25 percent. A 5.25 percent twenty-five-year mortgage for $11,400 will require monthly payments of $68.31. At 4 percent, a monthly payment of $68.31 for twenty-five years pays back a mortgage of $12,300. So the bank writes a mortgage for $12,300 to be repaid over twenty-five years at 4 percent interest, but gives the homeowner only $11,400. The $900 difference is about 7.5 percent of $12,300, so it is called "seven and a half points." It is another way of expressing the discount that would have to be applied to the mortgage to sell it on the market (and a good deal for the lender, because if the mortgage is repaid prior to maturity, as most mortgages are, his interest rate in effect is higher). It means that the interest rate limitation in the law is a fraud; but this is not, after all, the only law on the books that has been reinterpreted to mean something other than what it says.

The fundamental restriction on this process is that the appraisers must be squared, because they have to write up the "value" of the property being mortgaged before the bank can increase the face value of the loan (when this inflation of value is done with specific government approval, the phrase is "the points are mortgageable"). Such write-ups may be feasible up to about ten points or so, but after that—well, even a real estate appraiser has *some* pride. If the bank won't write up the mortgage, the points mean that the buyer must put down a larger initial payment. On the example of the previous paragraph, the bank would write a 4.5 percent mortgage on $11,400, but give the buyer only $10,575 for his monthly payment of $63.36; to pay $12,000 for the house, the buyer would have to come up with a $1,400 rather than a $600 down payment.

To the extent that American housing is financed from depository institutions, moreover—banks and S&Ls—the only money that can be lent is what the public deposits. In the years right after the war, this was no problem at all: there was a decade of accumulated savings temporarily at rest in government bonds, waiting to be invested on the private market. As the bonds were sold off and the loan-to-deposit ratios increased, the saving institutions increasingly could make housing loans only by selling off their old loans (at a discount, which now means a loss, if interest rates had been rising) or by drawing new deposits through the doors.

The natural way to draw new deposits would be to raise the interest rate, but that was easier said than done. As a trade-off for deposit insurance, the commercial banks and the savings banks had accepted federal controls on the interest they could pay. (The rationale for the ceilings was and is that in their absence the most troubled banks that needed the money worst would bid it away from the better banks, which would be bad for both the depositors and the insurance fund.) Until 1966, the S&Ls were free from such controls, paid higher interest on deposits, and grew more rapidly, but even they were subject to a significant systemic restraint: interest rates could not be raised on new deposits alone, but would have to be raised on all the accounts already in the bank. Moreover, the people who ran the S&Ls were among the least adventurous businessmen in the country. "This was a three-six-three business," says Maurice Mann, president of the San Francisco Home Loan Bank. "You got the money at three percent, you put it out at six percent, and every afternoon you were on the golf course at three o'clock."

In spring 1966, the Franklin National Bank of New York began offering "consumer certificates of deposit" at interest rates higher than those the California S&Ls had been paying their depositors (the beauty of the certificate being that you didn't have to raise the rates on your normal accounts). It turned out that the Californians really needed that steady flow of funds across the continent. One of the largest institutions in the state failed, and several others just staggered through the summer. Lawmakers saw that the S&Ls in the worst shape were mostly those that had been paying the highest interest rates on deposits, and in September Congress authorized the Home Loan Bank Board to set interest ceilings for the S&Ls. (It should be noted, by the way, that Franklin itself later went belly-up in the biggest bank failure in history: the feeling that interest rate ceilings must be kept for safety's sake shows not fuddy-duddyism

but a keener appreciation of realities than economists permit themselves.) Because they invest in housing to a much greater extent
than the commercial banks, and the government wants to encourage
investment in housing, both the savings banks and the S&Ls are
permitted to pay a slightly larger interest rate on deposits.

As the 1960s proceeded, the "interest differential" in favor of
the "thrifts" eroded from 1 percent (which was 40 percent of the
earnings on savings when banks paid 2½ percent and thrifts paid 3½
percent) to ¼ of 1 percent (which is less than 5 percent of the earnings on savings when the banks can pay 5¼ percent and the thrifts
can pay 5½ percent). The tiny margin became less and less effective
as a competitive weapon. The number of commercial bank branches
grew, and the banks were able to offer convenience and services
(including consumer loans) that savings institutions could not
match. In the war for savings accounts, the banks gained an increasingly dominant share, and they did not put their growing deposits
into housing.

Meanwhile, the life insurance companies stopped expanding their
mortgage holdings; and the pension funds, the most rapidly growing
bundle of savings in the country, had been lured away from their
early interest in mortgages by the siren songs of the stock market.
The system of housing finance carefully built in the Depression and
postwar years began to malfunction. In 1955–56, the proportion of
GNP that went to residential construction had been about 6 percent,
and 40 percent of all new money raised by nonfinancial sections of
the economy had gone to residential mortgatges; in 1966, in the first
big money scare in a generation, the value of new residential construction dropped to about 3 percent of GNP, and only about 20 percent of the new money raised by nonfinancial sectors went to
mortgages on housing.

The financial system that supported the housing cornucopia of the
postwar years had failed. The failure did not produce as much soul-
searching, or even informed thinking, as one might expect. The
Johnson Administration said, in effect: No matter. We will build a
whole new system. We are the government.

Chapter 16

The Inflationary Binge and the Morning After

This is the essential fact: *The government did not know what it was doing.*

—Senator Daniel P. Moynihan (writing of the urban poverty programs as a whole)

The crisis in housing finance came at the high noon of optimism in the Johnson Administration, when America could have guns and butter and all other good things, all at once. The "problems" of society could be "solved." For each there was a magic bullet, that would strike this target and no other, and such bullets could be forged in the bright retort of intellect. The nation had a "housing problem." A committee of businessmen, labor leaders, and politicians (ready to take guidance from the right academics) was put to work to solve it. The Congress by and large enacted the committee's recommendations; and presently we were in the final third of this century, when objectives and their constituencies would become so much more important than mere accomplishment, and government intervention in housing finance would become incompetent.

The President's Committee had called for the construction of 26 million dwelling units in ten years, and the existing institutions were in trouble financing more than about 1.5 million a year. If the "goals" were to be accomplished—and they had been legislatively declared to be "feasible"—new financial institutions would have to

377

be created to draw more money to housing. Four such—some of them modifications of existing institutions—were summoned from the bottle by 1971:

(1) A private Federal National Mortgage Association, five of its fifteen directors appointed by the President, which could operate more efficiently in the secondary market and could also support lending for homes when other institutions failed by purchasing mortgages from the originating bankers. Initially, the new Fannie Mae was restricted to FHA and VA mortgages, but ultimately all mortgages were made acceptable. Fannie Mae would have a stock issue listed on the New York Stock Exchange, but would get 95 percent of its money by selling bonds. (Really, 100 percent—because the stock would be purchased initially by those who sold mortgages to the agency.)

Though the bonds are the issue of the private corporation, they are sold as federal securities through official marketings and are interchangeable with government bonds in all the accountants' boxes where government securities are required. Moreover, FNMA has a call on the Treasury to buy $2.25 billion of its bonds on demand if the market dries up. Bank deposit insurance itself rests on no firmer a foundation. But Fannie Mae bonds are not part of the national debt; Fannie Mae mortgage purchases are not on the budget and are not included in the deficit.

FNMA grew enormous almost immediately after it went private; in the last quarter of 1969 and the first quarter of 1970, the agency accounted for *half* of all new-home financing in America. In 1977, the agency held almost $40 billion of residential mortgages, and still counting. Profits run about 15 percent of investment, and the feel of the place is definitely private—President Oakley Hunter in 1977 was getting a well-publicized $130,000 a year, and the furniture in his office, highlighted by a dart board and a huge stuffed lion, was anything but GI.

Because Fannie Mae bonds are so safe, they sell to yield considerably less than corporate mortgage bonds of the same duration, which means Fannie Mae can profitably purchase mortgages at yields below what the market would otherwise require. At first, the agency purchased mortgages over-the-counter at a fixed price, but soon it moved to a monthly auction at which mortgage originators bid for the right to sell mortgages to FNMA in specified amounts at specified interest rates at any time during the next four months. In this system, FNMA operates essentially as a primary lender (like the Danish mortgage credit funds), not as a supporter of the secondary market. The "commitment fee" of ½ of 1 percent, paid by the winners of these auctions, is a prime source of profits to the agency. The value to builders of knowing the terms and source of a guaranteed mortgage "take-out" is of course considerable.

(2) A new Government National Mortgage Association, GNMA, quickly called "Ginnie Mae," which would take over the support of government-subsidized mortgages that had been carried by Fannie Mae under a "special assistance" program while it was still a government bureau. Since the early 1960s and the "BMIR" (Below Market Interest Rate) programs that subsidized miltifamily housing, the government had been helping nonprofit apartment house owners charge low rents by reducing the interest they had to pay on their mortgage. After receiving the GNMA commitment and FHA insurance on the property, a bank would write a mortgage for the full cost of the apartment house at an interest rate of 4 percent (later 3 percent). If market interest rates were higher than that, the value of this mortgage for resale would of course be considerably less than its face value, but the government would buy the paper at par. Under the new Section 235 and 236 programs put in place by the 1968 Housing Act—which could subsidize interest rates on multifamily down to 1 percent—a great deal more of this "intermediation" would be required.

How Ginnie Mae should be funded to perform this service was a hard question. The planners of the agency had liked the idea of bonds similar to the old FNMA (or HOLC) bond—fully insured government obligations—the proceeds of which could be invested in the mortgages. But Treasury Department lawyers had ruled that such bonds were part of the national debt (this was why Fannie Mae had been spun off to be a private corporation); and in any event the Treasury hated to see too much encroachment on the hunting preserves set aside for the sale of government bonds. So GNMA prepared "pass-through" certificates, each representing the right to a share of the monthly repayment of principal and interest in a big bunch of FHA or VA insured mortgages.

This document looked to investors surprisingly like the old-fashioned "Participation Certificates" that had got people in a lot of trouble a generation before—what happened to the certificate-holder, for example, if some of the homeowners didn't make their payment that month? After the first "straight" pass-through offering failed, Ginnie Mae developed the "modified" pass-through, in which, essentially, GNMA guarantees with the majesty of the federal government that every penny of principal and interest due on a given date of a given month will be paid on that date. (In practice, the GNMA pass-throughs are actually issued, in minimum quantities of $1 million, by the lenders who write the mortgages; by contract with GNMA, they agree that if any mortgage in the package goes sour another one with identical terms will be substituted to maintain the flow to investors. As the mortgages in the bundle are paid off, each investor gets his pro-rata return of capital and the total value of the package is gradually reduced.) For this paper, there was an enormous market, demanding yields only ½ percent greater than the yield on a long-term government bond.

"It wasn't always easy to figure out," says Oliver Jones of the Mortgage Bankers Association, whose members originate about 85 percent of these GNMA mortgages, "why the Ginnie Mae guarantee meant so much when the mortgages were already insured by FHA. But it did." From the beginning, "Ginnie Maes" were heavily traded in the money market; as the highest yield federal government security, they competed against Tressury Bills as a place for corporations and banks to keep overnight money. They were bought and sold by the Federal Reserve system as part of its "open market" operations to control interest rates and bank reserves. By 1977, more than $3 billion of Ginnie Mae pass-throughs were being traded back and forth every day.

GNMA really is a government agency, atypical because it runs so lean with so professional and so small a staff (less than three dozen people, as against 1,200 at Fannie Mae); it is housed on one of the endless curving corridors at the Department of Housing and Urban Development. GNMA and FNMA can work together, with FNMA purchasing the paper. From early on, "tandem plans" were organized by which GNMA would commit itself to buy reduced-rate mortgages at par and then sell them at a loss (which *was* part of the federal deficit) to FNMA. These arrangements do not have to benefit lower-income families, by the way: under the Emergency Home Purchase Assistance Act of 1974, GNMA committed $7 billion for conventional home mortgages to be written 1½ percent to 2 percent below the market. Comparing the purchasers of unsubsidized homes with the beneficiaries of this tandem plan, GNMA estimated that in 1975 the buyers who did it on their own had an average income in the range $21,800–$23,600, while the average subsidized purchaser had an income of $21,500. But it is probably true that the tandem plan generated a good deal of homebuilding that otherwise would not have happened. In 1976, thanks to the tandem plans, GNMA was responsible for almost $14 billion of housing finance, 20 percent of all the new mortgage money in the country that year. By mid-1977, outstanding GNMA paper totalled $41 billion.

(3) A mortgage association (officially, Federal Home Loan Mortgage Corporation, FHLMC; unofficially, "Freddie Mac") for the S&Ls, to be administered by the Federal Home Loan Bank Board. In theory, the regional Home Loan Bank had always stood ready and able to advance funds to its members on the security of their mortgages when they needed money. But this looked bad on the balance sheet, and the regional Banks were not always as sympathetic as the members wanted them to be. Freddie Mac, like Fannie Mae, would sell bonds on the market and use the proceeds to buy mortgages—but from S&Ls rather than from mortgage bankers. Unlike Fannie Mae, however, Freddie Mac would not keep mortgages in a portfolio of investments: when the pressure went off the market, it would sell the mortgages it had acquired. Its purpose was neither to subsidize mortgages nor to feed new permanent money into the mortgage market—just to give the S&Ls a feeling of security, that somebody up there loved them, that they continued to have a role in the government's grand strategy of 2.6 million new housing units a year, that they would not be placed at a disadvantage in their competition with the mortgage bankers because the mortgage bankers could feed at the federal trough with their FHA mortgages while the S&Ls remained dependent on depositors.

This does not mean FHLMC was a minor factor: in the 1974

crunch, Freddie Mac sold $22 billion of paper and bought $22 billion in mortgages from the S&Ls. Norman Strunk, executive director of the U. S. Savings and Loan League, says that the avail-ability of this backstop facility made a great difference in his mem-bers' attitudes: ''The first flicker of increase in savings now triggers an increase in lending, because they're confident now that the Home Loan Bank will bail them out if it turns.'' There are elements of risk-free boondoggle in this situation, because the S&Ls can use the proceeds of their sales to Freddie Mac as a source of funds to purchase higher-rate GNMA paper.

(4) The Real Estate Investment Trust (REIT, pronounced ''Reet''), an entirely private corporation that would give small in-vestors the chance to participate in leveraged investments and tax shelters of the sort that were making the rich richer. The model was the mutual fund for securities investment; like the mutual fund, the REIT would pay no taxes as long as it passed on to its shareholders virtually all its earnings. But the purpose of promoting the REIT was less to lure new equity money into land development than to increase real estate lending by commercial banks and to give hous-ing a piece of the rapidly growing ''commercial paper'' market in which corporations lent short-term millions to each other. REITs came in all varieties—some of them bought and held property for income, some wrote mortgages, some specialized in construction lending. All were engaged in borrowing for short terms (supposedly at lower interest rates) to lend at longer terms (supposedly at higher rates). In theory, they could borrow from the banks more cheaply than the developers could themselves, because the cushion of the stockholder's equity made the loan more safe. In 1973, HUD es-timated that REITs were doing 25 percent of all the apartment house construction financing in the country. At their peak in 1974, the REITs controlled an alleged $16 billion worth of real estate assets.

''I had them coming down from your part of the country,'' said Billy Satterfield in Winston-Salem, leonine head over turtleneck shirt. ''New York? Chase-Manhattan Trust? They'd come down and they'd say, 'Billy, you-all are the best builder in North Caro-lina.'

"And I'd say, 'I know that.'

"And they'd say, 'We want to be your partner.'

"And I'd say, 'Gentlemen, I'm right flattered at that, your wanting to be partners with a country boy like me. What did you have in mind?'

"And they'd say, 'Anything you want to build, Billy. We'll put up all the money, and when we're finished, we'll split the profits fifty-fifty.'

"And I'd say, 'Gentlemen, that's a very generous offer you're making me. But while we're partners, I suppose you'll expect me to pay some interest on that money.'

"And they'd say, 'Well, of course.'

"And I'd say, 'How much interest do you-all think you'd have to charge me while we're partners and all?'

"And they'd say, 'Well, Billy, you know, money's tight; we'll need fohteen percent.'

"And I was very sad to hear that. I said to them, 'Gentlemen, that's too bad. But, you see, I never even finished *high* school, and I'm just not smart enough to make a living on half the profits after I pay you fohteen percent.'

"You'd be surprised how many people down here took deals like that," Billy Satterfield added, with a look of mourning former colleagues. . . .

Someday someone will write the book about the REIT movement of the late 1960s and early 1970s; this is a book about something else, and besides, I'm squeamish.

"Florida developers came up here looking for one hundred percent financing," says Agnes Nolan of New York reminiscently, "and we had REITs lined up eager to meet them. There were some men building a condo of concrete slabs, they thought it would be cheaper, found out later it would cost twice as much. Their plan was to charge separately for all utilities, make it look like cheap maintenance. Everything was electric—electric heat, even an individual electric hot-water heater in each unit. Some REIT had put up $5 million, went to $6 million; they needed $2 million more. One of the insurance trusts put it up. I remember that the first $50,000 was to go to the model apartment designer, a girl the builder was living with; he was very concerned that the insurance company shouldn't know. . . ."

It will be years before Florida tears down the unfinished, shoddily-started condominiums and unwanted shopping centers built with money mediated through REITs. What is odd about them in retrospect is that there were no leaders in these trusts even when they were flying high: the presidents were faceless men seconded by the banks that organized and presumably advised the organizations. They lent money recklessly: Pat Crow of the Florida Apartment Association reports that "progress payments" triggered by the completion of various parts of a building were routinely released (especially by Chase) on receipt of a letter from the builder casually asserting that he was ready to spend some more money, without any inspection of any sort on the spot.

Once it became evident that the ship was sinking, there was a most unseemly spectacle of banks elbowing women and children out of the way to get to the lifeboats. Bob Cochran in Melbourne, Florida, remembers the day he discovered that an REIT that was financing a condominium on which he was general contractor had used the final draw of its stand-by credit at the bank not to pay part of his outstanding bill but to prepay some of the bank's interest charges. "They showed me the fine print in the contract, and it was legal. Twenty years I'd been a builder, I'd never heard of anything like that. They make contractors get a bond; what they ought to do is bond the lenders. . . ." REITs organized by insurance companies were less irresponsible, less eager to "establish" tax losses to make profits tax-free, more likely to be investors for the long term. But the fact is that these "trusts" had been started with government blessing to pour into real estate money that would not otherwise have gone in that direction, with no thought by the government about how much money might be misallocated.

Misguided private ventures get themselves in trouble; misguided public ventures get everybody in trouble. It has not been customary to consider the REIT phenomenon in parallel with the bloating of Fannie Mae, Ginnie Mae, and Freddie Mac—the defaulted condo loans and busted apartment deals at the REITs together with the 20 percent delinquency rate on Section 235 mortgages, the 40 percent failure rate on Section 236 projects at HUD. But in truth all these things are part and parcel of the same failed government policy, the drive toward additional housing production at whatever price, dictated by the Housing Act of 1968.

And the visible costs of this policy—subsidy commitments with a present value of perhaps $30 billion pledged to what is often rapidly deteriorating housing, out-of-pocket losses of perhaps $8 billion for the REITs and the FHA insurance programs together—are peanuts next to the costs that remain concealed in more general statistics. For the new housing finance schemes misfired in ways that damaged literally everyone in the country.

The basic proposition behind the creation of the new agencies was that the illiquidity and apparent risk of mortgage instruments drove away potential investors and raised costs to households. Before 1968, there was typically a "spread" of 1.5 or 2 percent between the interest rates on new triple-A corporate bonds and the interest rates on new mortgages. Providing fully insured bond-type securities to support mortgage investments, and improving the performance of secondary markets through which mortgages could be bought and sold, the new government agencies would reduce mortgage interest rates—and thus housing costs—by eliminating the spread. It does not seem to have occurred to the technical experts who designed these programs—first-class people, too, some of them friends of mine until just now—that the spread might be closed *not by reduced interest rates on mortgages but by increased interest rates on everything else.*

If the activities of the Federally Sponsored Credit Agencies, as they are called, were going to involve the sale of $46 billion of government guaranteed bonds between 1970 and 1975, as they did, then surely the rest of the bond market was going to react. Henry Kaufman of Salomon Brothers predicted the result even before the process began: "to intensify the competition for scarce resources, to lift costs, to sustain inflationary expectations and to temporarily immobilize monetary restraint." As money grew tighter in 1973–74, corporations began to slow their investments in response to higher interest rates, but the FSCAs kept pumping paper for sale into the market—$15 billion worth in 1973, $13 billion worth in 1974.

"More than any other demand factor," banker Sidney Homer said in 1976, "this agency borrowing was responsible for the crisis yields of 1974."

Worse: the dollar bill in your pocket is nothing but a federally guaranteed piece of paper. The bonds and notes issued by the federal credit agencies were "near money," that could substitute in a pinch for the real thing—as demonstrated by the use of Ginnie Maes in the trading markets previously reserved for Treasury Bills. The commercial banks used them effectively in their struggle to keep expanding credit against the snaffle bit the Federal Reserve System was jamming in their mouths; there can be little doubt that points were added to the inflation rate by the blind printing of paper to finance housing construction that the market would otherwise have deemed unworthy.

HUD had assured the Congress airily in 1967 that there would be plenty of savings to finance housing in the 1970s. Banking consultant Carter Golembe, however, reporting to the Kaiser Committee, had shown a severe shortage of funds for housing purposes at a production of 2.2 million units and an inflation rate of 3 percent (his worst-case analysis!). But the Congress went for 2.6 million units. By 1971, more than two years before OPEC jacked the price of oil, a President who had sworn he would never, *never,* NEVER do such a thing had imposed price and wage controls; but the new mortgage-financing institutions kept printing paper, and in 1972 housing production reached the Grail of 2.6 million units. When price controls were removed, predictably, inflation made up for lost time and soared to double digits. And housing starts dropped to 1.1 million a year.

If investment in housing is to be expanded greatly, it must be matched either by an expansion in savings or a reduction in other kinds of investments. Expanded savings means, in real terms, reduced consumption. That reduction can be accomplished either by putting aside for investment purposes money that might otherwise be spent—or through an inflation that forces spenders to reduce their consumption whether they like it or not, because their dollars buy them less. At the end of any period, as the first-grade economics course teaches, savings must equal investment: it is an "accounting identity." But the price level at which this equilibrium is achieved will be a function of the demand for resources by the investing sector compared to the supply of resources unused by consumers. Abusing their authority to compile and analyze statistics, official advocates of enormously expanded housing production misled the Congress about alleged "excess capacity" that would permit increased investment without reduced consumption. The damage done by the resulting inflation was most severe in the later production and costs of housing. The government did not know what it was doing.

Housing finance came out of the big boom and the big bust of 1971–75 in a thoroughly demoralized condition. A side effect of the elimination of the spread between mortgage yields and triple-A yields had been the virtual disappearance from the mortgage market of both the mutual savings banks and the life insurance companies (in 1976, the mutuals showed an increase in deposits of $13.5 billion, but an increase in mortgage holdings of only $4.3 billion; the life insurance companies showed an increase in assets of $31 billion, an increase in mortgages of only $2.2 billion). The entire burden now fell on the S&Ls and the Federally Secured Credit Agencies. In early 1977, worried technicians at Ginnie Mae saw no source of credit at all outside their agency for multifamily housing. Congressional committees were pushing on the agency to seek the release of additional tandem funds from the Office of Management and Budget (which must approve GNMA allocations), but a Ginnie Mae spokesman said, "We've looked at the pending applications for FHA insurance on multifamily projects, and so far as we can see we're already funding all of them. We don't think it's healthy for the government to be the financial intermediary for all the apartment house construction in the country." Especially not when the means of that intermediation is a pass-through security that the market treats as a short-term note.

There are proposals afoot to beef up the thrift institutions by giving them a more appealing product to draw depositors and a stronger lending instrument as a shield against inflation. In New England, first savings banks and now all banks have been empowered by

Congress to offer interest-bearing checking accounts; in New York
State, savings banks can draw deposits their way with no-interest
conventional checking accounts of the kind once restricted to commercial banks. Bills are pending to permit all federal S&Ls to offer
what would amount to "full service banking," including consumer
loans and credit cards. The homebuilders, the labor unions, and the
Congress, however, have not yet been convinced (nor have I) that
empowering the S&Ls to make consumer loans would increase their
supply of money to housing.

Most economists advocate the removal of ceilings on savings account interest rates to allow the institutions that lend to housing to
complete for savings when money is tight. Because the people who
buy houses have a higher average income than the people who put
their money into savings institutions, the ceiling on interest rates
can be condemned as a subsidy from lower-income households to
higher-income households. Unfortunately, high interest rates are
compatible with accelerating inflation in a modernized society (as in
Britain in 1975) and passbook holders are the people least able to
defend themselves if perpetually higher interest rates should make it
more profitable for banks and "cash management" experts to open
the throttle on the velocity of money.

On the other side of the balance sheet, bankers have called for the
removal of mortgage interest rate ceilings in the three dozen states
that have "usury laws" which impose them. Unlike the FHA and
VA ceilings, which can be evaded through the "points" mechanism, many of these laws permit no escape, and there seems no
question that they direct some mortgage money out toward states
like California where there are no ceilings or where ceilings are constantly adjusted by administrative action to keep them above market
figures. The question is whether the redirection involves significant
quantities of money, because a bank or savings association may
have good community and political reasons to make mortgage loans
in its immediate area even if there are more profitable investments
available far away. And usury laws may in fact have held down interest rates on a lot of mortgages in 1966–1974. Timothy Cook of
the Federal Reserve Bank of Richmond has argued that the appearance of a "negative spread" between mortgage interest rates and

corporate bond rates in 1970—when the yield on mortgages briefly and for the first time ever fell below the yield on bonds—was an artifact of low usury ceilings in several large states.

Figures compiled by the New York State Banking Department indicate that New York experienced a drop in its share of single-family mortgages (which is what the ceilings cover) from 1966 to 1970, but that thereafter, though the worst of the crunch, its status was stable. In the 1970–1974 period, however, the gap between the revised New York ceiling and the market was smaller than it had been in the previous period. The authors of the study concluded "that the New York housing market was not as fully served as it might have been and that the usury ceiling was a contributory factor in the depressed state of that market." Perhaps more serious is the indication that lenders to the New York market demanded higher downpayments and shorter terms, making lower-income purchasers bear the burden of the law.

The major thrust in the industry has been toward the creation of what is called "an alternative mortgage instrument." This comes in three different varieties, for three different purposes.

The simplest is the annuity mortgage, for older people who would like to continue to live in the home they have bought and paid for, but can no longer afford because taxes, insurance, fuel and maintenance bills are too great to be paid from a retirement income. Under the terms of the annuity mortgage, their expenses would be paid for them by the bank, and their debt would gradually grow; and when they died or decided to move away the total principal and interest on that debt would be deducted from the proceeds of the sale of the house. This is simple, practical, no serious problem for anybody, except that most older people hate to incur debt and would greatly prefer to see their bills reduced by senior-citizen tax exemptions and subsidies.

The second different variety of mortgage is the Graduated Payments Mortgage, or GPM, which would reduce monthly payments in the early years of a mortgage at the price of an increased level payment after the first five or ten years. Young people—or people who think they can count on increased income with the passage of time (as most do these days)—could then purchase homes for less current sacrifice, giving up in return part of what would presumably be discretionary income later. The problem is that the principal on the mortgage rises through the early years. People will owe the bank more than they paid for the house, and if it turns out that home prices do *not* go up they may find themselves still owing money on the mortgage after they sell the house. Under those circumstances, the temptation to mail the bank the key and walk away, treating the house the way construction workers in 1974–75 treated their mobile home, might become irresistible.

The GPM in a sense institutionalizes inflation, like the Danish mortgage bond for an over-appraised house. (Senator Edward Brooke has proposed a fancy GPM that expresses this relationship precisely—each year the homeowner would automatically have the principal of his mortgage increased by the percentage figure which is one-half the previous year's rise in the Cost-of-Living Index, and the amount of this addition to principal would be deducted from his payments. Unfortunately, the mathematics of this won't work: the principal rises not only at the beginning but—at an ever increasing pace—at the end; and the homeowner winds up owing the bank about twice as much as the original mortgage if the inflation rate is 6 percent a year.) It must be pointed out, however, that at 9 percent interest the conventional thirty-year mortgage is itself a kind of GPM: the purchaser's equity builds so slowly that after five years of monthly payments he still owes 96 percent of what he borrowed; and if home prices do not rise he may well find that he can't cover the repayment of his debt and his selling costs from what a buyer pays him.

Since early 1976, FHA has been experimenting with GPMs on a small scale, and in summer 1977 a group of private mortgage insurers announced a willingness to underwrite a version of GPM they called a FLIP mortgage, by which the purchaser's 10 percent down payment (plus interest credited to him for the down payment, which would be held in escrow) can be used to reduce the first five years' mortgage payments. Under this arrangement, at a 9 percent interest rate, the owner's mortgage bill in the first year would be reduced by about 25 percent, but his bill after the five years would be about 10 percent higher than what he would have paid on a straight mortgage for the purchase price less the down payment. (The instrument was presented deceptively in the initial advertising, which implied that the mortgage payment after the first five years was identical to what it would have been in conventional arrangements.) The principal on the mortgage would rise for the first four years, but would remain below the purchase price of the house. The sponsors claimed that the new instrument would reduce from $17,000 to $13,400 the annual income needed by a new home buyer to qualify for a mortgage on a $41,000 home. The willingness of five large mortgage companies to take the risk may make the FLIP mortgage attractive to lenders, and the probabilities of continuing inflation may also make it attractive to large numbers of home buyers.

The Graduated Payments Mortgage, does not, of course, address the problem of the *lender* who signs a contract for a fixed interest rate to run through a thirty-year period when his price of money may rise above what the homeowner is paying him. ("What other businessman," economist Donald M. Kaplan of the Home Loan Bank Board asks rhetorically, ". . . contracts to deliver goods and services for thirty-years on a fixed-price contract?") Bowery Savings Bank in New York reported in 1977 that it still had in its vaults $900 million of mortgages written in the 1950s at rates of 5½ percent or less. What the S&Ls want—what the California S&Ls have got, and the federal S&Ls won from the Bank Board for a heady few weeks, before Congress forbade it—is a third variety of new mortgage instrument, a Variable Rate Mortgage (VRM), which periodically changes the homeowner's monthly payments to reflect the lender's cost of funds.

"Actually, we already have a *one-way* variable rate," says Morris (always "Rusty") Crawford, chairman of Bowery, New York's largest savings bank. "When rates go down, homeowners come in with a gun at the bank's head and you lower their interest rate for them. We bought some loans up in Connecticut, where they've never had a prepayment penalty [or a usury ceiling]. Those loans were at nine, nine-and-one-half percent. Every one of them has been brought back to eight, eight-and-a-half percent level."

An unpublished study done by Wat Tyler of McKenzie & Co., a bank consultants group, tracked early redemptions of mortgages as

a function of changes in mortgage interest rates for the period
1965–75. The two graphs were a mirror image of each other: when
interest rates rose, mortgages were not paid back early; when interest rates fell, they were—and presumably were then refinanced at
the lower rates. FHA and VA regulations, and a number of state
laws, prohibit prepayment penalties, so the refinancing can be inexpensive for many homeowners. (It is not, of course, inexpensive for
the FHA/VA borrower who started out paying "points.") John
Heimann, now Comptroller of the Currency, estimated while New
York State Superintendent of Banks that the trigger gap was 1 percent—whenever prevailing mortgage rates fell 1 percent below what
a homeowner was paying, he felt a tug toward paying off the old
mortgage and taking on a new one. To put a number on it, the
reduction from 9 percent to 8 percent on a thirty-year $30,000
mortgage with 25 years to go is about $19 a month.

Then when interest rates turn up again, the bank is stuck: only the
borrower can initiate a change in the mortgage. Ceilings on interest
payments to depositors are not much protection: the authorities
willy-nilly lift the ceilings, in fear that otherwise people will take
their deposits out of the banks and S&Ls and invest them directly in
some form of government or corporate bond or note, in the process
known elegantly as "disintermediation." Nobody can quarrel
seriously with the bankers' plaint that if they have to pay more on
the savings, they also have to get more on the mortgages. Even
those who are in no way patsies for the banks—a category that
includes Heimann and I hope myself—cannot see how depository
institutions can survive a long inflationary period without some opportunity to split with their debtors the agony of depreciating currency.

Variable Rate Mortgages are not that new an idea in the United
States. The Rand study of Green Bay, Wisconsin showed that most
homeowners there had mortgage contracts that permitted the banks
to raise the interest rate at stated intervals (the mortgage could be
paid off without penalty if the borrower found someone who would
take it on at a lower rate); Gerald Levy of the Guaranty Savings and
Loan Association of Brookfield, Wisconsin told a meeting of a
Home Loan Board committee that in his part of the country 75–80

percent of all the banks and S&Ls wrote such mortgages, with interest rates adjustable at any time after the first three years on four months' notice to the homeowner. Joseph Fahey of the State National Bank in Bridgeport, Conn., commented that before the Second World War the banks in New England wrote virtually all their mortgage loans that way. Indeed, such "roll over" loans resemble what was the standard procedure pre-New Deal, pre-FHA. Canadian mortgages have always been and still are written to "roll over" every five years.

The modern Variable Rate Mortgage when first proposed would have lengthened the term of the mortgage (in effect, reducing the amortization component of the monthly payment) rather than increased the payment. The Committee for Economic Development called it a " 'variable maturity' instrument" when recommending it in 1973. But this gimmick had already been overtaken by 9 percent interest rates: on a 9 percent variable maturity thirty-year mortgage with a fixed monthly payment, an increase to 9⅝ percent stretches the term out to sixty years, and at anything over 9⅝ percent the mortgage doesn't work at all because the level monthly payment no longer covers the interest and the principal starts rising instead of falling.

What has been put in place in California is a variable rate that changes the borrower's monthly payment, up or down, to make it more representative of market interest rates. The reference scale is the Federal Home Loan Bank semiannual report of the average interest rate the California S&Ls are actually paying for the money they lend—on deposits, on borrowings from the Home Loan Bank itself, certificates, etc. If that index figure goes down one-quarter of one percentage point, monthly payments on a VRM *must* be reduced by .25 percent; if the index figure goes up a quarter of a percentage point, the monthly payments on a VRM *may* at the S&L's discretion be raised to reflect an interest rate .25 percent more. The interest rate cannot be changed in the same direction more than once a year, or more than a quarter of a percentage point on any move— or more than two and a half percentage points over the life of the mortgage.

It was considered that this would be a hard sell: all the public opinion polls and some violently negative reactions when VRMs had been tried in the 1960s argued that people wanted level-payment mortgages. (An attempt by Vermont banks to offer only mortgages with variable rates had provoked near riots and a quick ban by the state legislature.) The two large commercial banks that decided to try California's VRM (Bank of America and Wells Fargo) both offered the mortgages at a starting rate a quarter of a point below their rate for conventional mortgages. The state S&Ls after the first months offered no bargains. ("I am concerned," Senator Brooke said in a speech, "that the interest rates being offered on variable rate mortgages . . . are not the rates which one would expect considering the fact that lenders are protecting their portfolios against market interest rate increases.") What they advertised instead was two sweetners: the right to transfer the mortgage to any buyer of the home when the borrower wanted to sell, or to take the mortgage along to the new house if the borrower wanted to move; and the right to take all the amortization out in cash at any time, restoring the mortgage to its original size.

Neither was much of a sweetener really (if you transfer your mortgage with the house, you remain responsible for the buyer's payments, which is not a good idea; and the very slow amortization of a high-interest loan means that the available cash reserve will be trivial for more than ten years); but both looked good in the ads. Automatic transfer was especially attractive to people who had been unable to buy (or sell) a house during one of the credit crunches and thought that with the transferrable VRM they had an insurance policy against any repetition of that unpleasant experience. For whatever reason, California customers bought the deal; the local banks and S&Ls wrote more than $8 billion in VRM mortgages from summer 1975 to summer 1977. By January 1977, Great Western—a $7 billion institution—had more than a third of its assets in variable rate mortgages, and would not make a fixed-rate conventional loan.

Unfortunately, the closer one looks at the question of alternative mortgage instruments, the more complicated it gets. Let us touch briefly on some of these complications, as an introduction to the argument of the final chapter—that when you tinker with just one part of a big machine it's hard to know what else may start shaking off its fastenings.

For example, neither VRM nor GPM fits conveniently into any of the boxes of the Truth in Lending Act, which requires lenders to supply borrowers with their Annual Percentage Rate. There is no APR that can be stated for the contract as a whole for either of these instruments. In the case of the GPM, the interest rate is entirely a function of when the homeowner pays back the mortgage. In California, where the average mortgage turns over in about six years, that interest rate will be quite different from what it would be in New England, where the average mortgage endures for about fifteen years.

By the same token, it's all but impossible for the secondary market to put a price on any of these pieces of paper. As 40 percent of the money for mortgages is now mediated through some sort of secondary market (mostly the federal agencies), the VRM or GPM market becomes separated out to a disconcerting degree. The California S&Ls are already much more dependent on advances from the Home Loan Bank—and much less able to take advantage of even their own Freddie Mac facilities—than they were in the past. In summer 1977, Home S&L, the largest in the state, tried to wriggle out of that dependency with a "mortgage-backed bond" (and Great Western followed in the fall), but the face value of the mortgages that had to be pledged was a lot bigger than the bond issue.

Thrift institutions have been able to keep making at least some mortgage loans through tight periods—and have not had to worry too desperately about deposit withdrawals—because the monthly mortgage payments by the customers have included amortization. With GPM, the bank will in effect be lending additional money rather than reducing the loans of a fair fraction of the borrowers. Thus a time of credit stringency might well hit even harder at institutions offering the new instruments.

Mortgages are registered in courthouses: that's what makes it possible for lenders to foreclose if the borrower defaults. In nearly every state, the amount of the mortgage must be part of the document. But that blank in the form cannot be filled in with the GPM. At a bad—an awful—moment, a lender might find himself on uncharted waters in the legal sea of "creditor's remedies."

Taxes: does the borrower get to take a tax deduction for the inter-

est added to the principal of a GPM mortgage in the early years?
And does the lender get to include in his income for profit-and-loss
purposes the accrued interest he has agreed to add to that principal?
If not, where does that bookeeping item fit? If the advances on an
annuity mortgage exceed what the owner put into the house years
ago, is that income for the owner, or capital gains, or just a loan?
Who is programming the computer?—and why has it gone down
again?

Fair Credit laws prohibit discrimination against elderly borrowers
and racial minorities. But an instrument tailored to the prospects of
a young couple may be most unsuitable for an older purchaser
whose income is not likely to rise much. Lending on inner city properties via conventional mortgages already requires a banker smart
enough to hold his nose and cross himself at the same time; how can
the bank offer a GPM-type mortgage that shows some years of increasing loan on a property that looks none too likely to appreciate?
If the bank offers a GPM in the suburbs, can it refuse to do so in the
city?

Is it in fact possible—or desirable—to achieve parity between the
fellow who borrows when money is easy (which means that fewer
people have the confidence or courage to incur debts) and the fellow
who borrows when interest rates are soaring (which means he's following the crowd)? VRM turns out not to answer this question.
Where once the bank wrote mortgages at lower or higher rates,
depending on the state of the money markets, now it writes
mortgages with higher or lower differentials between the borrower's
interest rate and the Home Loan Bank index. Confronted with the
need to put money out, the S&L is still likely to shade requirements
and sharpen the lending officer's pencil real good; and it is still
likely to hold up (genteelly) the borrower who comes around when
money is tight. The customer who signs up at 2.5 percent over the
index rate will still come back for refinancing when new mortgages
are being written at 1.5 percent over the index rate.

And nobody has even begun to think through the money supply
and inflationary implications of new mortgage instruments.

Ollie Jones of the Mortgage Bankers Association insists it's really
a simple story: "They've reduced the downpayment about as far as

they can go, and they've lengthened the loan about as far as it can go. Now all that's left is to screw up the amortization.'' He likes the Canadian five-year roll-over, mostly because it's the simplest device, easy to understand and to explain to lending officers, customers, and Congressmen. Assuming something can be worked out to guarantee that the customer (by definition an amateur in these matters) does not have to bear *all* the risk of interest rate fluctuations—and the Home Loan Bank rate, which is sticky and moves rather slowly, does in fact limit the customer's risk—Dr. Jones may have seen the future.

Chapter 17

Taxes and Subsidies

The result which the legislator has produced is the reverse of
beneficial; for he has made his city poor, and his citizens greedy.
—Aristotle (ca. 340 B.C.)

In a market economy, government is likely to have its most pervasive impact through the taxing system, for taxes do not fall like
rain equally upon the parched corn and the picnic. Decisions about
what will be taxed, and what will not be taxed, reverberate through
the systems of production and consumption—especially in a modern
society where taxes are sure and everything else is uncertain. Housing is such a central element in the economy that virtually every tax
imposed touches it somewhere, heating oil and gasoline taxes, corporate and individual income taxes, sales taxes on building materials and (in some states) homes themselves, estate taxes, mortgage
taxes (inherited, sometimes visibly, from the ''stamp taxes'' against
which our forebears rebelled)—and especially property taxes. The
total real estate tax on residential property in the United States probably reached $35 billion in 1977—one sixth of all the money Ameri-
cans spent for housing.

Real estate taxes in America are imposed mostly by municipalities, counties, and special-purpose geographically restricted districts that operate schools, sanitary facilities, and occasionally some
other public services. Usually the grant of taxing authority specifies

a maximum rate *ad valorem*—on value. A "tax assessor," some-times separately elected and sometimes appointed, is empowered to establish the value of the property to be taxed, subject to limited right of appeal by the taxpayer. The first call on the receipts from the tax is usually for the payment of principal and interest on debt instruments the taxing body has issued to build its facilities. Taxes assessed for operating expenses (especially school expenses) can usually be increased with the approval of the voters in a special elec-tion. On the average across the country, real estate taxes run about 2 percent of the market value of residential property, a result that can be achieved by a 2 percent tax on full value or an 8 percent tax on one-quarter of the value, according to taste or the demands of the state constitution. But the burden may vary from less than 1 percent in towns with lots of industry and few children to something like 5 percent in some hard-pressed residential suburbs, and in Boston, Newark, New York, and Berkeley, California.

When Lou Harris asked the public in 1973 which taxes were too high, 68 percent of his respondents pointed the finger at the real es-tate tax. "It is simply inconceivable," economist Dick Netzer argued in a report to the National Commission on Urban Problems, "that if we were starting to develop a tax system from scratch, we would single out housing for extraordinarily high levels of con-sumption taxation." Property taxes are widely held to be a major cause of slums in the cities: "Extremely high tax rates," George Pe-terson of The Urban Institute writes, "drain small owners of cash flow which they normally use to make annual repairs and mainte-nance. In this way, high tax rates lead to a gradual but cumulatively significant disinvestment in the housing stock."

Running for President in 1972, Richard Nixon said that "the property tax is one of the most oppressive and discriminatory of all taxes", George McGovern suggested that "the federal government may have to step in to allow for reduction of property taxes"; Hubert Humphrey described the fight for property tax reform as a battle by "the people" against "the international financiers, the credit companies, the giant banks, the bastions of the wealthy, and the giant corporations." There is all but universal agreement (the Milton Friedman school of economists politely dissents) that the real estate tax is "regressive," lying more heavily on low-income than on high-income families. It is clear enough that nobody ever got elected by calling for an increase in real estate taxes. But when the Governor of New Jersey promised to refund through real estate tax reductions some of the bundle the state would take in through his spanking new income tax, the proposal was supported in most of the high-income suburbs and most strongly opposed by the blue-collar communities. They seem to have known some things that the writers of the newspaper editorials didn't know—for there is in fact a great deal to be said in favor of the real estate tax.

Its most obvious contribution—totally neglected in the modern literature—is its power to hold down the price of land. Americans buy their homes through level monthly payments that include not only interest and amortization on the mortgage but also a pro-rata share of the annual real estate tax. The bank likes to handle the tax bill itself, because a lien for unpaid taxes takes precedence over a mortgage; trusting the owner looks to the bank like an avoidable risk. Take now a $45,000 home with a 9 percent, thirty-year $40,000 mortgage and $800 in annual taxes. The monthly bill to the owner is $388.50, of which $66.67 is tax. Take off the tax, and the monthly payment would be enough to support a $48,000 mortgage—and sure as shooting the price of the house will rise, probably quickly, to $53,000. The use value and (for argument's sake) the replication cost of the house having remained the same, all the increased price must be increased land "value." Conversely, an increase in the tax to $1,200 a year would mean that a family with $388.50 a month to pay the bank could carry only a $35,000 mortgage, and unless the increased taxes purchased some amenity that might draw new customers to this area (more visible police patrols, for example, of reduced class sizes at school), the market for this house would shrink, reducing its price.

Especially if vacant land is constantly reassessed to reflect an increasing market price, the real estate tax acts as a brake on land speculation. "There's no more speculative purchasing of land around here," says assessor Clerk Maxwell in Florida's Brevard County, "because I assess at the price they're asking for it. They scream bloody murder." In the great inflation of this decade, land prices in America have indeed risen faster than the prices of most other assets, but the rate of increase has been only a fraction of that experienced in Europe and in Japan. In France in 1976, the price of land for housing, having tripled in less than ten years, approached 50 percent of the total cost of a new dwelling unit. Pierre Schaefer of the federation of land developers inquired about the percentage in America, and could not understand how it had been held below 25 percent. The role of the real estate tax (which barely exists in France) was explained. Schaefer was puzzled for a moment, then exploded: "Real estate tax!" he said. "Real estate tax! Why, that 's a capital levy!"

And so it is—by European standards, the most politically radical of all taxes, a tax on wealth. To argue that the tax is by nature regressive in the United States is to assert that in this country the poor possess a share of the property wealth (or, in the case of renters, enjoy the use of a share of property wealth) greater than their share of income. This is almost certainly true across the population as a whole, because retired people have little income while controlling considerable residential wealth. Eliminating the senior citizenry, however, the tax almost certainly does not lie more heavily on the poor: as the economist Mason Gaffney argues, "Enterprise and labor are worth more to a nation than inert wealth"—i.e., it's sound public policy to tax wealth.

In rental situations, Alfred Marshall theorized in the nineteenth century that landlords bear the portion of the tax that lies on the land value while tenants bear the portion that lies on the building value. That's hard to prove but probably correct: Henry Aaron while at the Brookings Institution (he moved on to become Assistant Secretary for HEW in the Carter Administration) concluded after an empirical study that "the property tax is borne substantially by the owners of capital."

Looking at the real estate tax as a "consumption" tax rather than a "wealth" tax, as Dick Netzer does, it becomes regressive to the extent that high-income people spend a smaller share of their incomes on housing than low-income people do. The data in this area tend to be poisoned by the virtual impossibility of defining either "income" or "housing expense," and by the time lag between a

change of income status and a change in housing: a man who loses
his job or takes a sick-leave continues to pay about the same for his
housing as he did before, skimping on other things; and a man who
gets a big raise takes some time finding a new house for himself.
Looking at housing through Milton Friedman's prism (which sepa-
rates out the "permanent" income on which people make decisions
from their "transient" income), Margaret Reid insisted that housing
expense actually rises as a share of "permanent" income when the
income itself rises—which would make real estate taxes inherently
progressive in their impact.

In any event, a tax that lies on real estate is by no means an unrea-
sonable way to pay for the basic services a community requires—
police, fire, sanitation, road maintenance, and schools—for these
enhance and sustain the values of the property being taxed. "In
most suburbs," Netzer writes, "the connection between property
tax payments and local public services . . . is a clear one." Except
for the improvements that can be paid for through user taxes (water
and sewage systems, airports, and bridges), construction for public
purposes *should* be paid out through the property tax, because the
later generations that will also enjoy these facilities will still be
located on the same land however much they vary in other ways. A
legislature commits the future when it approves borrowing as it does
with no other act (laws can always be repealed, but debts have to be
paid off; that's why so many state constitutions and city charters
require a referendum vote before debts can be incurred); and what's
permanent, of course, is the land.

The tax burden on a piece of land grossly influences its use. The
tract developers who have moved out to the boondocks have been
driven there not only by the greater immediate cost of land (and of
zoning and enivronmental controls) near the center city, but also by
the much lower taxes in the countryside, which reduce the homebuyer's monthly bill from the bank. In principle, the local nature of
real estate taxes is an advantage, giving people a choice between
highly taxed, richly served communities and low-tax areas where
the quality of public service is lower. In reality, taxable real estate
wealth is inequitably distributed, and people who live in one township must tax themselves much harder to achieve the revenues collected with ease in another township.

The downward spiral of the cities results from loading a steady or even increasing burden of costs—fixed debt service charges, tenured employees who can't be fired as population shrinks, extensive pension funding obligations—on the city's diminishing share of an area's industry and commerce and upper-income homes. Worse: during the lush days of the 1960s when municipal revenues were growing very rapidly, a number of cities began to use the receipts from their wealth tax for purposes of income redistribution—increasing welfare and medical help payments, housing subsidies, day-care centers, assorted social services—thereby breaking the perceived link between property tax payments and the services received by the taxpayers. Among the reasons for the departure of the middle class was their observation that they got a much better buy for their tax money in the suburbs.

For thirty years, municipalities have been creating land-use incentives by real estate tax abatements (in New York, the worst case, literally nothing gets built without real estate tax concessions by the city—an apartment house on Park Avenue at 79th Street, where the lowest rent is over a thousand dollars a month, and a new ABC-TV studio begun at a time when that company had by far the largest profits in its history, are among the recent beneficiaries of the city's tax incentive program). The result is development by crazy-quilt, ignoring the practical elements of transportation and municipal service cost as well as the apparent rationalities of the land-use planners.

As the taxes mount, the value of the city's land falls—but the city dare not reassess to show the declines, because it is the assessed value of the real estate that supports the bonds the city has issued. There are blocks in New York where apartment houses cannot be sold for more than one and a half years' rent roll, and the real estate tax as a proportion of market value is more than 15 percent. For the borough of the Bronx as a whole, according to a study by the Center for Local Tax Research, the *average* real estate tax actually paid on apartment properties in 1976–77 was 7.12 percent of value. But it takes three years to get an assessment reduced in New York unless you go to the right lawyer, who (with his friends in city government, of course) keeps up to half of what you save, merely shifting that part of the burden on your books from taxes to legal fees.

In most parts of the country, sunshine works effectively as a disinfectant to an assessment process that obviously offers great scope for bribery. Developers with large tracts of vacant land, a few favored banks and stores, always the local newspaper—and the homeowning public, which notices its taxes more than the rent-paying public—are beneficiaries of lower assessments; and new industries that must be persuaded to move may receive tax breaks denied to sitting industries that would incur costs by leaving. Still, outside of a handful of big cities, the assessment process seems relatively clean, because the rolls must be exposed for at least a few days every year.

Once the city is so big that nobody can make sense of the assessment rolls, however, all bets are off. A man with a complicated medical clinic in Brooklyn, part private enterprise and part charity, received in 1971 an announcement from the tax department that his assessment had been roughly doubled. When he called to see what could be done about it, he was given the name of a lawyer, who said he would take care of the matter and did—for a fee of 40 percent of the doctor's savings. The next year, just before the tax bills were to arrive, the lawyer called again, to tell the doctor that his assessment had been boosted again, and to ask if the doctor wanted the matter taken care of "in the usual way." Every year for the next three years the same scenario was played out. The sixth year, the doctor said, "To hell with it—I'd rather pay the city." The lawyer said, "You don't understand. They can raise your assessment *more.* . . . ." It will be noted that because the call came and the adjustments were made *before* the arrival of the bill, there is a record in this matter only for the first year's reassessment. And the doctor must process his Medicaid bills through city officials who may well be friends of this same lawyer. In Chicago this sort of work is apparently done by relatives of highly placed people in the city administration.

Despite the hanky-panky and the economic distress, however, it was still true coming into the end of the 1970s—contrary to the beliefs of the public-interest law firms and the judges who have heard their cases—that assessed property wealth per capita is much greater in the center cities than in their suburbs. Quite apart from questions of bigotry, suburbs resist the intrusion of low-income apartment developments because they simply don't have the tax base to pay the bills for serving the people who would live in them. (Objections to elderly housing are less strong because the elderly don't bring with them children who must be educated and, increasingly, policed.) The cities still do have such resources but are losing them. Conceivably, the recent court cases requiring states rather than municipalities to assume the costs of education will reduce suburban resistance to the arrival of lower-income families, but in the meantime, carried out along the lines of their rhetoric, the decisions will simply take money out of the cities and distribute it to the suburbs.

Brainless application of mechanical assessment and tax formulas forces the development of farming land that stands in the path of urban expansion and cannot from farming revenues pay the taxes imposed on it as a potential homesite. Mechanical tax assessments obviously hinder redevelopment in the cities—the effort to make improvements in what looks like a declining neighborhood is fragile enough without loading on it a burden of knowledge that once the improvements are in place the city will come around and whomp up the tax bill. (It should be noted that what is being said here is that in urban renewal areas property taxes must not be allowed to perform their natural function of holding down land prices, because the first requirement of a successful rehab program is an increase in the valuation of the land. George Peterson of the Urban Institute says that in fact even legally mandated tax increases do not occur in these areas: in his study of 192 significant private rehab projects, only fifteen— none in a blighted area restoration—had actually been reassessed.)

Manipulating the real estate tax to retard or encourage development appears attractive, but it is subject to corruption, inflation (you have to give a bigger subsidy every year to achieve the same result), and simple stupidity. Peterson estimated that from 1968 to 1972, the Lindsay heyday, New York City paid out $275 million in property tax reductions to achieve $220 million of private rehab investment—and then the investors, of course, had the benefit of the price appreciation.

Economists who look at the problem almost always decide that taxation should fall most heavily on "site-value," on the land itself. Australia and New Zealand, and to a lesser extent Canada, moved in that direction in the 1910s and 1920s; so did one American city— Pittsburgh, which (perhaps not by coincidence) has had the most generally successful urban renewal experience. Land values, after all, are frequently created by government activity (the construction of roads or harbors or transit facilities, schools, museums, parks, universities); and they are, as Mason Gaffney insists, the most inert form of wealth. Unfortunately, the real estate tax is now too great a revenue source. As urban land economist James Heilbrun wrote in 1966 (and things have got worse since), "Unless we are prepared virtually to end private ownership of rights in land by taking almost its whole rent, it is no longer feasible to substitute a land tax for the real estate tax at the present level of yield."

What lies ahead, unless inflation can be controlled, is a painful period for both taxpayers and their governments. Because the monthly mortgage payment includes taxes as well as interest and amortization, rising land costs and interest rates place direct pressure on people's willingness (and ability) to pay real estate taxes. With the proliferation of "economic development commissions," corporations are able to drive increasingly hard bargains with states and municipalities prepared to give tax concessions to increase the nonresidential tax base. As the drain of people and enterprise across the border from Massachusetts to New Hampshire vividly illustrates (and so, on a different scale, does the success of Fox & Jacobs), localities and states have begun to derive disproportionate benefits from *not* providing public services. There may or may not be a "tax revolt"; there certainly is a tax-induced migration.

The disappearance of industry from the big cities means that drawing poor people to the metropolis to provide a pool of low-wage workers is no longer advantageous to the cities—or, for that matter, to the poor people. Economic rationality now demands that in the conflict between traditional "municipal" services and modern "human" services, the cities must concentrate their resources on the "municipal" sector. The drive to persuade the federal government to undertake the entire burden of welfare and health care should be seen in this context—and supported, though the immediate result would be a decrease in the public charity extended to the urban poor.

The English economist Ralph Turvey distinguished between "onerous" and "beneficial" real estate taxes, defining the latter as "those which were spent in providing services from which ratepayers benefited." Wealth taxes can be used to redistribute wealth—to increase the use-value of urban land occupied by poor people, providing schools, roads, water and sewage, police and fire protection, recreational services that enhance the attractiveness of the area. But they are irrelevant to the purpose of redistributing income, and the attempt to use them for that purpose will bankrupt the cities.

"Taxing the poor to provide services to the poor," Dick Netzer writes, ". . . is surely nonsense." But to the extent that the taxes are merely means of purchasing services collectively, it is no greater nonsense for poor people to pay for their municipal services than it is for them to buy their own food and clothing. The problem then is that the property they occupy cannot pay enough tax to provide vital services that in law and justice must be more or less equitably distributed. The solution to *this* problem is conceptually simple—a subsidy to make up the short-fall between the costs of property-related municipal services and the revenues available from the property of the poor—and it would not be politically so difficult if the problem were properly conceived, without the haze of poverty-program rhetoric and judicial decision that now enshrouds it.

Most discussion of the impact of taxation on housing deals with federal income tax laws and the rewards they give to homeowners. Heilbrun states the issue precisely: "On the one hand, owner occupants are not taxed on the imputed net annual rental value of their home. On the other hand, they are allowed to deduct from taxable income the interest and property tax cost of owning the home, even though these are really costs of generating the imputed rental income on which they are not being taxed." Homeowners in America—a large majority of the country, an especially large majority of those in their most productive years—unquestionably benefit by a national policy in favor of homeownership, and this policy is reflected in the tax laws. But the numbers being put on this "tax loophole"—especially by reformers in the Treasury Department, who publish an annual estimate of "tax expenditures"—are wildly high; and as even the most ardent reformers admit, there isn't much that can be done about it anyway.

Still, if only because the question surfaces periodically in political debate, it's worth getting the story straight. With only a handful of exotic exceptions (borrowings to carry tax-exempt securities, for example), *all* interest is tax-deductible. The reason is that the interest payments by the borrower instantly become income to the lender, so the government gets its tax on this money in any event. Taking two bites from a single income stream—though it is done in the taxation of dividend income paid from post-tax corporate earnings—is generally considered bad manners by the government.

Moreover, denying deductibility to just one kind of interest—that relating to home ownership—would merely shift the borrowing habits of people who can get credit on other assets. Taking deductibility away from second homes will increase borrowing on primary residences. Taking deductibility away entirely would lead people to borrow against stocks, bonds, insurance policies, and savings passbooks—on all of which interest is tax deductible—instead of real estate. Small businessmen who often have to pledge their equity in their home to get credit would simply put the whole house on a large commercial loan. Millions of Americans live in a more-than-one-family building and rent part of it; deciding what proportion of their interest payments should be tax deductible would make a fine ten years' work for a select committee. In the end, only a lower-income fraction of the homeowning community would seek mortgage loans, and risks and rates for such lending would rise. Cui bono?

The idea that state and local taxes should be paid out of income after federal tax would destroy the autonomy of state and local government and punish people for living under a system of government with divided jurisdictions. In economics if not in politics, the government is the government and taxes are taxes; it doesn't make sense to say that because there is a division of function and taxing power each level of government should be empowered to take from the taxpayer's entire income stream as though the other layers didn't exist. (Such procedures would imply the possibility—realized in Sweden shortly before the socialist government was voted out—that progressive tax rates separately imposed by several elements of government would take more than 100 percent of a taxpayer's increase in earnings.) The deduction of state and local taxes from income for federal tax purposes does not put money in the taxpayer's pocket—he has paid that money out, in taxes. The effect of the deduction is to reduce, by no means to eliminate, the out-of-pocket costs of state and local taxation. A family in a 50 percent federal tax bracket which owns its home in New York City will be paying well over 20 percent of its income in state and local income, property, and sales taxes; what the deduction does is to reduce that bite to a little more than 10 percent of income. The subsidy really is a kind of revenue-sharing to the states and cities, which would have infinitely more trouble imposing their taxes (they have enough trouble already) if people had to pay them after federal taxes.

The "reform" case is strongest when its advocates criticize the failure to include imputed rent in the definitions of taxable income: here there is indeed a bald-faced favoritism to investment in owner-occupied housing. The occupant of a $30,000 house purchased ten years ago probably has invested $10,000 in downpayment, amortization, and improvements. If he had that $10,000 in the bank in long-term savings, it would earn him $750, on which he would have to pay tax. By sinking it in the house rather than in an income-earning investment, he saves himself the tax on $750. And he also is getting the use of what he financed with the mortgage, without paying taxes on the benefit though he deducts the cost—the interest portion of the monthly payment—from his taxable income.

There are some big holes in this argument, for one could also impute a rental income from the free use of the family car. ("Perhaps you have neglected to be thankful for your good fortune," the socialist Paul Porter writes scornfully, "but if you own the bed you sleep in you are getting a government subsidy equal to the tax that is not charged on the income you earn (but do not report) from renting your bed to yourself instead of to someone else.") The same logic would tax "imputed income" from any money you kept in the mattress because you didn't trust banks, or a checking account because you never got around to opening a savings account. And to eliminate this tax benefit to housing would be to discourage a kind of investment vigorously promoted by a number of other government programs.

Still, taxing imputed income is an attractive idea, because it tells a truth: there *is,* in the present tax structure, a tax-free income for homeowners that is denied to tenants. Nothing else the federal government can do would be so likely to retain for the cities that fraction of the middle class that now rents urban housing. By taxing away some of the windfall gain homeowners now receive from inflation—for the imputed rent would have to rise with the valuation of the home, which has been going up faster than the Cost of Living Index—such a tax would hold down house prices by making people analyze what they can afford (and lenders analyze the debt they can carry) in a different way.

But now the vice of the tax becomes apparent. What is being proposed here is a deliberate reduction in the American standard of housing (or an increase in the proportion of their incomes people pay for housing); and that's hard to sell. Taxing the unrealized capital gains on the house is, like any tax on "paper profits," a dubious policy: among other things, it eventually forces a ponderable fraction of owners to sell. (This is one of the ways it holds down prices.) It would hit at local real estate tax collections, which are already immensely unpopular in large part because rising assessments are also a form of taxing unrealized capital gains; confronted with an inescapable federal bite from this apple, people would never tolerate a local one too. And unless the federal income tax schedule were "indexed"—that is, unless the numbers on the tax brackets rose in step with inflation—an imputed income from rent that did rise with inflation would be near enough to monstrous to guarantee the defeat of any legislator who voted for it. (Seeking greater fairness in the tax structure through the elimination of the "tax break" on capital gains, the Carter Administration prudently made an exception to assure the 70+ percent of the voters who own their own homes that they would not be asked to pay income tax on the profits from the sale of a home.) Moreover, *query:* would the President pay tax on the imputed rent from the White House and Camp David?

In any event, the revenues derived would be much lower than the proponents of imputed rent seem to believe. Because it would increase income to be reported, this tax would strike at all homeowners, not just those who itemize deductions. Congress would immediately and substantially raise the standard deduction to make sure that people of average income did not have their tax bills increased—or would give a special credit for homeownership, which removes the policy reason for imposing the tax. Depreciation allowances, insurance premiums, and maintenance costs, maybe even heat and utilities, would have to be added to tax deductions once imputed rent made homeownership a business rather than something consenting adults did in private. The net gain to governments as a group, after deducting the diminution in real estate tax collections, would probably be trivial.

Most European countries do tax imputed rent from home ownership, but that rent has been "controlled" for a generation, as though the house really were rental property, and probably averages about 5 percent of market value (against the 10 percent that would be necessary here). Once the imputed rent is reduced to that level, depreciation, insurance, and maintenance will pretty much eat it up, leaving a situation not unlike our own. The British taxed imputed rent until 1961—the hated "Schedule A" that occasionally appeared in the fiction of those times—but gave it up as more trouble than its yield. What the British have done recently that might be looked at in the United States is to subsidize the mortgage payments of people who do not itemize their deductions on their tax returns, giving lower-income homeowners the benefits that now accrue only to higher-income homeowners. I'd go for that. It would reduce the tax receipts of the government; but if there's one subject of general agreement in America outside the District of Columbia, it's that the government ought to learn to run a little more lean than it does today.

What makes the figures look so large in the Treasury's "tax expenditure" statements, by the way, is that the reformers include in their "loophole" *both* the failure to tax imputed rent and the deduction for interest and real estate taxes. This way of handling the figures traces back to Henry Aaron's brilliant initial analysis of the situation in 1972, when he claimed that "Homeowners paid $7 billion less in taxes in 1966 than they would have if they had been governed by the rules applicable to investors in other assets.'' But that's not so: in terms of Helbrun's more accurate formulation at the head of this section, the vice is that home ownership is *not* treated as a normal investment because the income part of the equation is left out. Once imputed rent is taxed, then the deductions for interest payments and taxes are a normal business expense. To tax them, too, as Aaron and the Treasury Department propose, is overkill, biasing the tax system *against* home ownership. It could be advocated only by people who believe in their hearts, as so many do in Washington, that all income really belongs to the government, which out of charity permits private persons to use some of it some of the time.

Much government subsidy of lower-income and ''moderate-income'' housing has also been arranged through the tax mechanism, and a surprisingly high proportion of it is borne by the states and localities rather than by the federal government. The 1.1 million units of public housing, for example, would have a present value of about $25 billion (maybe more). Since 1949, they have been exempted from local real estate taxes, and not many of them pay the ''10 percent of shelter rent'' that is permissible as an exaction by local government under the law. That's $500 million of real estate tax revenue foregone by the localities. Probably two million more units are at least partially exempt from real estate taxes under state and local laws (in New York State, 160,000 units of housing built via the Housing Finance Authority and the Urban Development Corporation paid a derisory $10 million in local real estate taxes in 1976). Without counting the costs of any special programs run in connection with such housing, the local subsidy runs well over a billion dollars a year.

The construction of public housing is financed by municipal bonds guaranteed by the federal government, and these are the lowest-yielding debt instruments on the market because they couple an exemption from income taxes with complete safety. The issuance of this paper fluctuates from year to year; there was about $12 billion outstanding in 1977, involving an interest subsidy through tax exemption of perhaps $300 million. Starting in the 1950s, and led by New York, the states themselves, without federal guarantee, have been issuing bonds to finance moderate-income rental projects (and more recently some home ownership programs). The interest rates on these bonds have fluctuated according to the credit of the state—New York and Massachusetts wound up paying full mortgage interest rates on their tax-exempt paper in 1974–75. The total outstanding in 1977 was about $7 billion, and the value of the federal tax exemption was probably about $175 million.

There is also a subsidy of unknown dimensions—say, $150 million a year and growing—from the state agencies to the housing they have financed with their bond issues. It is an interesting comment on the intelligence with which housing policy was made in the 1960s and early 1970s that state and local agencies—and churches—were drawn into the sponsorship of housing by the federal government at precisely the time that capacious tax shelters were being erected for private venturers (at a cost to the Treasury which the Commissional Budget Office estimated in spring 1977 at $1.3 billion a year) because it was clear that rental apartment projects for middle-income people could no longer pay their way.

By far the biggest housing subsidy is the $12 billion or so of welfare and Old Age Assistance payments that the recipients spend for housing—though this money is almost never considered part of the "housing subsidy" program because it appears in the HEW and state public assistance budgets rather than in the HUD appropriations. In the HUD budget, counting everything—annual contributions contracts with public housing authorities, absorbed losses on FHA foreclosures, the interest rate subsidies, the rent supplement programs, the leasing programs, the Section 8 fiasco described in Chapter 9—the publicly admitted subsidy for 1977 appeared to be about $6 billion. But that's like saying the national debt is only the $35 billion or so the government must put out in interest payments every year. Most of the twenty or so subsidy programs HUD supports are based on annual contributions contracts of one sort or another, and their present value (assuming a 7.75 percent federal interest rate on forty-year paper) approaches $50 billion. Thanks to the housing programs, in other words, the national debt is really $50 billion larger than the government admits.

   Worse: the commitments are open-ended. Under the Section 236
program, the government brought down owners' mortgage interest
rates to 1 percent through a two-step process. The first step was an
agreement to buy at par a 6 percent mortgage written by a private
lender, even if the interest rate in the market was above 6 percent.
Let us take a 300-unit apartment house built in 1971 for $7.5 mil-
lion. The market rate was 8.5 percent. Reselling the mortgage purchased at par, GNMA took a loss of about $1.9 million. That went
on the budget for 1971. Thereafter, HUD was committed to pay the
owner an annual subsidy sufficient to bring his out-of-pocket costs
on the mortgage down to a 1 percent rate. This subsidy came to
$267,000 a year, which would dutifully be included each year for
forty years in HUD's appropriations. Over the course of the forty
years, the interest-rate subsidy would cost the government an additional $10.7 million.

   Now it turns out that even with the mortgage rate subsidized to 1
percent—and with tenants receiving separate rent supplement allowances from HUD, and with real estate tax concessions in many
cities—the owners can't pay the operating expenses of these expensive new buildings. Under the terms of recent legislation, HUD is
obliged to make additional subsidy payments to keep the owners
from losing money (and, incidentally, to make their properties salable so they can reap the rewards of their tax shelters; one group of
owners has already sued HUD for $50-odd million extra subsidy
payment that they say the Department has so far failed to pay). And
Section 8 was set up on these terms from the start.

   In spring 1976, HUD and the Senate Banking Committee staff
got into a fight on the relative cost of subsidizing new public housing projects and subsidizing the Section 8 program. The figures
were not wildly different in the two sets of estimates; what was
striking about them was their size—for the committee staff and the
HUD experts agreed that for either public housing or Section 8, the
government in 1976 had to figure an average subsidy of more than
$300 a month per unit. When the annual subsidy necessary to create
new housing for poor people becomes greater than the total housing
costs of the average American family, something is seriously
wrong. To date, nobody has made much of a fuss about it; but the

Congress if not the public knows the numbers. ("We worked out the full costs to the end," said Representative Ashley of the $400 million added Section 8 authorization his committee gave Patricia Harris as a welcoming present to HUD; "it was a therapeutic exercise.") The programs envisaged in the Housing Act of 1974, like those laid out in the Housing Act of 1968, have aborted: the costs are too high.

Again, it should be noted that only part of these costs can be measured by the subsidy figures, for the evil of the Housing Acts has been to push *everybody's* housing costs—land costs, building costs, money costs—far higher than they would have been in the absence of the national "commitment" to 2.6 million new units a year. The burden has fallen mostly on families of average and just-below-average income, for it is demand for their sort of housing that the government has overstimulated and overfinanced.

There are limits to what subsidy can accomplish—limits that have been exceeded in the preamble, the "statement of legislative intent," in every recent piece of housing legislation. "Without the goods and services needed to achieve the goal, and sufficient to satisfy the demands generated by this housing activity," Roger Starr wrote with distaste, looking back on the Housing Act of 1968, "establishing a ten-year housing goal of 26 million units is hardly more practical an enterprise than tacking mailboxes on palm trees of a tropical paradise and expecting air mail letters dropped into these boxes to be placed on airplanes without having made any provision for recruiting mailmen from among the happy islanders who are busy fishing, dancing, and posing for anthropologists."

Chapter 18

What Can
the Governments Do?

One cannot help being impressed by the amount of ingenuity and
energy that has gone into housing legislation already on the books
and by the almost uniformly disappointing results. . . . Nevertheless, past frustrations cannot excuse our present inertia.

> —Arthur P. Solomon, director,
> Harvard-MIT Joint Center for Urban
> Studies (1974)

Faulty or misconceived control may impede the flow of investment. . . . When such an impediment to the desired flow of investment does appear, the government is less likely to remove the ob-
stacle than it is to introduce new devices of regulation or
stimulation, and in this way to broaden the area of intervention.
Thus, the increase of governmental control introduces a new element of uncertainty into the market—uncertainty as to what government policy will be. As a result, private decisions come more
and more to wait upon the judgment of public officials, and there
follows a tendency for initiative to shift from the private to the
public source.

> —Miles Colean (1950)

I told them in Washington that I didn't know all there was to be
known about housing, but I certainly knew more than any of them
did.

> —William Levitt (1948)

Shortly after he took office as New York State Housing Commissioner, John Heimann gave a talk advocating that for the future all

housing construction subsidies, federal, state, or local, should be put "up front"—the costs budgeted immediately, as though the government were building a dam or an office building—before development begins. A shudder not far off horror went through the housing community. Washington lawyer Philip Brownstein, former head of the VA loan program and of the FHA, said that "if you put these huge capital contributions up front there's no way any administration would propose it or any Congress would approve it."

Nevertheless, the path toward a rational housing policy lies in the prior assessment of its costs. This does not make housing unique— weapons systems, health care, highway projects, dams, environmental protection programs, occupational safety rules could all use greater honesty in the presentation of their costs. Housing becomes a special problem because the product lasts so long, because rising standards mean that what is sound today will be substandard tomorrow, because demand is insatiable (by definition, there will always be more customers than dwelling units when housing is subsidized to rent or sell below the market price) and because inflation imposes a double burden on housing, an increase in the cost of building the unit multiplied by the higher cost of money as lenders seek to protect themselves against the depreciation of the currency.

Housing is a special problem, too, because of a kind of sweet fierceness that characterizes professional housers: there is an enormous human satisfaction to be derived from watching people who have been living in bad conditions move to the shiny new comfort you have built for them. Even Ed Logue of New York's catastrophic Urban Development Corporation—a difficult man who will tell you without being asked what scoundrels all his former colleagues are— begins to glow with a pleasure any companion must share when he talks about the homes his work supplied. Philip Brownstein, who has given his whole life to housing (he started with FHA when it began in 1934), displays a serenity rare in veterans of government service, even when he is discussing old wars. That's why this unusually candid man can openly advocate trying to hide from the public the costs of the programs he supports. For the professional houser there is *no* public purpose so important as the provision of more dwelling units.

The mission has been misconceived for a decade. The "housing problem" in America is not one of supply but of prices. Usually an increase in supply reduces price, but what happens in housing once the gross supply of units is adequate to the population is that government-stimulated production of new homes eliminates "substandard" (i.e., low-priced) units, maintaining or even shrinking the total supply while increasing the average price. When production is pushed to the limits of the capacity of this complicated industry (which includes limits on the supply of capital funds), prices rise rapidly.

Though a generation of shortage, when the poor were driven from pillar to post by urban renewal projects, housing advocates yearned for a day when there would be excess units to which people could be moved while neighborhoods were upgraded. But this sort of social engineering could not possibly work in high-density cities in a society where people are free to choose their own residence. A multifamily building does not immediately empty out because a city has an excess supply of apartments; it shows an increase in the percentage of vacancies. Reduced income from these vacancies makes maintenance impossible, and leads to rapidly deteriorating structures still housing people who now live much worse than they did before.

In a suburban nexus with a density of two or three houses to an acre, an occasional vacant house does not create a major neighborhood problem. The row houses of a city street, however, are linked much more closely through the shared use of public spaces (if only the sidewalks), and the opening of gaps in the row weakens social cohesion as a knot in the grain weakens the wood. Once the production of new housing is largely for replacement purposes— and half the 26 million units demanded by the Housing Act of 1968 were for replacement or a desired increase in vacancies—abandonment occurs in ways and in areas not predicted by the planners.

Though individuals have benefited hugely by the government housing programs of the past decade (especially older people, whose housing needs had never been specifically addressed before), there is a strong case for the proposition that the net impact of the programs has been negative for the population as a whole—that there would be fewer run-down neighborhoods, fewer blighted urban apartment houses, and certainly a lesser allocation of family income to housing costs if HUD had been kept on a tight leash. Consciousness of this developing crisis produced the program changes written into the Housing Act of 1974, with its emphasis on lifting effective demand rather than increasing supply. But as the steady escalation of "fair market rents" demonstrates, Section 8 is even more certain than the programs it replaced to generate an inflation of housing costs—costs borne not only by the government but by all the neighbors of the Section 8 recipients, who receive few of the benefits.

As the story has unfolded, cutbacks have been imposed in precisely the wrong place: the provision of basic municipal services. What has provoked the decline in the physical condition of so many neighborhoods is a process of *public* disinvestment—a failure to provide adequate police protection, to keep the streets and sidewalks clean, to preserve the behavioral (let alone academic) standards of the schools, to maintain the roads, to supply adequate public transportation, even, increasingly (this is the hidden crisis of our big cities) to preserve standards of care in water and sewage services. Municipal services, to restate an obviousness that gets forgotten, are most essential in crowded communities; they are labor-intensive, and thus increasingly expensive in a society where technology constantly increases the productivity of labor in capital-intensive activities; and they cannot be supported on the receipts of a wealth tax except in areas where property value per capita is high.

If the government in a neighborhood does not perform properly the tasks that only the government can accomplish, the people who live in that area will scramble to leave—and then the "housing problem" in that area becomes insoluble, however hard the legislators push on the subsidy pedal, however firmly the "code requirements" are enforced. For reasons that relate to the professional competencies and loyalties of the people who wind up planning national social programs, the money allocated from far away goes to what are in this context irrelevancies—to social workers, day care centers, health clinics, legal services, the sociological leaf-raking of community action. ("Client and professional groups at the national

level become the experts and the shapers of local policy," wrote
Stephen Berger, executive director of the Emergency Financial
Control Board supervising New York City finances. "Thus we soon
find these programs shaped, not around the needs of the populations
they were designed to serve, but rather around the particular needs
of the professionals and bureaucrats who dominate the field.")
Meanwhile, the natural symbiosis of sociology and journalism
builds a wall of misinformation between politics and reality.

What, then, are the parameters of a potentially effective housing
policy?

Mrs. Patricia Harris, making her first appearance as Secretary of
HUD before the National Housing Conference early in 1977, announced that history had demonstrated the incapacity of private enterprise alone to supply good housing to the poor. The truth in that
statement is not what it would normally be taken to mean (and what
Mrs. Harris probably meant); the truth is, in Paul Porter's words,
that "the main route to a decent home has been, as it should be, a
rise in family income that makes public assistance unnecessary." It
was the rise in the average absolute income of poor people, not any
government program, that reduced Baltimore's proportion of homes
without indoor plumbing from 18 percent to 2 percent between 1960
and 1970. One improves the housing of the poor by making them
not-poor—and for that purpose, private enterprise has been spectacularly effective. Housing poverty as we have known it in the last
two hundred years is in fact scarce in America. The government
programs that have *really* improved the housing of the lower half of
the income distribution are not the subsidy programs but the unsubsidized mortgage-insurance and mortgage-guarantee programs. To
the extent that poverty is reflected in housing conditions (and, of
course, *relative* poverty we have with us always), the truth in Mrs.
Harris' remark is that neither the government nor private enterprise
in housing alone can supply good housing to the poor.

Because of the peculiar economics of housing, it can be said that
we are dealing not with a supply problem *or* a demand problem, but
with a price problem. Most poor people are decently housed in
America, but too many of them must pay for their housing too high

a proportion of their income (it ranged not uncommonly up to 50 percent in the Rand study of Green Bay, for families with older children). Government policy to increase supply will not control price because it provokes the abandonment of low-priced units; and government policy to increase effective demand inevitably raises prices. The pressure must therefore be exerted on the factors of production that tend to increase the price of housing:

(1) Money. Assuming that inflation is with us for the foreseeable future—a horrible but inescapable assumption—depository institutions will not be able to carry as much of the weight of housing finance as they have in the past. Fannie Mae, Ginnie Mae, and Freddie Mac can take up the burdens the depository institutions drop, but only at a cost of worsening the inflation and driving interest rates still higher. Because the loans are so long and so small a portion of each payment goes for amortization, housing costs rise intolerably as interest rates go up. Housing needs a limited, protected pool of capital that does not have to be bid for. The obvious source is the group of institutions already receiving tax exemptions for which they do nothing—life insurance companies, pension funds, and mutual savings banks. As the price of tax exemption, each of these should be required to invest in housing-related paper a predetermined portion of each year's increase in assets. A more general mandatory housing finance program like the French payroll "tax" (which requires all companies employing more than ten workers to place 0.9 percent of their payroll in low-yield housing-related investments) would wind up less expensive for everybody than the inflationary activities of the Federally Secured Credit Agencies. And it might not be so unpopular with businessmen as first glance suggests. Commenting on a Chilean system of housing finance by corporations, Robert Alexander noted that "most businessmen prefer to carry out housing programs themselves rather than to throw money down what they conceived to be the government's bottomless well."

FNMA should be renationalized (or made entirely private, with the presidential directors removed and the Treasury's bond-purchasing obligations cancelled); and *all* federal housing investment should be duly appropriated and included in the national debt; long-term subsidy contracts should be capitalized. A national usury ceiling, perhaps one percentage point above the yield on long-term government bonds, should be imposed on mortgage loans to make sure the protected pool does indeed hold down money costs. The mildly countercyclical implications of the usury ceiling—the fact that housing production will run out of money at those rare times when a superheated economy pushes interest rates on private borrowings well over those on public borrowings—should be accepted calmly. No doubt it would be best to maintain a steady investment in housing through good times and bad, but that won't happen; the danger that housing will become a pro-cyclical force (making the booms louder and the busts deeper—which is what happened in 1972–75) far exceeds the damage that could be done by permitting housing to remain to some degree a residual user of funds.

A national usury ceiling on mortgages and the degree of credit allocation advocated here do not seem to me intrinsically desirable programs. But they will be the least costly responses to persistent inflation.

(2) Land. This is the hardest one, and in truth one of the most grievous wounds imposed by inflation on modern society. The right to own, use, and dispose of land is, as the Founding Fathers knew and so many of us have forgotten, one of the great bulwarks of an independent citizenry. But even with the brake of real estate taxes, land prices rise much more rapidly than other prices during an inflation, because usable land almost alone among assets is sure to hold its value. And the effects of rapidly rising land prices are chilling: as Edwin Mills and Katsutoshi Ohta observe in an article about urbanization in Japan, "People consume relatively little housing where one of its most important inputs, land, is expensive."

As the emerging California experience indicates, the imposition of land use controls without some check on the prices of land left for development will be violently inflationary, a fact of which environmentalists take no notice. No public body should be permitted to institute land use controls, including zoning, without preparing a defensible estimate of their impact on land costs for housing. The effects of land use planning are likely to be extreme, because the planners to a man have a vision of the American future very different from what the American people want. Certainly, land use planning cannot be carried out in an inflationary period without a heavy tax on land sales, to take away the windfall profits produced by government action.

With or without land use controls, some sort of special tax on land sales is going to be necessary in the United States to discourage the abuse of investment in land as a hedge against inflation, with the resulting price run-up. In general, the taxation of capital gains at income tax rates is probably unwise, because it works to lower the time horizons of businessmen; but profits on land should certainly have their tax break taken away. This will be a distasteful political struggle in America, as it has been throughout Europe (the Dutch government fell on this issue in spring 1977; and the French chamber at the last minute in 1976, left and right linked in distaste for all the advanced thinking, refused the *Loi Fonciere* that was at the heart of Giscard d'Estaing's land use policy). But we cannot control housing costs if land speculation is highly profitable.

(3) Materials and labor. Obviously, one Operation Breakthrough per generation is enough. But rigorous policing of competition among the suppliers of cement, lumber (especially plywood), gypsum, and roofing materials would seem in order: the line between "price leadership" and collusion seems thin in those industries.

Restrictive union work rules would be best combated by repeal of the Davis-Bacon Act, which now requires that all federally subsidized construction live by the union contracts of the locality. AFL-CIO itself has specifically *not* called for the imposition of the Act, or even permission for situs-picketing, where what is being built is one-to-four-family housing (which is what AFL-CIO members buy); to burden urban development and rehab programs with these rules further diminishes the prospect that the cities can compete effectively with the remote suburbs as a place for working-class housing. If Davis-Bacon is retained (which is like saying *if* inflation persists), serious efforts should be made to persuade the construction unions to swap make-work rules for guaranteed annual employment contracts. In the north, costs are needlessly raised because men, bosses and machinery are kept idle during the winter; there is something to be learned from Canadian systems for building in cold weather.

(4) Government—but this whole long book turns out to be, not

by design, virtually a compendium of the ways that government actions have inflated housing costs. The plea must be that the government start doing its job well, and letting other people get about their
business. If the government could sufficiently control its own appe-
tites, and effectively manage the money supply to prevent inflation,
every problem associated with housing would be infinitely more
manageable.

Among the government's key functions is to be the houser—not
of *last* resort, which is the destiny of the slumlord, who deserves the
tenants he gets—but of late resort, for that large fraction of the low-
income community that really wants, and will care for, better housing than it can buy. A first call on subsidies should be home ownership programs for this group. Applicants should be selected on criteria which emphasize the likelihood that this particular family will
be able to take care of this house. In dealing with urban neighborhoods, it is of the first importance to avoid the shift of single-
family homes from owner-occupied to rental status. Both rehab and
new construction of four-to-six-family structures, which should become a prime focus of federal effort, ought to offer "ownership"
(subject to dynamic contractual conditions) to poor people who in
the American tradition would become the landlord and super for the
other units. There could be no more suitable employment for a capable welfare mother—and there are many such—than serving as
owner-caretaker of a four-to-six family building. NOW could be
subsidized to run the handyperson training program.

For others, there should be renewed and expanded construction of
low-rise public housing under government auspices—probably
through private building contractors, but eliminating the great gaggle of fee-takers, profit-makers, and tax-shelterers who now drain
so much money out of our housing subsidies. In the cities, it is
crucial that governments pay low prices for low-value land—urban
landowners and bankers will never be willing to swallow their
losses and make land available at prices that encourage redevelopment so long as there is a realistic hope that government will bail
them out at a profit. (And so long as the properties remain in the
hands of people who show losses on them, reinvestment cannot
occur.) Public housing construction should be deliberately counter-

cyclical: the jobs should be bid on a stand-by basis, to be put in work (land acquisition and design phases completed long before) when other demands on the industry run slack.

In a society where so high a fraction of behavior and attitude relates to consumption patterns, there must be grave doubts whether income mixing is a plausible social policy. (Lou Winnick questioned it as long ago as 1960: "We do not really know whether economically diverse groups truly mix or merely live side by side. And casual observation indicates that many exclusively high-income or middle-income neighborhoods seem to have withstood neighborhood decline extremely well while many economically mixed neighborhoods have proven quite vulnerable. The social gain of mixture and the social losses from homogeneity have yet to be demonstrated.") Increasingly, our neighborhoods and suburbs are self-segregated by income stratum—which is, at that, preferable to the old self-segregation by ethnic origin.

"Housing improves markedly as one goes up the economic hierarchy," Margaret Reid noted in 1962, "—much more than does food and clothing and probably even more than automobiles." We are a lot richer since and the housing gaps are greater. Sometimes it turns your stomach. One day in Christmas week 1976 I spent the morning in a black slum in Tampa and the afternoon at the Countryside development in Clearwater. I know no valid reason why the old black lady scrubbing her stoop to the jeers of a crowd of bejeaned young punks in the one place should be so disfavored by comparison to the old white lady protecting her hair against the wind on the golf course in the other. Those who argue that you remedy this condition by transporting the black lady (and the punks) to Clearwater, however, have not thought through the problem.

Neighborhoods can bridge income differentials of 50 percent or so without strain, provided that the people in the less well paid group are there voluntarily, making sacrifices to buy better housing that they deeply care for. Government efforts to promote income mixing should concentrate not on making higher-income neighborhoods available to all lower-income people, but on supplementing a greater than usual degree of effort for people who place housing high in their value-preference ranking. Public housing worked well when the people in it were paying a greater share of their income for housing than their neighbors in the surrounding slum. If we could create a minimal but intelligent income maintenance system that did not reward sluttishness, we would be in a position to offer rent supplements on the basis of effort—to offer what might be called challenge grants.

There is no reason why every family should have to pay 20 percent of its income for housing, and good reason for the government to extend additional help to those who would wish to spend more. Historically, even during periods when all the savants were proclaiming that three-quarters of American families could not afford to buy a new house, up to a third of the home buyers were drawn from the lower half of the income distribution. If income mixing is the national policy, it is those families, priced completely out of the market by inflation, who will make the most use of the help. In the last analysis, housing is an area where the government, like God Himself, can help only those who help themselves. There remains a secondary obligation to help people learn how to help themselves. The first assumption of a social housing program should be that poor people, like those with higher incomes, will some day be able to afford a "better" home, and are looking forward to it.

To help poor people who are willing or able to make no more than normal efforts for their housing, new public housing units must be built and the tenancy of all public housing must be upgraded. If public housing is "the housing of last resort," then nobody who doesn't absolutely have to live it in will remain. Roger Starr while New York City housing administrator stated the central, painful fact: there are a certain number of people who want to live in a slum, they will make a slum of any place where they do live, and the only sound public policy is to provide a slum for them. That slum, for every imaginable reason, must not be public housing.

Few human rights are more absolute than the right not to be locked by a combination of economic circumstance and government action into a situation where you must bring up your children in a nest of multiproblem families. The people who run public housing projects—and it can be tenant organizations, under some municipal control—must have the authority to police the admission and behavior of tenants. Some time after my visit to the successful public housing projects in Winston-Salem, I found myself at dinner there with a young lawyer from the poverty program, who was bringing suit to force on the authority a complicated and time-consuming process (in effect, a severe reduction of standards) before an applicant could be rejected. Did I, he asked indignantly, want David Thompkins to have all but unchecked power to turn people away from public housing? To which the answer was: Damn right, I do. David Thompkins cares about public housing and the people who live in it; you and the judge who may well rule in your favor care about nothing but pieces of paper.

Lawyers in and out of Congress have made the problem seem much more difficult than it is. The people who run the redevelopment program at Jeff-Vander-Loo in St. Louis visit the homes of applicants for apartments; the people who run the Section 235 home-buying program in San Francisco and the Housing Development Corporation of the New York Council of Churches require applicants to take and pass a fairly formal course in home maintenance, and to take it again if subsequent inspection indicates the lessons haven't been learned. "We explain to them," says HDC's Enoch Williams, "that if they don't develop some measures of control we're going to have accelerated rent increases because of the things we as tenants allow to happen in the buildings." (But it should be noted that the Section 8 rent-subsidy program empties the content from this threat, for the Section 8 tenant pays the same fraction of his income however high the rents may rise.)

Demonstrated desire and reasonable capacity to maintain the unit assigned to them should be reimposed as criteria for applicants for admission to public housing. The current New York City Housing Authority rule by which anybody who torches an apartment in a privately owned building can get priority for public housing (plus a cash grant from the welfare department to compensate for ruined belongings) will damage not only a great public investment but the life chances of hundreds of thousands of people who want nothing more from their housing than the chance to live decently, quietly, and safely. It is appalling that public policy should deny that opportunity to the tenants of government-sponsored housing.

There is another side, of course. The arrogance of office ranked
equal with the law's delays in Hamlet's arguments for suicide.
Another lawyer in Winston-Salem said that the maid who had
worked in his family for twenty years had decided against moving to
public housing when her apartment building was demolished "because she didn't want those inspectors poking around her home."
That's legitimate; presumably she will have to pay some price for
the preservation of her privacy to rent equivalent accommodations
to the private market. But if there is enough public housing, that
price should not be high, and the costs imposed on a greater number
of others by the absence of inspection powers are much higher.
Where order and dignity are in conflict, as they can be toward the
bottom of the income or status distribution in all societies, public
policy must begin from the recognition that order is a precondition
of dignity. Not end there, but begin.

For the rest, the government like everybody else has to pay for
what it buys. The use of "regulation" to achieve public purpose
against private wishes can be effective in those situations where a
faulty legal order has permitted people to impose on others the costs
of their activities (i.e., industries that pollute the air or water or
create hazardous conditions for employees or consumers); but as the
history of housing code enforcement shows, an unrealistic regulation creates costs without benefits. It is fair enough for government
to insist that state-chartered lending institutions—not to mention
government's own insurance underwriters—look upon applications
individually and not simply save their own time by "redlining" an
area. But to pass a law forbidding lenders to consider the location of
a property is King Canute stuff, the arrogance of office in its silly
season—for a rehabilitation loan is, as Charles Abrams pointed out,
primarily (not exclusively) a loan to a neighborhood. To pass laws
forcing private loans into areas where there is public disinvestment—where municipal services are inadequate to the needs of the
residents and there is no budget to improve them—will eventually
devastate public confidence in government more than any number of
building inspectors with their hands out.

In housing policy as in energy policy, the unanswered question of
this time and place is whether the government is *serious*—or, more

precisely, in Nat Rogg's formulation, whether the government understands what it is being serious about.

Free market, for-profit rental housing cannot survive a prolonged inflation. From an investor's point of view, rentals must support not only the increased cost of the construction and the financing but also a continuous rise in the market value of the property to match the continuous fall in the value of the dollar. The significance of cooperative or condominium ownership is that it enables the apartment dweller to gain the advantage of the hedge against inflation as compensation for the increased ''rental'' payments. For the next few years, while the rate of household formation puts pressure on our housing stock, it would be sensible public policy to subsidize the creation of true tenant cooperatives, probably on the basis of pre-existing affinity groups (members of a labor union or a credit union, employees of a company or a public body, a church congregation if this can be squared with the Constitution, maybe even a PTA). Resale formulas indexed to the rate of inflation could prevent the abuse of subsidies without depriving the residents of what is, after all, the prime reason for moving to cooperative rather than landlord ownership. As inflation continues, the fact that the occupants' incomes rise while the mortgage payments remain stable should permit a phasing down of the initial subsidy.

By the late 1980s, the squeeze will end: the huge age cohorts from the baby boom will be settled in their homes and the new young adults from the reduced birth rate of recent years will demand many fewer new housing units every year. Intelligent economic planning for the next decade requires an awareness that homebuilding toward the end of the century could suffer the economic depression that afflicts education today. Certainly it does not make sense to strain our sinews and distort all our economic priorities now for the purpose of deepening the housing depression of the 1990s. Part of the attractiveness of the current emphasis on rehabbed housing is the likelihood that these homes will want replacement at about the time the housing industry begins to suffer from excess capacity.

But all planning should start from the recognition that governance, however necessary, however well-intentioned, is a painfully simple-minded decision-making system by comparison with the variegated economic market, which measures so many more purposive inputs. A large number of the world's problems have to be handled as they arise. As James Heilbrun writes, "the character of the housing stock in a particular place frequently changes under the influence of market forces in ways that neither city planners nor housing policy administrators can readily control. Probably the greatest weakness in U.S. housing policy has been its failure to acknowledge the power (and therefore to anticipate the consequences) of the adaptive process in the urban housing market." Or, as Elbert Hubbard once put it, "Keep away from that wheelbarrow! What the hell do you know about machinery?"

In the upper five-sixths of the income distribution in America, people can pretty well take care of themselves, provided the long-established policies that support homeownership are not gutted in the eager search for alleged equality. If the regulatory apparatus can be shrunk back closer to its legitimate size, and the money cost and land cost factors are kept under control—ideally, by a halt to inflation, though there are as noted some palliatives that can work for a while—the great beast of the marketplace will bestir itself to provide for the large majority of residents of a rich country. There will still be a housing shortage—there will always be a "housing shortage"—but there will not be the crisis we can create for ourselves if we tolerate increasingly frivolous behavior by an increasingly pervasive government.

Appendix

A Glossary of the Numbers

In 1975, the staff of the Subcommittee on Housing and Community Development of the Committee on Banking, Currency and Housing of the House of Representatives developed a summary list and tables of federal programs in housing from 1892 to 1974. The list alone occupies more than 220 pages, 200 of them abstracting programs developed since World War II, 100 of them devoted to programs adopted since 1967. In 1972, a staff at The Brookings Institution supplied Henry Aaron with an heroically reduced schema of "Data on Selected Federal Housing Programs, Early 1970s." It filled twenty-three pages of agate type.

Housing people lisp in numbers, because the numbers come. The hope in this book was that only a relative handful of federal programs would have to be examined, and that their purposes could be expressed in English clear enough so that the later use of the numbers would be a shorthand comprehensible to all readers. But the gap between hope and reality must be recognized by writers as well as by policy-makers. What follows, then, is a handy reference guide for those who may feel there are more important things in their lives than remembering the dates of the Housing Acts and the meanings of enumerated federal programs. Please remember that this is only a smattering, chosen for utility to readers of this book; there are many, many other Housing Acts, Sections, and Titles—not to mention some tens of thousands of pages (literally) of HUD regulations.

HOUSING ACTS:

of 1934: Ur-papa: the foundation stone of all federal housing programs. Established the Federal Housing Administration and the mortgage insurance programs for single-family and multifamily houses, new and old. Insured deposits in Savings & Loan Associations. Authorized the creation of national mortgage associations to provide a secondary market for home mortgages.

of 1937: the Public Housing Act. Invited the establishment of Local Housing Authorities to be subsidized by the federal government through Annual Contributions Contracts.

of 1939: required payment of union wages by contractor as precondition of insurance of multi-family mortgages.

[The Servicemen's Readjustment Act of 1944 created the VA Mortgage Guaranty Program]

of 1949: declared "the goal of a decent home and suitable living environment for every American family." Authorized the Urban Redevelopment (later "urban renewal") program.

of 1954: established special insurance programs for housing in urban renewal areas and in areas impacted by people thrown out of urban renewal areas. Rechartered the Federal National Mortgage Association to invite private money and remove the expenditures from the federal budget. Required localities to prepare "workable programs" of slum clearance to qualify for federal housing assistance.

of 1961: created first program to subsidize and insure mortgages written "Below Market Interest Rate." Permitted subsidy to local housing authorities beyond Annual Contributions Contract if necessary to "maintain solvency" of public housing projects.

of 1964: stressed rehabilitation rather than new construction in urban renewal areas, help to localities in code enforcement.

of 1965: created the cabinet Department of Housing and Urban Development, which absorbed (among other bureaus) the FHA. Authorized rent supplement programs to make it possible for some poor people to rent housing on terms profitable to their landlords.

of 1968: set the goal of 26 million new housing units in ten years. Vastly extended the interest-rate subsidy programs to increase home ownership by poor people and both construction and rehabilitation of low-rent housing. Authorized a "special-risk insurance fund . . . not to be actuarially sound." Ordered large-scale use of "new and advanced technologies." Created the Government National Mortgage Association to purchase mortgages written below market interest rates, and spun off the Federal National Mortgage Association to private ownership. Established the "new communities" program, neighborhood-development programs, rehabilitation loans and grants program, "national housing partnerships" program to tap corporate funds for low-income housing. Funded a great gaggle of programs to support planning and pay planners. In its entirety, the Housing and Urban Development Act of 1968 could define the phrase "catastrophic optimism."

[The Emergency Home Finance Act of 1970 established the Federal Home Loan Mortgage Corporation, authorized subsidy of loans by the Federal Home Loan Bank Board to Savings & Loan Associations to enable them to hold down interest rates, gave FNMA authority to purchase noninsured mortgages, and expanded the subsidy powers of GNMA.]

of 1974: created a "guaranteed rent" program as the central subsidy device for low- and moderate-income housing assistance, assuring revenues to landlords for construction, rehabilitation, and occupancy of existing structures. Established community development fund to consolidate previous assistance programs, and sought (unsuccessfully) to compel HUD to allocate money by formula rather than by *ad hoc* bureaucratic policymaking. Authorized GNMA "tandem" plan to reduce interest rates; also "passthrough" sales of GNMA mortgages to reduce apparent cost of programs.

SECTIONS:

8: The "guaranteed rent" program immediately above. Authorizes HUD to enter into contracts with housing owners—up to twenty years for private owners, forty years for state-sponsored owners—to make up the difference between the "fair market rent" of a housing unit and roughly 20 percent of its occupants' income. From Housing Act of 1974. (The number 8 had previously been used—from 1950—for a program insuring higher-risk mortgages on "very low cost" single-family homes in outlying suburban and exurban areas.)

23: the first "leased housing" section, authorizing local housing authorities to rent as well as build and own housing for low-income families; forerunner of Section 8. From 1965.

115: grants to low-income homeowners for rehabilitation of housing in urban renewal areas. From 1949.

203: the basic federal home-mortgage insurance program. From 1934.

207: the basic insurance program for mortgages on multifamily housing, available only to sponsors organized on a nonprofit or (allegedly) limited-profit basis. From 1934.

213: insurance for mortgages on cooperatively owned multifamily housing. From 1950.

220: insurance of mortgages representing a higher fraction of costs, for housing in urban renewal areas. From 1954.

221: insurance of mortgages representing at least 100 percent of developers' costs for low- and moderate-income housing in urban areas. From 1959. Separate sections of Section 221—notably d(3) and d(4)—permitted federal subsidy to reduce interest rates on such mortgages to 3 percent. From 1961.

233: promotion of new technologies. From 1961. Later supplemented by Section 108, to promote "experimental housing." Led to "Operation Breakthrough."

235: insurance (100 percent less $200 downpayment) of mortgages on new or rehabilitated moderate-cost single-family housing to be bought by lower-income households. Interest rates down to 1 percent were possible through federal subsidy. From 1968.

236: insurance (at 100 percent of cost-plus) of mortgages on new and rehabilitated multifamily housing for rent to low- and moderate-income households; 1 percent interest rates through federal subsidy. From 1968.

312: direct federal loans for housing rehabilitation in urban renewal and "code enforcement" areas; interest rate, 3 percent. From 1949.

502: the charter section for the Farmers Home Administration: direct loans or insurance of mortgages on low-cost rural housing, with interest rates subsidized to 1 percent. From 1949, replacing prior program established in 1937.

608: insurance at 100 percent of "replacement cost" for mortgages to finance multifamily veterans' housing. First use of "acceptable risk" rather than "economically sound" as criterion for writing insurance. First big corruption stories, too. From 1943.

TITLES:

I: insurance on loans for home improvement. From 1934. Same Title number later used for the loans and grants under the Urban Renewal Act.

II: the umbrella for the mortgage-insurance sections. From 1934.

III: the umbrella for the institutions created to operate a secondary market in mortgages. From 1934.

IV: rules for federal savings and loan associations. From 1934.

V: rural housing. From 1949.

VI: the umbrella for defense and veterans' housing other than VA insurance programs. From 1949. Same Title number used in 1970 for crime insurance programs. Military base, NASA, and AEC (now NRC) housing now come under Title VIII.

VII: New Communities. From 1970. Previously used for military housing (1953).

. . . Last I looked, we were up to Title XIV (flood control), but none of the others has figured significantly in this book.

Acknowledgments

This book was suggested by the Lavanburg Foundation, and commemorates its fiftieth anniversary. The foundation has been devoted exclusively to housing matters from its beginnings. Its best known project is still the original Lavanburg Houses on New York's Lower East Side, currently in use as an indispensable way-station for impoverished families in process of relocation. I am grateful to Roger and Oscar Straus (nephews of Fred Lavanburg), to Roger Starr (a member of the foundation's board, who proposed the idea that I should do a book he might have done himself had he not been about to assume the job of Housing and Development Administrator for the City of New York), and to executive secretary Ruth Glover.

I owe an extraordinary debt, unique in my experience as an author, to Theodore B. Lee of San Francisco, lawyer, quondam civil rights advocate and developer, who made available to me not only the education of his experience and the contents of his invaluable library (from which perhaps one-fifth of the ensuing footnotes derive) but also the delights of his splendid home and the company of his wife Doris and his sons Greg and Ernie. Rarely have my wife and I spent so engaging—or so useful—a week.

A number of people were helpful to me far beyond the call of duty. Louis Winnick arranged appointments for me *ex cathedra* from the Ford Foundation. Nathaniel Rogg, the New Deal economist who became executive vice-president of the National Association of Home Builders, managed for me a number of introductions that could have worked no other way. John Heimann, an old friend who was a partner in E. M. Warburg & Co. when the work on this book began, moved during the course of its preparation to be New York State Banking Commissioner, New York State Housing Commissioner, and U. S. Comptroller of the Currency. In each capacity, he was helpful. Arthur Holden, who was an adviser on housing matters to Governor Al Smith more than half a century ago, brought the unique perspective of an architect who is also a monetary theorist. These four offered an enthusiasm for which I am most grateful, significant guidance, and (if I may) wisdom; but none of them, of course, is to be held responsible for anything in these pages.

The great majority of those helpful on this book have been quoted in the preceding chapters. None was obliged in any way to give me his time, and many undoubtedly expected to be the losers by their cooperation (now that I get interviewed myself once in a while, I know the quality of resigned concern a subject feels). But with even fewer than usual exceptions, they tried to understand questions not phrased in their lingo, to answer them honestly by their own perceptions, and to steer me right. Those who enjoy being interviewed and contributed some fun to the book may now, I know, rather regret having been so outgoing; let me express a very special gratitude to them.

A number of friends contributed generously to the economy of my time in their bailiwick, especially B. Frank Brown in Melbourne, Florida; Bill Straughan and Russell Brantley of Wake Forest University in Winston-Salem; Nils-Erik Enkvist in Åbo (Turku), Finland; and Gerald T. Dunne in St. Louis. Martin Feinstein's hospitality at Kennedy Center and Bernice Feinstein's welcome to their home made me much more willing than I would otherwise have been to spend time in Washington.

Among others whose names have not appeared in these pages, I am grateful to the press representatives and executive organizers: to Larry Beckerman and George Norris of the Department of Housing and Urban Development, Don Alexander at the Federal Home Loan Bank of San Francisco, Jerrold Hickey of the Boston Housing Authority, Val Coleman of the New York Housing Authority, Susan Rezksick of the San Diego City government, W. Scott Ditch of the Rouse Company, John Cassado of the Jim Walter Company, Torben Egede of the Danish Housing Ministry. Thomas Stamm of Buowcentrum and J. M. W. Droog of the Netherlands Housing Ministry, and Mrs. Tuma of the French Cultural Consulate in New York.

During the first year of work on this book, I was also helping the late Louis G. Cowan with his memoirs, and Lou's ardent curiosity and practical questions were a source of both pleasure and assistance; I feel an ache of regret that he is not here to read the results.

I owe special thanks to William Schauer, who runs the apartment house where I live and attempt to function as president of the tenant cooperative. Willy taught me more than I can remember about how an apartment house runs, protected and still protects me from any number of nuisances that would otherwise wind up on my desk. For similar assistance on a higher level of abstraction, I am grateful to Mark Harris of Whitbread-Nolan, Inc.

My sons are now old enough to participate in the final boarding process of literary venture, and both read my manuscript. Tom dissuaded me from several (not all) of what he called "snottinesses," and Jim saved me from a number of technical gaffes in Chapters 10–12. As suggested by the dedication, my wife Ellen Moers read my drafts with close attention, and told me with whatever firmness was necessary where the sentence had floundered, the page failed to track, or the argument was off the rails. Murray Rossant generously, carefully, and helpfully read the final manuscript.

Evan Thomas of Norton saw that there was a book here when other editors did not; I thank him for that, and for some dozens of valuable suggestions as the manuscript staggered in its sections over his desk. It is scary to think that this book will be published almost exactly twenty-five years after Evan was the only editor to see that there might be a book in my just-curious approach to Wall Street.

The admirable index was prepared by Sydney Wolfe Cohen, to whom I am most grateful.

Finally, a kow-tow to the Texas Instruments Business Analyst,

for $38 (you can get it now for less than $30), the best toy a boy
ever had. When I found out I could work up the figures myself, and
wasn't dependent on other people's parameters and examples, it
was like the prisoners coming up from the dungeons in *Fidelio*.
Beautiful music. Liberation. And truth: now *they* (be it business or
governments or ideological advocate) can't lie with numbers to us
poor scribblers, ever again.

Martin Mayer

Notes

Chapter 1

4. *Harrington quote:* Michael Harrington, *The Other America*. Baltimore, Penguin, 1964, p. 166.
4. *AFL-CIO figures:* "Survey of AFL-CIO Members Housing," Washington, D.C., AFL-CIO, 1975, p. 3.
4. *Urban Institute estimates:* Frank deLeeuw, Anne B. Schnare, and Raymond J. Struyk, "Housing," in Nathan Glazer and William Gorman, eds., *The Urban Predicament*. Washington, The Urban Institute, 1976, p. 136.
4. *von Eckardt quote:* Wolf von Eckardt, "Architecture," in *The New Republic,* Aug. 6 & 13, 1977, p. 33.
5. *Marshall quote:* Alfred Marshall, *Principles of Economics*. New York, Macmillan, 1948, p. 107.
5. *James quote:* Henry James, *Washington Square*. New York, Modern Library, 1950, p. 32.
6. *Bryce quote:* Lord Bryce, *Modern Democracies*. New York, Macmillan, 1921, Vol. II, p. 108.
8. *Winnick quotes:* Louis Winnick, *Rental Housing*. New York, McGraw-Hill, 1958, p. 1, p. 12.
9. *think-tank quote:* TEMPO Professional Staff, United States Housing Needs 1968–1978, in *The Report of the President's Committee on Urban Housing, Technical Studies,* Vol. I, p. 14.
9. *Aaron quote:* Henry Aaron, *Shelter and Subsidies*. Washington, The Brookings Institution, 1972, p. 55.
9. *BLS statement:* Peter Kihss, *Taxes Found Crucial to Rise in Living Costs,* New York *Times,* July 5, 1977, p. 47.
9. *Meyerson quote:* from Introduction to Winnick, *op. cit.,* p. xiii.
    Wood quote: Robert Wood, *Suburbia*. Boston, Houghton-Mifflin, 1959, p. 210, pp. 215–16.
11. *Bellow quote:* Saul Bellow, *To Jerusalem and Back*. New York, Viking Press, 1976, p. 146.

12.   *Sternlieb quote:* from James W. Hughes and Kenneth D. Bleakley, *Urban Homesteading.* New Brunswick, N.J., Rutgers University Press, 1975, preface.

12.   *Talbot and Horsley quotes:* Carter B. Horsley, "In Suburbs, Quest for Space Pits 'Have-Nots' vs. 'Haves'," New York *Times,* Feb. 6, 1977, section 8, pp. 1, 8.

12.   *Kristof quote:* from Henry J. Aaron, *op. cit.,* p. 41.

13.   *land cost figures:* from Elsie Eaves, *How the Many Costs of Housing Fit Together.* A report to the National Commission on Urban Problems, Washington, 1969, p. 33.

15.   *Wood quote:* Edith Elmer Wood, *Recent Trends in American Housing.* New York, Macmillan, 1931, p. 46.

15.   *Weinfeld quote:* from Margaret G. Reid, *Housing and Income.* Chicago, University of Chicago Press, 1962, p. 392.

15.   *Friedin quote:* Bernard J. Friedin, "Housing and National Urban Goals," in James Q. Wilson, ed., *The Metropolitan Enigma.* Cambridge, Mass., Harvard University Press, 1968, pp. 187–88.

15.   *"about a third":* see William Grigsby, *Housing Markets and Public Policy* (Philadelphia, University of Pennsylvania Press, 1963), pp. 207ff.

16.   *President's Committee comment:* from *A Decent Home.* The Report of the President's Committee on Urban Housing, Washington, 1969, p. 119.

17.   *Kaplan quote:* Donald M. Kaplan, *The Alternate Mortgage Investment Research Study,* speech (mimeo), Federal Home Loan Bank Board, Washington, Sept. 15, 1976, p. 2.

17.   *Conference Board quote:* Business Bulletin, *The Wall Street Journal,* Aug. 18, 1977, p. 1.

18.   *tenant turnover rates:* from Institute of Real Estate Management, *Income/Expense Analysis 1976, Chicago,* p. 27.

18.   *Pitt quote:* from *Blackstone's Commentaries,* Ewell ed., 1889, p. 26.

20.   *Boston Figures:* from *Housing in Boston: Background Analysis and Program Direction.* Boston Redevelopment Authority, June 1974, pp. 11–20.

20.   *Sternlieb quote:* George Sternlieb, *The Tenement Landlord.* New Brunswick, N.J., Rutgers University Press, 1966, p. xvii.

22.   *Colean quotes:* Miles L. Colean, *The Impact of Government on Real Estate Financing in the United States.* New York, National Bureau of Economic Research, 1950, p. 16, p. 3.

24.   *Riis quote:* Jacob Riis, *How the Other Half Lives.* New York, Dover (paperback), 1971, pp. 115–16.

24.   *segregation laws:* see John Hope Franklin, *From Slavery to Freedom,* 3rd ed. New York, Knopf, 1967, pp. 436–37.

25.   *Abrams and FHA quotes:* Charles Abrams, *Forbidden Neighbors.* New York, Harper, 1955, pp. 229–30.

26.   *Sternlieb study:* Sternlieb, op. cit.

Chapter 2

29. *Safdie quote:* Moshe Safdie, *Beyond Habitat.* Cambridge, Mass., MIT Press, 1970, p. 47.

29. *Abrams quote:* Charles Abrams, *The City Is the Frontier.* New York, Harper & Row, 1965, p. 40.

29. *Arlen quote:* Michael Arlen, *Passage to Ararat.* New York, Ballantine Books (paper), 1976, p. 141.

32. *Barnett quote:* from John Portman and Jonathan Barnett, *The Architect as Developer.* New York, McGraw-Hill, 1976, p. 7.

32. *Nichols quote:* J. C. Nichols & Co., *Your Dream House.* Kansas City, undated, p. 1.

32. *Wright quote:* Frank Lloyd Wright, *The Natural House.* New York, NAL (paper), 1970, p. 14.

33. *FHA regulation:* cited in Winnick, *op. cit.,* p. 206.

34. *Gruen quote:* from Laurence B. Holland, ed., *Who Designs America.* Garden City, N.Y., Doubleday-Anchor (paper), 1966, pp. 238–39.

39. *Downing quote: Ibid.,* pp. 55–56.

39. *Kennedy material:* Robert Woods Kennedy, *The House.* New York, Reinhold, 1953, pp. 116–25.

40. *"is seen and sees": Ibid.,* pp. 175–76.

40. *"precarious balance": Ibid.,* p. 144.

41. *Wright quote: Ibid.,* p. 128.

42. *"medicine cabinet": Ibid.,* p. 291.

42. *"conventions": Ibid.,* p. 143.

42. *LeCorbusier quote:* LeCorbusier, *The City of Tomorrow* (originally, *Urbanisme*), trans. Frederick Etchell. London, The Architectural Press, 1971 ed., p. 76 fn.

43. *Whyte quote:* William H. Whyte, *The Last Landscape.* Garden City, N.Y., Doubleday-Anchor, 1970, p. 281.

43. *Wright quotes:* Frank Lloyd Wright, *op. cit.,* pp. 68, 26, 37.

43. *Whyte quotes:* Whyte, *op. cit.,* p. 246.

45. *Safdie quotes:* Safdie, *op. cit.,* p. 114.

48. *Kennedy quote:* Kennedy, *op. cit.,* p. 27.

49. *McMichael quote:* Stanley C. McMichael, *McMichael's Appraising Manual.* Englewood Cliffs, N.J., Prentice-Hall, 1954 ed., p. 7.

49. *Mumford quote:* Lewis Mumford, *Sticks and Stones.* New York, Dover (paper), 1955 (originally published 1929), p. 50.

50. *Goldfeld quote:* Abraham Goldfeld and Beatrice Greenfield Rosen, *Housing Management.* New York, Covici, Fried, Publishers, 1937, p. 34.

51. *Veiller quote:* from Roy Lubove, *The Progressives and the Slums.* University of Pittsburgh Press, 1962, p. 1952.

51. *Riis quote: Ibid.,* p. 181.

52. *Starr quote:* Roger Starr, "The New York Apartment House," in Bern Dibner and Murray Rubien, eds., *Concepts, Critiques and Comments.* A Festschrift in honor of David Rose, New York, 1976, p. 304.

52. *Tenement House Commission quote:* from Lubove, *op. cit.,* p. 95.

52. *Lubove quote:* from *Ibid.,* pp. 95–98.

53. *Gribetz and Grad quote:* Judah Gribetz and Frank P. Grad, *Housing Code Enforcement: Sanctions and Remedies,* 66 Columbia Law Review 1290.

53. *Woodlawn comment:* Winston Moore, Charles P. Livermore, and George Galland, "Woodlawn: The Zone of Destruction," in *The Public Interest,* #30, Winter 1973, p. 47.

53. *Downs quote:* Anthony Downs, *Opening Up the Suburbs.* New Haven, Yale University Press (paper), 1973, p. 7.

53. *Boston quote:* Boston Redevelopment Authority and Boston Urban Observatory, *Housing Policy Considerations,* 1975, p. 108.

54. *Fitch quote:* James Marston Fitch, *American Building: The Historical Forces that Shaped It,* 2nd ed. New York, Schocken Books, 1973, p. 178.

CHAPTER 3

57. *Juvenal quote:* from Winnick, *op. cit.,* p. 60.

57. *Maurois quote:* from *Your Dream House,* p. 2.

61. *Meyerson quote:* Martin Meyerson, Barbara Terrett, and William L. C. Wheaton, *Housing, People and Cities.* New York, McGraw-Hill, 1962, pp. 5–6.

62. *Gans quote:* Herbert J. Gans, *The Levittowners.* New York, Vintage ed. (paper), 1967, p. 167.

62. *British Housing quote:* Ingrid Reynolds, Roy Ince, and David Davies, *The Quality of Local Housing Authority Schemes.* London, Housing Development Papers, Department of the Environment, p. 4. An exhaustive study of convenience as a source of residential location in England can be found in R. Baxter, *A Model of Housing Preferences,* Urban Studies 1975, Vol. 12, pp. 135–49. Baxter concludes it's trivial.

62. *Boston quote:* Robert Earsy and Kent Colton, *Boston's New High-Rise Apartments.* Boston Redevelopment Authority, 1974, p. 12.

62. *Meyerson quote:* from Meyerson *et al., op cit.,* p. 88.

63. *"mathematically stable solution":* see Thomas Schelling, "On the Ecology of Micromotives," *The Public Interest,* #25, Fall 1971, p. 61 @ 87.

63. *Downs comment:* Downs, *op. cit.,* pp. 70–71.

63. *Zeisel quote:* John Zeisel, *Sociology and Architectural Design.* New York, Russell Sage Foundation, 1976, p. 28.

63. *Donaldson quote:* Scott Donaldson, *The Suburban Myth.* New York, Columbia University Press, 1969, p. 1.

63. *Wood quotes:* Wood, *op. cit.,* p. 16, p. vii.

64. *Banfield quote:* Edward Banfield, *The Unheavenly City.* Boston, Little, Brown (paper), 1970, pp. 32–33.

64. *Milwaukee lady:* from Alvin L. Schorr, *Slums and Social Insecurity.* Research Report #1, HEW, no date, p. 9.

64. *Abrams quote:* from Bernard Taper, "A Lover of Cities," *The New Yorker,* Feb. 11, 1967, p. 86.

65. *Schorr quote:* Schorr, *op. cit.,* p. 71.

66. *Woodruff quote:* from M. A. Woodruff, "Recycling Urban Land," in

C. Lowell Harriss, ed., *The Good Earth of America,* Englewood Cliffs, N.J., Prentice-Hall, 1974, p. 31.

66. *Fitch and Mack quote:* Lyle Fitch and Ruth P. Mack, Land "Land Banking," in *Ibid.,* p. 148.

67. *New York State legislative report:* from Banfield, *op. cit.,* p. 25.

68. *Babcock quote:* Richard F. Babcock, *The Zoning Game.* Madison, Wis., University of Wisconsin Press, 1966, p. 5.

68. *New Jersey court:* Gabrielson v. Glen Ridge, 13NJMisc 142.

68. *Supreme Court:* Village of Euclid v. Amber Realty Co., 272 US 365 @ 388 and 394–395.

69. *Lutheran high school:* State ex rel Lutheran H.S. Conference v. Sinar, 65 NW2nd 43 @ 46.

69. *Douglas opinion: Berman v. Parker,* 348 US 26 @ 33.

70. *architectural appeal:* State ex rel Saveland Park Holding Corp. v. Wieland, 69NW2nd217 @ 220, 219; italics added.

70. *Mumford quote:* Lewis Mumford, *The City in History.* New York, Harcourt, Brace & World (paper), 1961, pp. 323–24.

71. *advice to zoning boards:* from *The Uses of Land,* A Task Force Report Sponsored by the Rockefeller Brothers Fund. New York, Thomas Y. Crowell, 1973, pp. 184–91.

72. *Minneapolis-St. Paul Council:* Robert Reinhold, *Move to Regional Planning Is Led by Minneapolis-St. Paul,* New York *Times,* March 8, 1977, p. 1, p. 18.

73. *Hagman quote:* Donald G. Hagman, "Windfalls for Wipeouts," in Harriss, ed., *op. cit.,* p. 114.

73. *Meyers quote:* from Herbert B. Lawton, *No Vacancy, Wall Street Journal,* Jan. 31, 1975, p. 1, p. 21.

74. *Pennsylvania case:* National Land Investment Co. v. Easttown Board of Adjustment, 419Pa504, 437Pa237.

75. *Babcock quote:* Babcock, *op. cit.,* p. 94.

75. *Guitar quote:* Mary Anne Guitar, *Property Power.* Garden City, N.Y., Doubleday, 1972, p. 92.

75. *Fischer quote:* from *Ibid.,* pp. 278–79.

CHAPTER 4

80. *Rouse quote:* James W. Rouse, *Great Cities for a Great Society.* Rouse Development Corp., April 8, 1965, mimeo. p. 5.

80. LeCorbusier quote: LeCorbusier, *op. cit.,* p. 130.

81. *Bell quote:* Colin and Rose Bell, *City Fathers.* New York, Praeger, 1969, p. 210.

83. *PEP Pamphlet:* Ray Thomas, *London's New Towns,* PEP Broadsheet 510, London, April 1969, p. 391.

85. *Roosevelt quote:* Charles Abrams, *op. cit.,* p. 240.

85. *Blumenfeld quote:* Hans Blumenfeld, *The Modern Metropolis,* Cambridge, Mass., MIT Press (paper), 1971, p. 42.

86. *Hertzen quote:* Heikki v. Hertzen, *Practical Problems of New Town Development.* Tapiola Garden City, Finland, 1973, p. 10.

87. *Hoppenfeld quote:* Morton Hoppenfeld, *The Columbia Process: The Potential for New Towns,* The Architects' Year Book, Letchworth; reprinted by Rouse Co., Columbia, p. 7.

88. *Downie quote:* Leonard Downie, Jr., *Mortgage on America.* New York, Praeger (paper), 1974, p. 200.

88. *Rouse quote:* James W. Rouse, "How to Build a Whole New City from Scratch," Rouse Development Corp., Columbia, May 13, 1966 (mimeo), p. 4.

91. *much Gananda material* from Richard Karp, "Building Chaos," *Barron's,* Sept. 6, 1976, pp. 3, 8, 10, 12.

93. *reviews and approvals:* Public Law 910609, 84 Stat 1796, Section 712(a)5.

93. *housing analyst:* from Scott Jacobs, "New Towns—or Ghost Towns," *Planning,* Jan. 1975, p. 16.

94. *Landauer quote:* from Mary Perot Nichols, "Faulty Blueprint," *Barron's,* Nov. 29, 1976, p. 6.

94. Eichler comments from Edward P. Eichler and Marshall Kaplan, "The Community Builders," Institute of Urban and Regional Development, University of California (Berkeley), undated mimeo, p. 130.

98. *Downie quote:* Downie, *op. cit.,* p. 228.

98. *Goldberg quote: Techniques & Architecture* #301, November–December 1974, Paris, p. 113.

102. *LeCorbusier quote:* from Philippe Boudon, *Lived-in Architecture: LeCorbusier's Pessac Revisited* (trans. Gerald Onn). Cambridge, Mass., MIT Press, 1972, p. 1.

Chapter 5

103. *Aristotle quote:* from *Politics,* Book VII, Chapter 4, Jowett translation. New York, Modern Library, 1949, p. 287.

103. *Blumenfeld quote:* Hans Blumenfeld, *op. cit.,* p. 43.

104. *Saarinen quote:* Eliel Saarinen, *The City.* Cambridge, Mass., MIT Press (paper), 1971, p. 41.

104. *Colin and Rose Bell quote:* Bell, *op. cit.,* pp. 13–14.

104. *Blumenfeld quote:* Hans Blumenfeld, *op. cit.,* p. 11.

104. *hides and squares: Ibid.,* pp. 7, 10.

105. *Lynch quote:* from Kevin Lynch, "Quality in City Design," in Holland, ed., *Who Designs America,* p. 123.

105. *McKelvey quote:* Blake McKelvey, *The Urbanization of America.* New Brunswick, N.J., Rutgers University Press, 1963, p. 13.

106. *Kelper & DeLeuw:* cited in Robert A. Futterman, *The Future of Our Cities.* Garden City, N.Y., Doubleday, 1961, p. 55.

107. *McKelvey quote:* Blake McKelvey, *op. cit.,* p. 12.

107. *Weaver quote:* Robert Weaver, *The Urban Complex.* Garden City, N.Y., Doubleday/Anchor (paper), 1966, p. 9.

108. *George quote:* Henry George, "More About American Landlordism," *North American Review,* February 1886, #142, p. 391; cited in Charles Albro Barker, *Henry George* (New York, Oxford University Press, 1955), p. 445.

110. *Blumenfeld quote:* Hans Blumenfeld, *op. cit.,* p. 24.

111. *Colean quote:* Miles Colean, *op. cit.,* p. 24.

112. *Vernon quote:* Raymond Vernon, *The Changing Economic Function of the Central City.* New York, Committee for Economic Development, 1959, p. 53.

112. *Minneapolis study:* Paul R. Porter, *The Recovery of American Cities.* New York, Two Continents Press, 1976, p. 55.

113. *Winnick quotes:* Winnick, *op. cit.,* pp. 44–45.

113. *Schlivek quote:* Louis B. Schlivek, *Man in Metropolis.* Garden City, N.Y., Doubleday, 1965, p. 31.

114. *McMichaels quote:* McMichaels, *op. cit.,* p. 159.

114. *Urban League quotes: The National Survey of Housing Abandonment,* National Urban League, Washington, 1971, p. 16.

115. *Sternlieb and Hughes quote:* George Sternlieb and James W. Hughes, *Post-Industrial America,* New Brunswick, N.J., Rutgers University Press, 1976, p. 2.

115. *Sternlieb quote: The Tenement Landlord,* p. 119.

CHAPTER 6

116. *LeCorbusier quote:* LeCorbusier, *op. cit.,* pp. 280–81.

116. *Jane Jacobs quote:* Jane Jacobs, *The Life and Death of Great American Cities,* New York, Random House, 1961, pp. 270–72.

116. *Anderson quote:* Martin Anderson, *The Federal Bulldozer.* New York, McGraw-Hill ed. (paper), 1967, p. xi.

117. *San Jule quote:* James San Jule, "Housing Comes into Its Own," Meeting Transcript, *Guidelines to Success in Multi-Family Housing,* April 9–11, 1969, Associated Home Builders of the Greater East Bay, San Francisco, p. 54.

117. *Olson quote:* Robert Olson, *URC: Renewal Gone Astray,* St. Louis, September 1975, p. 11.

119. *Anderson quote:* Anderson, *op. cit.,* p. 8.

119. *Gans quote:* Herbert Gans, *The Urban Villagers.* New York, The Free Press, Macmillan, 1962, p. 105.

120. *House Committee quote: Evolution of the Role of the Federal Government in Housing and Community Development,* Committee on Banking, Currency and Housing, 94th Congress, October 1975. Washington, Government Printing Office, p. 26.

120. *Weaver quote:* Robert Weaver, *op. cit.,* p. 105.

120. *Starr quote:* Roger Starr, *Urban Choices,* pp. 44–45.

120. *Hartman quote:* Chester Hartman, "The Politics of Housing: Displaced Persons," in *Society,* July/August 1972, p. 61.

121. *University of Chicago quote:* from Brian J. L. Berry, Sandra J. Parsons,

Rutherford H. Platt, *The Impact of Urban Renewal on Small Businesses.* University of Chicago Press, 1968, p. xv. This report caused considerable offense because the director of the University's center for urban studies had been the planner who laid out the renewal project for the city. The authors replied that the attack and the fact that the study was conducted on federal funds "made us constantly aware of the need for objectivity and accuracy. . . . We take it as a mark of courage on the part of the public agency that they were willing to support research on these terms." (Fn., p. 78.)

122. *Jane Jacobs quote:* Jacobs, *op. cit.,* p. 300.

123. *Section 608 descriptions:* from *Evolution of the Role of the Federal Government,* p. 11.

124. *Abrams quote:* Abrams, *The City Is the Frontier,* p. 88.

124. *Winnick quote:* Winnick, *op. cit.,* pp. 180–81.

129. *Milgram quote:* Morris Milgram, *Good Neighborhood,* New York, W. W. Norton, 1977, in press.

129. *Lefrak quote:* "A Lecture Delivered by Samuel Lefrak . . . At the Harvard University Graduate School of Business Administraton," 1971. The Lefrak Organization, Forest Hills, N.Y., unpaginated.

132. *Heilbrun quote:* James Heilbrun, *Urban Economics and Public Policy.* New York, St. Martin's Press, 1974, p. 239.

CHAPTER 7

133. *Hughes and Bleakly quote:* James W. Hughes and Kenneth D. Bleakly, Jr., *op. cit.,* p. 48.

133. *Forrester quote:* Jay W. Forrester, *Urban Dynamics.* Cambridge, Mass., MIT Press, 1969, p. 11.

140. *Boston quote:* from Boston Redevelopment Authority and Boston Urban Observatory, *Housing Policy Considerations,* 1975, pp. 110–11.

146. *Benitez book:* A. William Benitez, *Housing Rehabilitation.* Washington, D.C., National Association of Housing and Redevelopment Officials, 1976, p. 18.

150. *Catholic priest:* Ann Crittenden, *Big Burden for Small Business: Government Rules,* New York *Times,* July 2, 1977, pp. 23, 27.

152. Brian J. L. Berry, "The Decline of the Aging Metropolis," in George Sternlieb and James W. Hughes, eds., *Post Industrial America,* p. 181.

152. *45,000 units: St. Louis Development Program,* St. Louis Plan Commission, 1973, p. S-8.

154. *tremendously impressed:* I went to see Strauss on Brady's suggestion.

157. *Mayor's letter: Central Hoboken Neighborhood Preservation Program,* Community Development Administration, Hoboken, N.J., 1976, p. v.

157. *Hoboken description: National Survey of Housing Abandonment,* p. 58.

161. *Weaver quote:* Robert Weaver, *op. cit.,* p. 111.

161. *1954 Act:* from *Evolution of the Role of the Federal Government,* p. 54.

162. *Weaver quotes:* Robert Weaver, *op. cit.,* pp. 109, 110.

162. *Radisch quote:* from *Profile of New York City's Municipal Loan Program,* Housing and Development Administration, New York, 1967, p. 1.

163. *New York quote:* from *Community Development Program and Application,* 1st Program Year, p. 36.

163. *"special risk insurance":* Evolution of the Role of the Federal Government, p. 130.

166. *Abrams quote:* Charles Abrams, *The City Is the Frontier,* p. 189, p. 190.

168. *Wein and Quirk quote:* William J. Quirk and Leon Wein, "Homeownership for the Poor," *Cornell Law Review,* July 1969, Vol. 54, #6, pp. 828–29.

168. *Newark quote:* from James W. Hughes and Kenneth D. Bleakly, Jr., *op. cit.,* p. 169.

169. *Moreland Act Commission quote:* from *Restoring Credit and Confidence,* A Report to the Governor by the New York State Moreland Act Commission on the Urban Development Corporation and Other State Financing Agencies, March 31, 1976, p. 41.

CHAPTER 8

173. *Fyvel quote:* T. R. Fyvel, *The Insecure Offenders.* London, Penguin ed., 1963, p. 11.

181. *Sachs quote:* Daniel Y. Sachs, *Handbook for Housing and Urban Renewal Commissioners.* Washington, D.C., NAHRO, 1972, p. 12.

181. *federal court: U. S. v. Certain Lands in the City of Louisville,* 78F2nd 684.

181. *Abrams quote:* Abrams, *The City Is the Frontier,* p. 266.

183. *Meyerson and Banfield quote:* Martin Meyerson and Edward Banfield, *Politics, Planning and the Public Interest.* Glencoe, Ill., The Free Press, 1955, p. 94.

183. *Ford quote:* James Ford, *Slums and Housing.* Cambridge, Mass., Harvard University Press, 1936, Vol. II, pp. 774, 775.

184. *McDougal and Mueller quote:* Myres McDougal and Addison A. Mueller, *Public Purpose in Public Housing: An Anachronism Reburied,* 52 Yale Law Journal 48.

184. *Salisbury quote:* Harrison Salisbury, *The Shook-Up Generation.* New York, Harper & Row, 1958, p. 75.

184. *Greeley quote:* Andrew Greeley, "The Ethnic Miracle," in *The Public Interest,* #45, Fall 1976, p. 23.

187. *Goldfeld quotes:* Beatrice Greenfield Rosen and Abraham Goldfeld, *op. cit.,* p. 64.

188. *Abrams quotes:* first, from *Forbidden Neighbors,* p. 306; second, from *The City Is the Frontier,* p. 35.

189. Wagner quote: from *Congressional Record* 81, August 3, 1937, p. 8099.

190. *Newman quote:* from Oscar Newman, *Defensible Space.* New York, Collier (paper) ed., 1973, p. 188.

190. *Boston figures:* from *State of the Development Report 1975,* Boston Housing Authority, pp. viii, xii.

190. *Gutman quote:* from Daniel P. Moynihan, ed., *Toward a National Urban Policy.* New York, Basic Books, 1970, p. 127.

191. *Newman quote:* Oscar Newman, *op. cit.,* p. 193.

191. *St. Louis quote:* from *Tenant Management Corporations in St. Louis Public*

Housing: The Status After Two Years, Center for Urban Programs, St. Louis University (mimeo), 1975, p. 47.

191. *Jephcott quote:* Pearl Jephcott, *Homes in High Flats.* London, Oliver & Boyd, 1971; cited in Charles Mercer, *Living in Cities* (London, Penguin, 1975), p. 118.

192. *"respiratory organs":* Gerhard Ulmman, "High-rise Flats—Source of Social Conflict?" in *Kulturbrief,* #12, 1975-E. Bonn, West Germany, Inter-Nationes, p. 20.

193. *53 to 83 percent:* from Marie McGuire, "Managing Public Housing for the Elderly," in Frederick A. Vogelsang, ed., *Public Housing Management in the Seventies,* Washington, D.C., NAHRO, 1974, p. 151.

196. *Solomon quote:* Arthur Solomon, *Housing the Urban Poor.* Cambridge, Mass., MIT Press, 1974, p. 85.

196. *Cox story:* James Cox, "Management of Section 23 Leased Housing," in *Public Housing Management in the Seventies,* p. 113.

200. *"leadership roles":* from *Tenant Management Corporations in St. Louis Public Housing,* p. 7.

200. *Baron's brief:* Richard D. Baron, *Tenant Management: A Rationale for a National Demonstration of Management Innovation,* St. Louis, McCormack & Associates, undated, p. 74.

204. *"disciplined, trained residents":* from *Ibid,* p. 8.

204. *Jean King quotes:* from *Tenant Management Corporations in St. Louis Public Housing,* Appendix, p. 6.

CHAPTER 9

206. *Austin quote:* from Solomon, *op. cit.,* p. 21.

206. *Solomon quote: Ibid.,* p. 142.

208. *Derthick quote:* Martha Derthick, *New Towns in Town,* Washington, D.C., The Urban Institute, 1972, p. 1.

208. *Johnson's distance: Ibid.,* p. 84.

209. *truckdriver quote:* from Meyerson & Banfield, *op. cit.,* p. 110.

209. *Shannon & Luchs:* Linda Kuzmack, "Private Management of Public Housing," in *Housing Management in the Seventies,* p. 143.

210. *Solomon quote:* Solomon, *op. cit.,* p. 14.

212. Quirk-Wein and Kennedy quotes: from William J. Quirk and Leon E. Wein, *op. cit.,* pp. 823–24; the Kennedy citation is *Hearings on Housing and Urban Development Legislation of 1968 Before the Subcommittee on Housing and Urban Affairs of the Senate Committee on Banking and Currency,* 90th Congress 2nd Session, pp. 642–43.

214. *FmHa figures* from *Brief History of the Farmers Home Administration,* Department of Agriculture, Washington, 1977.

217. *Glazer quote:* from Moynihan, ed., *Toward a National Urban Policy,* p. 58.

218. *HUD sales literature: 8 Facts About Section 8* Washington, HUD, 1976.

218. *Solomon quote:* Solomon, *op. cit.,* p. xiv.

225. *$73:* The Housing Assistance Supply Experiment, RAND Research Review, Santa Monica, Cal., Summer 1977, p. 5.

225. *Floor quote:* from *Rents, Subsidies and Dynamic Costs.* The Hague, Ministry of Housing and Physical Planning, 1972, p. 8.

226. *HUD quote: Housing in the 70's,* HUD, Washington, 1974, p. 1.

CHAPTER 10

234. *Dietz quote:* Albert G. H. Dietz, *Dwelling House Construction,* fourth edition. Cambridge, Mass., MIT Press, 1974, p. 74.

235. *Kennedy quote:* Robert Woods Kennedy, *op. cit.,* pp. 192, 193. "An Old Pro Goes Back to the Basic House—Starting at $16,490," in *House & Home,* July 1975, p. 74.

240. *accountants quote:* from Felix Pomeranz and William J. Palmer, *Managing Construction Projects.* New York, Cooper & Lybrand, 1975, p. 36.

241. Benjamin Shimberg, Barbara F. Esser, and Daniel H. Kruger, *Occupational Licensing: Practices and Policies.* Washington, D.C., Public Affairs Press, 1973, p. 96.

242. *Ann Arbor/Bay City comparison:* from Alan B. Mandelstamm, "The Effect of Unions on Efficiency in the Residential Construction Industry," in *Industrial and Labor Relations Review,* July 1965.

243. *Urban Land Institute quote:* from Jack Newville, *New Engineering Concepts in Community Development.* Washington, D.C., Urban Land Institute, 1967, p. 3.

244. *Canadian quote:* from *Costs in the Land Development Process.* Ottawa, Housing and Urban Development Association of Canada, 1975, p. 166.

244. *Maryland figures: Public Improvement Costs for Residential Land Development,* Regional Planning Council, Baltimore, Md. 1973, p. 13.

245. *San Diego quote: A Growth Management Program for San Diego,* San Diego Planning Commission, December 1976, pp. 24, 29–30.

CHAPTER 11

253. *LeCorbusier quote:* from Philippe Boudon, *op. cit.,* pp. 194–95.

253. *Safdie quote:* Moshe Safdie, *op. cit.,* 1970, p. 111.

254. *Giedion quote:* Siegried Giedion, *Space, Time & Architecture.* Cambridge, Mass., Harvard University Press, 1941, p. 271.

254. *Boorstin quote:* Daniel Boorstin, *The Americans: The National Experience.* New York, Vintage (paper) ed., 1965, pp. 149–50.

254. *anonymous observer:* in Giedion, *op. cit.,* p. 273.

255. *Woodward quote:* G. E. Woodward, *Woodward's Country Homes,* New York, 1869; quoted in Giedion, *op. cit.,* p. 274.

255. *Sims quote:* Christopher A. Sims, "Efficiency in the Construction Industry," in *The Report of the President's Committee on Urban Housing.* Washington, D.C., Technical Studies, 1968, Vol. II, p. 149.

256. *1948–1966 figure,* from Elsie Eaves, *op. cit.,* p. 52.

256. *Johnson list and comment:* Ralph Johnson, "Housing Technology and Housing Costs." Technical Studies, Vol. II, pp. 60–61.

258. *Larrabee quote:* Eric Larrabee, "The Six Thousand Homes that Levitt Built," *Harper's Magazine,* September 1948, p. 84.

261. *index descriptive material:* from *Uniform Construction Index,* Construction Specifications Institute, 1150 17th St. NW, Washington, 1972, p. 4.

264. *"a high-technology housing industry":* A Decent Home: The Report of the President's Committee on Urban Housing,* Washington, D.C., 1968, p. 191.

264. *Rosskam quote:* Edwin Rosskam, *Roosevelt, N.J.* New York, Grossman Publishers, 1972, p. 22.

265. *Committee quotes: A Decent Home,* pp. 196, 205.

267. *Eaves quote:* Eaves, *op. cit.,* p. 100.

269. *HUD standards: Federal Register,* Dec. 18, 1975, 58752-42.

271. *MHI publication:* N. G. Asbury, *A Formula for Determining the Feasibility of Mobile Home Developments.* Chicago, Mobile Homes Manufacturing Assn., 1971, p. 39.

272. *Danish architects quote:* from Maris Kjeldsen and W. R. Simonsen, *Industrialised Building in Denmark,* Copenhagen, Skandinavsk Bogryk, 1965, p. 10.

274. *Committee quote:* from *A Decent Home,* p. 212.

274. *Jespersen quote: The Jespersen System,* Copenhagen (undated), p. 7.

275. *Berry quote:* Fred Berry, *Housing, The Great British Failure,* London, Charles Knight, 1974, p. 86.

277. *Kristof quote:* Frank S. Kristof, "Housing," in Robert H. Connery and Gerald Benjamin, *Governing New York State: The Rockefeller Years,* New York, Proceedings, The Academy of Political Science, Vol. 31, #3, 1974, pp. 196–97.

279. *Moreland Act Commission quote:* from *Restoring Credit and Confidence,* p. 266.

CHAPTER 12

283. *Beechers quote:* from Catherine E. Beecher and Harriet Beecher Stowe, *American Woman's Home.* Hartford, Conn., Stowe-Day Foundation, 1975, p. 36.

283. *LeCorbusier quote:* from Boudon, *op. cit.,* p. 35.

283. *NHBC quote:* from *Over Three Million People Now Live in Houses Covered by NHBC Ten Year Insurance.* London, National House-Building Council, 1973, p. 14.

284. Details from *Qualitel,* 60, Rue de la Chausée d'Antin, 75541 Paris, France.

288. *NHBC quote: op. cit.,* p. 19.

289. *Ibid.,* p. 11, p. 15.

293. *Sternlieb quote:* George Sternlieb and Stephen Seidel, "Government Regulations and Housing Costs," mimeo executive summary, May 20, 1977. Center for Urban Policy Research, Rutgers University, New Brunswick, N.J., p. 27.

295. *MacDonald quote:* John D. MacDonald, *Condominium.* Philadelphia, Lippincott, 1977, pp. 21–23.

296. *Committee quote:* from Leland S. Burns and Frank G. Mittelbach, *Efficiency in the Housing Industry.* The President's Committee, Technical Studies, Vol. II, p. 101.
301. *Masonry contractor:* "Builders Look for Quality, Dependability, Reputation in Masonry Subs," *NAHB Journal-Scope,* July 4, 1977, p. 23.

CHAPTER 13

308. *"happiness":* *Real Estate Salesmen's Handbook,* National Association of Realtors, 6th edition. Chicago, 1974, p. 44.
311. *"How much can you afford to invest?"* Anthony Wolff, *Unreal Estate.* San Francisco, Sierra Club, 1973.
320. *cautionary quotes:* from *Handbook on Multiple Listing Policy,* National Association of Realtors. Chicago, 1975, pp. 14, 16.
323. *empathy and sympathy: Real Estate Salesmen's Handbook,* p. 40.
324. *sex breakdown:* from *Profile of the Realtor and Realtor-Associate,* National Association of Realtors. Chicago, 1975, pp. 1, 2.
325. *schedule:* from *Handbook,* p. 30.
328. *New Orleans lawyer:* from Alfred A. Ring, *Real Estate Principles and Practices.* Englewood Cliffs, N.J., Prentice-Hall, 1972, p. 115.
331. *reports:* Mary Clay Berry, *The Real Estate Settlement Procedures Act* and *The Magical Power of Letters from Home.* The Alicia Patterson Foundation, 535 Fifth Avenue, New York 10017; undated.
331. *Senate Report: 1974 US Code, Congressional and Administrative News* 6546 @ 6547; cited in Morton P. Fisher and Gregory L. Reed, *The Real Estate Settlement Procedures Act of 1974 (as Amended in 1975),* in *RESPA Under the New Regulations,* ALI-ABA Course of Study Materials, Philadelphia, 1976, p. 14.
333. *regulations: Federal Register,* Vol. 41, no. 109, June 4, 1976, p. 22704.

CHAPTER 14

334. *Ring quote:* Alfred A. Ring, *op. cit.,* p. 232.
334. *Starr quote:* in *Concepts, Critiques and Comments,* p. 303.
337. *Thiel quote:* from *Guidelines to Success in Multi-Family Housing,* p. 17.
337. *committee quote:* from *Tax Shelters,* Joint Committee on Internal Revenue Taxation, CCH Standard Federal Tax Reports, Chicago, Vol. 63, #7, March 31, 1976, Part II, p. 12.
341. *Rand study: Second Annual Report of the Housing Assistance Supply Experiment,* Rand Caorporation, R-1959-HUD, Santa Monica, Cal., May 1976, pp. 75–80.
342. *costs: Income/Expense Analysis,* p. 24.
343. *Goldfeld quote:* from Beatrice Greenfield Rosen and Abraham Goldfeld, *op. cit.,* p. 64.
345. Fortune *quote:* "Business Roundup," *Fortune,* April 1977, p. 14.
345. *Gans quote:* Gans, *op. cit.,* p. 13.

347. *Bendiner quote:* Robert Bendiner, *Just Around the Corner*. New York, Dutton (paper), 1967, p. 16.

348. *Turner quote:* from Colin Ward, Introduction, in John F. Turner, *Housing by People*. London, Pantheon (paper), 1977, p. xxxiii.

348. *New York statistics:* from Frank S. Kristof, *Housing and People in New York City. City Almanac, Center for New York City Affairs, The New School, Vol.* 10, #5, February 1976, p. 4.

348. *Community Development quote:* from *Community Development Program and Application, 1st Program Year*, New York, City of New York, 1975, p. 51.

349. *Community Development quote: from Community Development Program and Application, 2nd Program Year*, New York, City of New York, 1976, p. 104.

CHAPTER 15

354. *Real Estate Research Corporation figures:* from Steven W. O'Heron and David L. Parry, *Orange County Housing Markets*, San Francisco, Federal Home Loan Bank, September 1977, p. 22.

355. *Sim Van Der Ryn quote:* from Jim Burns, "A Talk with Sim Van Der Ryn," *Urban Design*, Winter 1976, p. 30.

356. *O'Heron and Parry quote: Ibid*, p. 25.

357. *112 percent: Ibid.*, p. 26.

360. For the author's views on the realities of money and credit generally, see Martin Mayer, *The Bankers* (New York, Weybright & Talley), 1975.

366. *report:* from *National Report Denmark,* Fifth European Congress of Building Societies, Vienna, 1976, p. 10.

372. *pessimism:* Larrabee, *op. cit.,* p. 85.

372. *Klaman comment:* Saul B. Klaman, *The Postwar Residential Mortgage Market*. Princeton, N.J., Princeton University Press, 1961, p. 139.

CHAPTER 16

377. *Moynihan quote:* Daniel P. Moynihan, *Maximum Feasible Misunderstanding*. New York, The Free Press, 1969, p. 170; italics in original.

381. *income figures:* from *Statement of David DeWilde,* President, Government National Mortgage Association, before Senate Banking Committee, Sept. 22, 1976, Table IV.

385. *Kaufman quote:* Henry Kaufman, "Discussion," in *Housing and Monetary Policy,* Conference Series #4, Federal Reserve Bank of Boston, October 1970, p. 105.

386. *Homer quote:* Sidney Homer, "The Great American Credit Market," Salomon Bros., Nov. 5, 1976, mimeo, pp. 7–8.

386. *Golembe quote:* from Carter Golembe Associates, *An Appraisal of the Availability of Funds for Housing Needs,* in President's Committee, Technical Studies, Vol. II, pp. 213–235.

388. *"negative spread":* Timothy Cook, "The Residential Mortgage Market in Recent Years," *Economic Review,* Federal Reserve Bank of Richmond, Sept./Oct. 1974. p. 3 @ 16.

389. *New York State:* Ernest Kohn, Carmen J. Carlo, and Bernard Kaye, "The Impact of New York's Usury Ceiling on Local Mortgage Lending Activity," New York State Banking Dept., 1976, p. 51.

391. *Kaplan quote:* Donald M. Kaplan, *op. cit.*, p. 7.

392. *Levy and Fahey comments:* from "Transcript of Proceedings," Initial Meeting of the Advisory Committee to the Federal Home Loan Bank Board's Alternative Mortgage Instruments Research Study, Federal Home Loan Bank Board, Washington, Sept. 10, 1976, pp. 45–46.

393. *Committee for Economic Development:* "Financing the Nation's Housing Needs," New York, CED, 1973, p. 17.

394. *Brooke quote:* from speech to AMIR Study Group lunch, Sept. 10, 1976; Federal Home Loan Bank Board, mimeo, p. 7.

CHAPTER 17

398. *Aristotle quote:* Aristotle, *op. cit.*, Book II, Ch. 9, p. 114.

399. *Harris poll:* from John Shannon, "The Property Tax—Reform or Relief," in George E. Peterson, ed., *Property Tax Reform.* Washington, The Urban Institute, 1975, p. 25.

399. *Netzer quote:* Dick Netzer, *The Impact of the Property Tax,* Joint Economic Committee of Congress, May 1968, p. 18.

399. *Peterson quote:* Peterson, ed., *op. cit.*, p. 10.

399. *Nixon, McGovern and Humphrey quotes:* from George E. Peterson and Arthur P. Solomon, "Property Taxes and Populist Reform," *The Public Interest,* #30, Winter 1973, pp. 60, 61.

401. *Gaffney quote:* Mason Gaffney, "An Agenda for Strengthening the Property Tax," in Peterson, *op. cit.*, p. 72.

401. *Aaron quote:* Henry Aaron, "What Do Circuit-Breaker Laws Accomplish," in *Ibid.*, p. 62.

402. *Reid analysis:* Margaret G. Reid, *Housing and Income.* Chicago, University of Chicago Press, 1962, pp. v–vi.

402. *Netzer quote:* Netzer, *op. cit.*, p. 29.

403. *Bronx taxes:* from "Effective Real Property Tax Rates in the Metropolitan Area of New York," Center for Local Tax Research, New York, Sept. 1976, p. 7.

405. *Peterson comment:* George Peterson, "The Property Tax and Low Income Housing Markets," in Peterson, ed., *op. cit.*, Chapter 7.

406. *Heilbrun quote:* James Heilbrun, *Real Estate Taxes and Urban Housing.* New York, Columbia University Press, 1966, p. 150.

407. *Turvey quote:* from *Ibid.*, p. 75.

407. *Netzer quote:* Netzer, *op. cit.*, p. 32.

407. *Heilbrun quote:* James Heilbrun, *Urban Economics and Public Policy*, pp. 332–33.

409. *Porter quote:* Paul R. Porter, *op. cit.*, p. 145.

411. *Aaron quote:* Henry Aaron, *Shelter and Subsidies,* p. 55.

415. *Starr quote:* Roger Starr, *Housing and the Money Market.* New York, Basic Books, 1975, p. 54.

CHAPTER 18

416. *Solomon quote:* Solomon, *op. cit.,* p. 181.
416. *Colean quote:* Colean, *op. cit.,* p. 34.
416. *Levitt quote:* from Larrabee, *op. cit.,* p. 87.
418. *half the 26 million:* see Tempo Professional Staff, *United States Housing Needs, 1968–1978,* in The Report of the President's Committee on Urban Housing, Technical Studies, Vol. I, p. 28.
419. *Berger quote:* from Stephen Berger, "Time for Government to Redefine Its Role, Limit Its Services," in *New York Law Journal,* June 22, 1977, pp. 1, 35.
420. *Porter quote:* Paul R. Porter, *op. cit.,* p. 133.
421. *French payroll tax:* for descriptions of mandatory schemes, see Morris L. Sweet and S. George Walters, *Mandatory Housing Finance Programs,* (New York, Praeger, 1975); the French section is pp. 32–61.
421. *Alexander quote,* from *ibid.,* p. 101.
422. *Japan quote:* Edwin S. Mills and Katustoshi Ohta, "Urbanization and Urban Problems," in Hugh Patrick and Henry Rosovsky, eds., *Asia's New Giant* (Washington, The Brookings Institution, 1976), p. 673 @ 706.
425. *Winnick quote:* in Robert Weaver (who, of course, does not agree), *op. cit.,* p. 70.
425. *Reid quote:* in Margaret Reid, *op. cit.,* p. 378.
430. *Heilbrun quote:* Heilbrun, *Urban Economics and Public Policy,* p. 245.

Index